TEXTILE TOWN

TEXTILE TOWN

Spartanburg County, South Carolina

by the Hub City Writers Project
Betsy Wakefield Teter, editor

2002

First printing, November 2002

Photo editors—Mark Olencki and Betsy Wakefield Teter
Front cover photographs—Spartan Mills spinner • Day nursery lunch kids in Pacolet Mills • Rising flood waters at Pacolet Mills #3 • 1930s Valley Falls Mill • Clarence Wilburn • Clifton weavers
Back cover photographs—Converse Mill, rebuilt after the 1903 flood • 1899 Lowell brochure • Crafted With Pride patch • 1931 Drayton Mills spinning room workers • Mayfair Mills spinning room • 1940s view down Liberty Street through Beaumont Mills, courtesy, George Mullinax • Mayfair Mills forklift operator.
Title page photographs—Clifton Mills #1 *(top left)* and Converse Mill, before the 1903 flood *(bottom left)*, courtesy, Mr. & Mrs. Justin Converse • Clifton weavers *(right)*, courtesy, Converse College
Scanning and photography—Christina Smith, Karen Waldrep, Katherine Wakefield, and Mark Olencki
Family support—Diana, Weston, Dave, Percysox, Sweety, Sammy, and all the singing hamsters
Printed by McNaughton & Gunn, Inc., Michigan

Library of Congress Cataloging-in-Publication Data

Textile Town : Spartanburg County, South Carolina / edited by Betsy Wakefield Teter.
p. cm.
Includes bibliographical references and index.
ISBN 1-891885-28-6 (softback : alk. paper) — ISBN 1-891885-29-4 (hardback : alk. paper)
1. Textile industry—South Carolina—Spartanburg County—History. 2. Spartanburg County (S.C.)—History. I. Teter, Betsy Wakefield.
HD9857.S6 T38 2002
338.4'7677'00975729—dc21
2002010860

Hub City Writers Project
Post Office Box 8421
Spartanburg, South Carolina 29305
(864) 577-9349 • fax (864) 577-0188 • www.hubcity.org

For the more than 100,000 men, women, and children
who labored in Spartanburg County textile mills

Contributors

Publication of *Textile Town* is made possible through
the generous contributions of the following:

The Jimmy & Marsha Gibbs Foundation

The Arcadia Foundation
The Arts Partnership of Greater Spartanburg
The Barnet Foundation
The Inman Riverdale Foundation
The Alfred Moore Foundation
The Phifer/Johnson Foundation

The Arkwright Foundation
Paula & Stan Baker
Carol & Jim Bradof
JoAnn, John & Brent Bristow
Ann Brown
Mr. & Mrs. M. L. Cates
Colonial Trust Co.
Paul & Nancy Cote
Mrs. & Mrs. William Gee
Marianna & Roger Habisreutinger
Agnes Harris
Don Fowler
Leigh Fibers
Stewart & Anne Johnson
Dorothy & Julian Josey
Sara, Paul & Ellis Lehner
Jill & John McBurney
Dr. & Mrs. Thomas R. McDaniel
Mrs. Roger Milliken
Mr. & Mrs. Walter Oates
Olencki Graphics, Inc.
The Palmetto Bank
Dwight & Liz Patterson
The South Carolina Arts Commission
Mr. & Mrs. George Stone
Bette & Katherine Wakefield
Tex-Mach Inc.

Margaret G. Allen
Dr. & Mrs. Mitchell H. Allen
Mr. & Mrs. Robert Allen
C. Mack Amick
Mr. & Mrs. Tom Arthur
Mr. & Mrs. Robert Atkins
Mr. & Mrs. John Bachman
Mr. & Mrs. Vic Bailey
Mr. & Mrs. W. D. Bain Jr.
Rebecca Barnes
William Barnet & Son, LLC
Dr. & Mrs. James S. Barrett
Michael Becknell
Mr. & Mrs. Philip Belcher
Linda & Victor Bilanchone
Jean Price Blackford
Shirley Blaes
Mr. & Mrs. Glen B. Boggs II
Bob Bourguignon
Don & Martha Bramblett
Will Brothers
Mr. & Mrs. Walter Brice
Pat Brock
Bea & Dennis Bruce

Duff Bruce & The Open Book
Jack & Myrna Bundy
Dr. & Mrs. William Burns
Mr. & Mrs. Dick Carr *(in memory of Richard Watkins Carr Sr.)*
Mr. & Mrs. Walter M. Cart
Mary Jo Cartledgehayes
Terry Cash
Ruth L. Cate
Elizabeth S. Chapman
Harrison Chapman
Mr. & Mrs. John Chapman
Marshall Chapman
Norman & Muffet Chapman
Margaret & Marshall Chapman
Mr. & Mrs. Robert Chapman
Robert Chapman
Michael & Jean Chitwood
Sally & Jerry Cogan
Marilyn Coltrane
Richard Conn
Mr. & Mrs. Richard L. Conner
Anna & Justin Converse
Helen & Ben Correll
Tom Moore Craig
Kristen & John Cribb
Daniel & Becky Cromer
Nancy Rainey Crowley
Mrs. Frank Cunningham
Mr. & Mrs. Reed L. Cunningham
Elizabeth Chapman Delorm
Magruder & Sara Dent
Georgie & Bill Dickerson
Chris & Alice Dorrance
Susan Willis & James Dunlap
Nancy S. Dunn
Mr. & Mrs. William Edwards
Anne Elliott
Mr. & Mrs. William C. Elston
Frankie Eppes
T. Alexander & Jennifer Evins
Dr. & Mrs. George Fields Jr.
Elsie Finkelstein
First National Bank of Spartanburg
First South Bank
Russell & Susan Floyd
Caleb & Delie Fort
Mr. & Mrs. William Fort
Lane Fowler Consulting
Dr. & Mrs. Sidney Fulmer
Betty & Theodore Gage
Mr. & Mrs. Theodore W. Gage
Mr. & Mrs. Samuel Galloway
Ralph Gillespie
Gerald Ginocchio
Billy & Judy Gossett
Margaret & Chip Green
James & Kay Gross
Lee & Kitty Hagglund
Louise L. Hagy
Benjamin & Tanya Hamm
Lou Ann & John Harrill
Peyton & Michele Harvey
Lawson & Eaddy Williams Hayes
Gary & Carmela Henderson
Mike & Nancy Henderson
Charlie Hodge
Max & Tom Hollis
Gary & Fergie Horvath
Jim Hudgens
Joe & Elsa Hudson
Col. John Hughes & Pic-a-Book
David & Harriet Ike
Lisa & Bob Isenhower
Mrs. James Ivey
Margaret Chapman Jackson
Sadie Chapman Jackson
Dr. & Mrs. Vernon Jeffords
Steve & Susan Jobe
Dr. & Mrs. Ralph H. Johns Jr.
Mr. & Mrs. Charles W. Jones
Lewis & Denny Jones
Frannie Jordan
Mr. & Mrs. Daniel Kahrs
Joe & Julie Kavanaugh
Ann J. Kelly
Polly Ketchum
Dr. & Mrs. Bert Knight
Mr. & Mrs. Paul D. Kountz
Dr. & Mrs. Cecil F. Lanford
Digit & Beth Laughridge
Mr. Jack Lawrence
Wood & Janice Lay
Dr. & Mrs. Joe Lesesne
George & Frances Loudon
Mary Speed Lynch
Susan & Ed Mabry
Tom & Pat Malone
Hal & Doreen Marshall
Zerno E. Martin Jr.
Marzoli International Inc.
Dan & Kit Maultsby
Byron & Linda McCane
Gail D. McCullough
Dr. & Mrs. Dean McKinney
Fayssoux Dunbar McLean
Bob McMichael
Ed Medlin
John Michener
Joann Miller
Mr. & Mrs. E. Lewis Miller
Milliken & Co.
Weston Milliken
Mr. & Mrs. Charles Minch
Karen & Bob Mitchell
Nancy & Lawrence Moore
George D. Mullinax
NBSC Bank
Mr. & Mrs. Douglas B. Nash
Woody Needham
Vivian Fisher & Jim Newcome
Margaret & George Nixon
Neely & Gibson Coal Company
Town of Pacolet
Alfred & Sally Gordon Page
Richard Pennell
Mr. & Mrs. Edward P. Perrin
Amelia Pettiss
Mickey & Nancy Pierce
Mr. & Mrs. Robert Pinson
Andrew Poliakoff
Gary & Anne Poliakoff
John & Lynne Poole
Dr. & Mrs. Jan H. Postma Jr.
Norman Powers
Elizabeth & W.O Pressley Jr.
Emory & Jeannette Price
Harry Price
Jim & Jane Proctor
Philip & Frances Racine
Eileen N. Rampey
Mr. & Mrs. William B. Ramsey III
Karen Randall
Allison & John Ratterree
Mr. & Mrs. John Renfro
Nancy & Robert Riehle
Elisabeth Robe
Mr. & Mrs. Steve Rush
Olin & Muffet Sansbury
Talmadge & Beverly Skinner
Curtis & Donna Smith
Rep. Doug Smith
Freeman & Laurel Weston Smith
www.Sparklenet.com
The Spartanburg County Foundation
The Spartanburg Development Council
Sally & Warwick Spencer
Mr. & Mrs. Jack Steinberg
Ginger Stephen
Anita Stoddard
Ben Stone
Mildred Dent Stuart
Christine & Bob Swager
Tom Swint
Symtech, Inc.
Eric Tapio & Erin Bentrim-Tapio
Allene & Jess Taylor
Nancy Taylor
Dr. & Mrs. Wallace Taylor
Betsy Teter
Ray E. Thompson Sr.
Michael D. & Janna M. Trammell
Robert & Judy Troup
Jack & Jane Turner
Mr. & Mrs. Harry B. Ussery *(in memory of William H. Carr)*
Sarah van Rens
Mr. & Mrs. Jay Wakefield
Mr. & Mrs. J.W. Wakefield
Mr. & Mrs. Lawrence Warren
Lindsay & Billy Webster
Mr. & Mrs. David Weir
Peter & Kathie Weisman
Dr. & Mrs. Fred Wenz
A. Marc White
John B. White
Mary K. Wilborn
Mary G. Willis
Jeffrey Willis
Janet Wilson & Sigmund Pickus
Cynthia & Stephen Wood
Bob & Carlyn Wynn
Jessalyn Wynn
Mrs. Thomas Young
Zimmer Machinery Corp.

Table of Contents

Introduction 11

1. Textile Town Pioneers: 1816 to 1879 15

Voices: Martin Meek 28
The Power of Shoals 30
Textiles in the Civil War 31
The English Manufacturing Company 33
Paving the Way 34
Song: Babies in the Mill 35

2. Boom Time in Textile Town: 1880 to 1909 37

Voices: James Chapman 60
Voices: David Camak 62
Voices: Carolyn Law 64
The Birth of Clifton 66
John H. Montgomery 67
Women and Children 68
Seth Milliken 70
Engineering the Mills 72
A War of Words 73
Mill Architecture 74
Newspaperman Charles Petty 76
The Flood of 1903 77
Vesta Mills 81
Migration from the Mountains 83
Pinto Beans and Cornbread 85
Song: Cotton Mill Colic 87

3. Improving Textile Town: 1910 to 1929 89

Voices: Mary Irene Gault 120
Voices: Lee Loftis 122
Voices: Ira Parker Pace 124
Voices: Ola Smith 126
Mill Village Religion 128
The Gift of Literacy 130
Power from the Hills 131
Company Stores 133
Mill Baseball 135
Olin Johnston 136
Marjorie Potwin 138
Pacolet Mills 140
Curing the Disease of the Poor 143
Lyman, South Carolina 145
The New England Textile Collapse 146
Song: The Mill Mother's Lament 149

4. Textile Town in Depression and War: 1930 to 1949 151

Voices: Boyd Israel 182
Voices: Winslow Howard 184
Voices: Annie Laura West 186

Voices: Laura Rodgers 188
Voices: Christine Gates 190
Spreading the Union Bug 192
Cotton Mill Music 193
Clipboards and Stopwatches 195
The Negro Textile Leagues 197
Blacklisting 198
Cotton Mill Poetry 199
Crowning a Queen 203
Women's Softball 205
Song: The Ballad of Spartan Mills 207

5. TEXTILE TOWN SETTLES IN: 1950 TO 1974 209
Voices: Alfred and Alaree Dawkins 232
Voices: Mac Cates 234
Voices: Rosalie Tucker 236
Voices: Clarence Wilburn 238
Heat, Noise, and Lint 240
The Last Strike 241
Walter S. Montgomery Sr. 243
A Fallout Shelter for 1,600 245
Making History 246
Funny Uniforms and College Stars 247
Working Class Hero 249
Selling Arcadia 250
Butte Knit 251
Andrew Teszler 252
An International Influx 254
Hoechst-Celanese 256
The Draper Story 257
Fred Dent 259
Song: Aragon Mill 261

6. TEXTILE TOWN IN TRANSITION: 1975 TO 2002 263
Voices: Beatrice Norton 280
Voices: Charles Sams 282
Voices: Mike Morris 284
Roger Milliken 286
Brown Lung 290
A Dynasty in Inman 292
Woodruff's Ambassador 294
Uprising of '34 294
Spreading the Wealth 296
Mill Villages 298
A Place Called Riverdale 301
Song: They Closed Down the Mill 305

TEXTILE TOWN APPENDIX 307
A TEXTILE GLOSSARY 319
TEXTILE TOWN AUTHORS 321
BIBLIOGRAPHY 326
ACKNOWLEDGEMENTS 333
INDEX 334

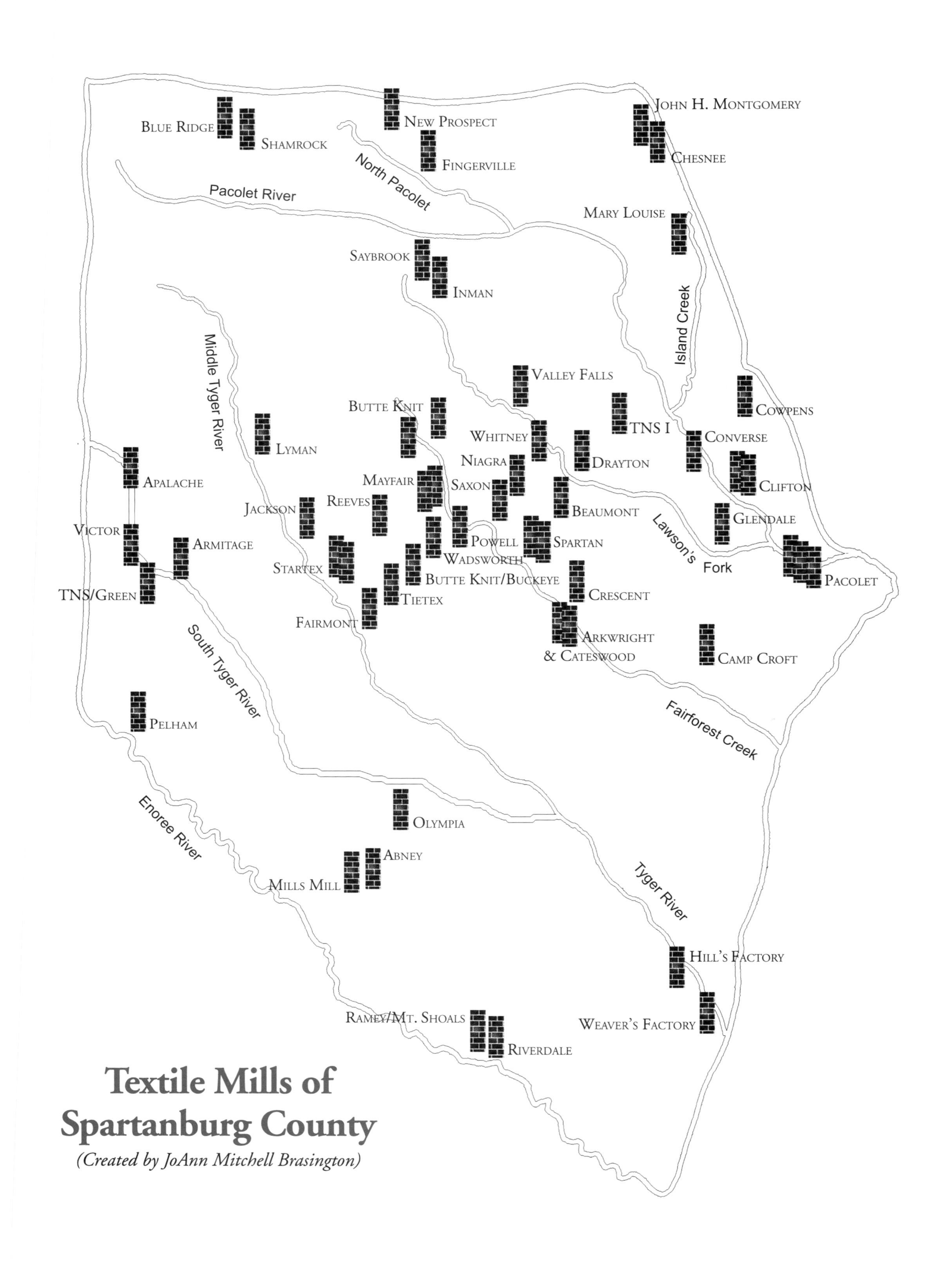

Textile Mills of Spartanburg County

(Created by JoAnn Mitchell Brasington)

INTRODUCTION

"Everything is historic, even ourselves."
—Franklin Burroughs, *The River Home*

Here at the Hub City Writers Project, we have deliberated for years about how to tell a tale with more than 100,000 characters, two centuries of history, and dozens of geographic locations. Would we write an encyclopedia of Spartanburg textile history? Would we tell the story of each individual mill village? Would we let the people who lived the textile life tell their own history? The story of *Textile Town* seemed almost too big to tackle.

There certainly was no question that the book needed to be written. It is the central story of our community. Spartanburg was, after all, "the Lowell of the South." Our northern namesake, a textile town in Massachusetts, is now home to the American Textile History Museum. And yet our own textile history is every bit as rich and deep and complex.

Two years ago, as we began this project, Spartanburg County's textile history began washing away with the fury of the 1903 Pacolet flood: the mills were not only going bankrupt, they were leaving town in container trucks bound for construction sites in Europe and Japan. The individual stories were passing away just as fast; abbreviated life stories of our mill village inhabitants were appearing on the *Herald-Journal* obituary pages daily.

From the beginning, we knew that this must be a "people's history," a book for the people who lived on the mill village and worked in the mills—a book for their children, grandchildren, and great-grandchildren. Because of their contributions and sacrifices, they deserved a thorough history of the textile experience. History tells us that our textile workers were (and are) people of tremendous spirit and endurance. To capture this story, that fact alone needed to be our central focus.

And so we assembled a distinguished group of historians and writers, some of them the best in their field on these subjects. Many of the authors were the sons, daughters, and grandchildren of textile workers. We deliberately chose not to write about the history of individual mill villages. Some of these villages, including Pacolet, Pelham, Clifton, and Glendale, already have fine published histories; others, including Drayton, Fairmont, and Tucapau, have written histories in the works. Our writers' mission was to concentrate on the shared experience of textiles in Spartanburg County.

Textile Town is one part historical narrative, one part scrapbook, one part encyclopedia, and one part oral history. We hope that by creating a book this way, there is something for everyone. Each long chapter is infused with factual material and excerpts from primary sources, such as letters, newspaper columns, and diaries. After each chapter, there are in-depth entries on a variety of subjects such as mill village music, food, and health issues. There is repetition inherent in assembling a book this way, but we hope that will not get in the reader's way of ingesting the culture of textiles.

Some entries are in here not because they are central to the history of textiles, but because they are curiosities, good stories worth preserving. Among them: the story of a cotton mill poem, the story of a labor newspaperman, the story of a gigantic fallout shelter. Hub City is in the business of preserving stories, and we couldn't resist including these among the weightier issues.

Eighteen oral histories, some collected with a microphone and others resurrected from long-buried documents, seem to tell this story the best. Read consecutively, they tell a remarkably epic saga of labor struggle and reconciliation. To reinforce these local stories, we have included a series of cotton mill songs. Interestingly, these songs, as they progress through time, illustrate the same story.

At the end of the book, we have compiled a brief history of every major textile business that has operated in Spartanburg County. This history is by no means complete and, chances are, inaccurate at places. Factual material was assembled from interviews and other publications, and we found that names and dates vary from source to source. A complete bibliography at the end can be used as a starting point for deeper research.

Finally, we hope that *Textile Town* is a document that announces a beginning for true appreciation and understanding of textile culture in Spartanburg County. There is an infinity of stories about this subject, just within the boundaries of this one South Carolina county. Every person who ever worked in a mill or lived on a village has his own set of stories. This volume is one collection point, and there need to be many others. *Textile Town* is our statement that we will not let this era pass without a serious attempt to understand it.

—Betsy Wakefield Teter
May 2002

1 Textile Town Pioneers 1816 to 1879

by Jeffrey Willis

Spartanburg's first spinning mill was a rustic factory built in a wilderness by a pair of Rhode Island brothers, Philip and Lindsay Weaver. Accompanying the Weavers on their voyage south were Thomas Slack, Leonard Hill, and Thomas Hutchings. Soon a third brother, John Weaver, joined the group. All of these adventurous souls possibly had gained experience in Rhode Island at America's first textile mill established by Samuel Slater in the 1790s. In 1816, on the Tyger River near the present-day community of Cross Keys, they acquired 60 acres and cleared away trees and chinquapin thickets to establish a small factory, which they grandiloquently named the South Carolina Cotton Manufactory.

Oddly enough, it was a depression in the American textile industry that gave Spartanburg County its start in manufacturing. In 1815, following the conclusion of the War of 1812, trade with Britain resumed and endangered the prosperity of cotton manufacturing in New England by competition from cheaper English goods. This first American textile depression, and the failure of some mills in New England, led Yankee entrepreneurs like the Weavers to seek new opportunities in the South.

The Weaver brothers and their associates had traveled from New England by water to the port of Charleston. From the Carolina coast they proceeded to explore inland for suitable mill sites. Their search eventually led them to the Piedmont region. It is unlikely that the remote village of Spartanburg impressed these travelers. The small settlement at the time had only 26 commercial buildings, including the shops of a blacksmith, a tailor, and a saddler and the offices of three lawyers and a physician. Perhaps the village's three houses of public entertainment (taverns) were more of an attraction. The district courthouse and jail, however, were shabby in appearance.

If the little village was not a great attraction, the surrounding country had much to offer a potential manufacturer. Spartanburg District (the term "district" was used from 1800 to 1868) had ample waterpower provided by its rivers and creeks. The Weavers may have been attracted to the Tyger River because of its unusual feature of having three branches, north, south, and middle, which flowed almost parallel to each other at a distance of only a few miles apart. Feeding the Tyger were Dutchman's, Ferguson, James, and Cane creeks. Falls and shoals along these waterways offered numerous sites for mills. The surrounding forests were thick with white, red, and Spanish oaks as well as pines, poplars, and maples. Not only was there a plentiful supply of wood for construction, granite and soapstone were available nearby.

Existing roads were in fair condition, and most crossings had bridges. Six bridges crossed the branches of the Tyger alone. In 1802 Governor John Drayton wrote: "a carriage and four may be driven from any part of this State to the other, and from the seashore to the mountains, without any

Remnants of Hill's Factory, here and far left, still stand on privately-owned property along the Tyger River. This was Spartanburg County's second cotton mill, built in 1819. (Courtesy, Mark Olencki)

James Leonard Hill, reading a book, was the second-generation Hill to run the Hill Factory on the Tyger River. He is sitting on the steps of his family home with other family members in about 1885. (Courtesy, Martin Meek)

other difficulty than such as naturally arises in long journeys." The journey from Charleston to the Spartanburg District took 10 to 14 days. Drayton seems to have given an unusually favorable picture of road conditions, but this could be expected from the state's governor. Perhaps Drayton was the first governor to recruit industry. Not all travelers through the district were as complimentary, however. Stumps had still to be cleared from some roads, and sections of rivers were without bridges and had to be forded.

Nevertheless, for the textile manufacturer, Spartanburg offered proximity to raw materials and lower taxes and property costs than in New England. Labor costs were also lower. In the 1820s the monthly wage for a common laborer was eight to ten dollars a month (about $100 to $130 today). This was lower even than in the coastal region of South Carolina. In addition to the low cost of labor, there was a growing supply. Between 1800 and 1820 the population of the district increased from 12,122 to 16,989, the majority of whom were of European origin. The 1820 population consisted of 13,655 whites, 3,308 slaves, and 26 free blacks. Spartanburg District included, at the time, what is both Spartanburg and western Cherokee counties today.

In 1822, Hill and Clark hired a man named John Craig for one year to weave and do other work at their factory "from Sunrise to Sunset" for $1 a day. In turn, they agreed to furnish him a home and to employ his children. His pay was to be one-third in cash and two-thirds in cotton yarn and other supplies.

—Author Ernest Lander

Not only were materials and workers to be found locally, but the necessary financing was available as well. The capital required to establish these early mills was usually contributed in small amounts by local merchants, bankers, and professionals. Among the local backers of the Weavers' enterprise was the Reverend Benjamin Wofford, future benefactor of the college that bears his name.

As well as offering advantages, the Upcountry presented the potential manufacturer with a number of difficulties. One was the problem of transporting machinery to the factory sites. The Weavers brought some of their equipment with them by ship from Rhode Island. They then faced the necessity of hauling this equipment the long distance from Charleston by horse-drawn wagons. It is likely that they also had some machinery constructed on site. The small factory that resulted had only 500 spindles.

In 1819, Spartanburg's second mill was founded by another group of Rhode Island brothers, George and Leonard Hill, in partnership with John Clark and William B. Sheldon. Leonard Hill had earlier accompanied the Weavers to the Piedmont. All four had experience in manufacturing and were master mechanics. A few miles upstream from the Weaver mill, in the present-day Hebron Methodist Church community, they set up the Industry Cotton Manufacturing Company with a similar number of spindles as the Weaver's enterprise. Robert Mills, in his *Statistics of South Carolina*, published in 1826, refers to the Weaver and Hill mills: "Two cotton factories are established on Tyger River, which do very good business."

Once established, these new mills still faced many difficulties. They had to deal with an inadequate number of local cotton gins to prepare the cotton fiber. In addition, distribution of their products was limited to the range of horse-drawn wagons. Both the Weaver and Hill operations suffered from instability. Partners withdrew, interests were sold, and funding was inadequate. Nevertheless, these two groups can be regarded as trailblazers who helped to prove that textile manufacturing was possible in the Piedmont.

These pioneers benefited from several factors that actually favored manufacturing. An increase in 1816 in the tariff on foreign imports of cotton goods provided a stimulus. Also, by the mid-1820s, over-production of cotton resulted in a sharp decline in the price. Investors, therefore, began to look for other outlets for their surplus capital. Perhaps this factor made it easier for entrepreneurs to find local financial backing for their ventures. Still, the limited quantity of local capital condemned the early mills to slow progress.

Land was the principal wealth-producing agent at the time. Those who owned enough land to produce surplus wealth were generally, and usually strongly, opposed to the development of manufacturing. Inextricably entwined with the land-based economy was the system of slave labor. Although the majority of farmers in the Spartanburg District owned no or few slaves, most of the white population was caught in a mindset dominated by agriculture and slavery. A threat to one part of the system was a threat to the whole system. Opposition from powerful political leaders in the state was an added hindrance to industrialization.

The absence of significant European immigration into the southern states was an additional factor that worked against southern manufacturers throughout the antebellum era. While Europeans streamed into the North, they avoided the competition of slave labor in the South. Not only did the southern states suffer from inadequate in-migration, the Piedmont region was plagued by out-migration as well. Cotton wore out the land rapidly. When the land declined in productivity, there was better soil available in Alabama, Mississippi, Louisiana, and Texas. In the 1840s alone the free population of the Piedmont declined 15 percent from out-migration. Between 1800 and 1850, the slave population increased by 124 percent for the Piedmont as a whole. In Spartanburg District, the rise in slave ownership was not as great as in other Upcountry districts. Potential planters considered that the frost-free season was too short in both Spartanburg and Greenville Districts. Nonetheless, the decline of the free population in the antebellum period worked against the success of manufacturing.

By 1830, the Weaver mill had closed. Philip and Lindsay Weaver returned to New England. In addition to coping with the hardships that faced manufacturers in the South Carolina Upcountry, the Weavers apparently had difficulty adjusting to social attitudes in the area. Philip Weaver wrote upon his departure: "I wish to leave this part of the country and wish to settle myself and family in a free state, where myself and family will not be looked down upon with contempt because I am opposed to the abominable practice of slavery."

Also in 1830, sole ownership of the Hill Factory passed into the hands of Leonard Hill. His brother George had returned to Rhode Island and the other partners sold their interests to Leonard, who operated the mill until his death in 1840. Twice the uninsured factory burned, and twice it was rebuilt. Upon Leonard's death, ownership passed to his four sons, two of whom continued its operation. In 1866 they closed the factory and sold the machinery to the firm of Nesbitt & Wright for the Barksdale Factory at Mountain Shoals on the Enoree River.

Another entrepreneur of the time was Thomas Hutchings. A Methodist minister as well as an industrialist, Hutchings had arrived from Rhode Island with the Weaver brothers and had been their partner in establishing the 1816 mill. He was a highly popular preacher and played a major part in raising funds to build Central Methodist Church's first building. After ending his business association with the Weavers, Hutchings was involved in building several small mills in both Greenville and Spartanburg Districts. In 1838, with financial support from fellow-minister Benjamin Wofford, and from Simpson Bobo and other investors, Hutchings established the South Tyger Cotton Manufactory at Cedar Hill on the South Tyger River about 18 miles northeast of Spartanburg. After selling his interests at Cedar Hill in 1840, Hutchings was accused of fraud for not fully revealing the liabilities of the factory. The Methodist Church expelled him from the ministry of that denomination. The Cedar Hill factory continued in operation, later under the name of the Arlington Mill and then the Apalache Mill.

> ***"I wish to leave this part of the country and wish to settle myself and family in a free state, where myself and family will not be looked down upon with contempt because I am opposed to slavery."***
>
> **—Spartanburg textile pioneer Philip Weaver, upon leaving the Upstate in the 1830s**

More insight into the difficulties faced by these early factories can be gained by an investigation into the early years of Upcountry settlement. It could be said that the manufacture of textiles is as old as human existence in the area. Certainly the Indians had to produce their own cloth. This was true of the early European settlers as well. In the New England colonies, the first European settlers were forced to develop manufacturing. For reasons of climate and topography, an economy depending primarily on agriculture could not sustain life. The same conditions were not present in the South Carolina backcountry. Land of reasonable fertility was plentiful and the climate allowed for a long growing season; thus, agriculture came to dominate the economy. There was, however, a necessity for backcountry farms to be self-sufficient in both agricultural products and manufactured goods. The colonists in the remote Piedmont region of colonial South Carolina lived in isolation from the centers of trade. Roads were either non-existent or inadequate and often impassable. Most of the rivers were not navigable. Manufacturing developed in the form of home industries to provide for the needs of the families. Most Upcountry farm families were clothed entirely in homespun. When the Irishman Michael Gaffney (for whom the town was later named) arrived in the Upcountry toward the end of 1800, he found that most farmers "dress generally in a homespun shirt and trousers of coarse cotton yarn. Every farmer or planter is his own shoemaker, tanner, tailor, carpenter, brazier, and in fact, everything else." Even in the centers of trade in coastal areas, the development of more extensive manufacturing was hindered by restrictive legislation passed by the British Parliament.

In the years prior to and during the Revolution, there is evidence of an interest in developing in South Carolina an economy that balanced agriculture and manufacturing. In the 1770s Henry Laurens chaired a committee whose purpose was to establish and promote industry. The purchase by the state of South Carolina of patent rights to Eli Whitney's improved cotton gin implies a similar motivation. One of the earliest mills in South Carolina was established in 1790 near Stateburg in the Sumter District. There is evidence that it was more of a plantation industry than an independent factory. For raw material, it had to import long staple cotton from the West Indies. The venture was short-lived and was not successful.

Black laborers—slaves—represented as many as 40 percent of South Carolina's 600 cotton mill operatives in the early 1840s. No Spartanburg County factories, however, appear to have had slaves. The company owners were Northerners who abhorred the practice of slavery.

In the period after the American colonies won their independence, Britain was anxious to keep to itself the new mechanical devices that were produced by the inventors of the Industrial Revolution. This effort was thwarted by Samuel Slater, an experienced machinist who disguised himself as a farm worker, left England, and brought his knowledge of spinning machinery to the new American republic. Around 1791 in Pawtucket, Rhode Island, he set up the first cotton mill in the United States that used the principle of spinning invented by Richard Arkwright in England. Soon other small cotton mills appeared in most of the other New England states. Some of these survived. Many did not. While New England was beginning a factory system, home manufacturing in the southern states was so prevalent that in 1810 the total value of manufactured goods in Virginia, North and South Carolina, and Georgia exceeded production in all of New England.

The Embargo Act of 1807 and the War of 1812 eliminated competition from English goods and stimulated growth in the New England operations. While manufacturing was growing in the Northeast, cotton manufacturing spread very slowly into the southern states. During this period the South Carolina Homespun Company opened in Charleston in 1808 under the ownership of John Shecut. When the mill closed in 1816, Shecut declared, "South Carolina is not prepared to become a manufacturing state."

In the Lowcountry of South Carolina, some planters hired out their slaves to work in small mills or set up small enterprises on their plantations using slave laborers. Because the cultivation of cotton was far more profitable than the manufacture of cotton yarn and cloth, plantation industries were limited in their operations. Slaves were an expensive source of labor, and many thought that labor of any condition could be more profitably used in agriculture.

Powerful leaders in the state, such as Langdon Cheves and John C. Calhoun, strongly opposed the development of manufacturing and thought South Carolina would be better off economically and morally by sticking to agriculture. They feared, in part, that the development of manufacturing would be accompanied by a pro-tariff attitude. Tariffs were thought to provide protection for northeastern industries and to increase the cost of living for the agricultural South. Langdon Cheves wrote in 1845: "manufacturing should be the last resort of industry in every country... [factories] serve no interests but those of the capitalists who set them in motion." For some South Carolinians, agriculture remained the only gentlemanly path to wealth. More powerful than tariff policies and social attitudes was the simple fact that more money could be made in agriculture than in manufacturing.

Two other factors operated in favor of agriculture and against the development of manufacturing in the South. The effects of the Industrial Revolution on the textile industry in England greatly increased the demand for raw cotton and the profits that could be made from growing it. A still more powerful influence was the invention of a practical cotton gin by Eli Whitney in 1793. Prior to this innovation, there was little potential for the cultivation of cotton as a cash crop in the South Carolina Upcountry. Long-staple cotton could be profitably grown on the coast and sea islands of the state. The climate and soil of the Upcountry was suitable for the growth of short-staple cotton only. Separating the fiber from the seed of short-staple cotton was labor intensive and, therefore, costly. The cotton gin changed this process by providing a means for easily separating the seed and fiber and was a boon to the advocates of an agricultural economy. In 1790 South Carolina grew 69,840 pounds of cotton. In 1800 the figure was 20 million pounds.

THE PERIOD FOLLOWING THE FOUNDING of the Weaver and Hill factories witnessed still more developments. As increased sectionalism emerged during the 1820s and 1830s, concern grew that the North was growing wealthy and powerful at the expense of the South. The South as a whole became concerned about its economic dependence on northern factories. Southerners bought cloth made in New England mills from cotton grown in southern fields. This situation inspired the founding of new and larger mills in the South.

At this stage, it was not Rhode Island but Lincolnton, North Carolina, that funneled leadership into the Spartanburg District. There, in the late eighteenth century, German settlers had established what was probably the first industrial center in the South. By 1790, four iron works were in operation. More important to Spartanburg was the founding in 1815 of the first successful textile mill in the South by Michael Schenck, a Swiss-German immigrant from Pennsylvania. Several of Spartanburg's future industrialists were connected, briefly or otherwise, with Schenck's Lincolnton Cotton Factory.

One of these was a medical doctor turned businessman, James Bivings. As a young man Bivings was engaged in business in Charleston, but by the time of the 1810 census he was resident in Lincolnton. When Michael Schenck expanded his original mill, Bivings became one of his partners. Benefiting from this experience, Bivings left Lincolnton, along with a number of skilled machinists, and arrived in Spartanburg District in 1834. He then proceeded to purchase 750 acres, including the site of the former Wofford/Berwick Iron Works east of Spartanburg on Lawson's Fork Creek, and established the Bivingsville Cotton Manufacturing Company in 1836. In addition to Bivings, the local investors were Simpson Bobo and Elias Leitner. Together the three invested $100,000 (about $1.57 million today). With 1,200 spindles and 24 looms, the enterprise became the largest cotton mill in Spartanburg District and represented a milestone in the development of the textile industry in Spartanburg.

The initial machinery had been acquired by Bivings in Paterson, New Jersey, and transported by sea and road to the new factory. Included in this equipment was an overshot wheel of 26-foot diameter and 12-foot breast. These first spinning machines and looms were operated by waterpower produced by a dam over Lawson's Fork Creek. The Bivingsville Factory spun both cotton and wool yarn and was one of the first to engage in weaving from the beginning. On high ground overlooking the mill, James and Susan Von Storre Bivings built for themselves and their family a large home with two-story porticoes on the front and back.

James Bivings took an interest in politics as well as business. In the 1828 presidential campaign, he backed John Quincy Adams against Andrew Jackson. In 1840, along with other Upcountry leaders such as General Waddy Thompson, Bivings supported the Whig Party and William Henry Harrison. In addition, he was a supporter of the temperance movement and urged his mill operatives to attend religious revival meetings.

Simpson Bobo, early investor in local mills (Courtesy, Louise Foster)

Except for the Bivingsville factory, the other early mills in Spartanburg District were primarily yarn plants that produced a coarse cotton yarn and linen warp destined for local household looms. In many ways they were little more than expanded plantation spinning houses. Falls and shoals provided the power source. A few were equipped with looms, weaving coarse cloth for local consumption. They sold their yarn and cloth themselves, sometimes by barter, or through local merchants. There was some trade by wagon with the western part of North Carolina, East Tennessee, and the South Carolina Lowcountry. Most employed between 10 and 25 workers, except for larger mills like that at Bivingsville. The labor force was white and local and included children. It was drawn from the poor farming population of the area. Skilled workers and managers had to be brought from the North.

After falling into disagreement with the other investors, James Bivings severed all connection with the Bivingsville Cotton Factory in 1844. Soon afterwards, he built a small factory on Chinquapin

Creek at the point where North Church Street now separates into Highways 9 and 221. The Chinquapin proved to be inadequate to operate the 264 spindles. An experiment with mule power did not solve the problem. The new mill closed.

Now in partnership with his son, James D. Bivings, Dr. Bivings acquired 650 acres on the Middle Tyger River, moved the equipment from the Chinquapin mill, and opened a new mill in 1846. The surrounding community, which included a school and cabins for the workers, became known as Crawfordsville after John Crawford, a local landowner. At the beginning, the Crawfordsville mill was engaged in spinning only. In the 1850s, James Bivings built an addition and set up looms as well.

The factory at Crawfordsville must have been successful. James and Susan Bivings built an impressive, Greek-revival home on Spartanburg's North Church Street in 1854. Nearby, Wofford College held its first classes in the same year. Two years later, in 1856, the Bivingses sold their interest in the Crawfordsville mill to the company of

Joseph Finger, founder of Fingerville (Courtesy, Spartanburg Herald-Journal)

Grady, Hawthorne, and Turbeyville. At the time it had 1,000 spindles and 20 looms. This mill later became Fairmont Mills.

Some accounts state that after selling his interest at Crawfordsville, Dr. Bivings left Spartanburg and moved to Georgia, where he had plans to establish a mill and where he subsequently died. Whether he had plans to build a mill in Georgia or not, Bivings did not die across the state border. Prior to his death in 1869 at the age of 82, Bivings spent his last years in Spartanburg District. The *Carolina Spartan* states that he died at his home near Crawfordsville having "been in feeble health for several years, which rendered him unable to engage in the active duties of life." A stone of white marble marks his grave at Fairmont Methodist Church. The *Carolina Spartan* gave him the credit he deserved: "He did more than any other individual to build up and promote the manufacturing interests of our District. He possessed a remarkable foresight and a discriminating judgment."

> **Many early mills prohibited alcohol on the factory premises. The proprietors of Fingerville even agreed to keep "no spirituous liquors in their private possession."**
>
> ***—Author Ernest Lander***

Descendants of James and Susan Bivings continued to reside in Spartanburg. In addition to a son, James D. Bivings, they had at least one daughter, Susan Elizabeth Bivings (1825-1862), who married John Blassingame Cleveland (1818-1870), the son of Spartanburg's first mega-merchant, Jesse Cleveland.

In addition to James Bivings, another entrepreneur active in the 1840s and 1850s was Joseph Finger. Finger also came to Spartanburg District from Lincolnton, North Carolina, and bought property on the North Pacolet River near the present-day McMillan community. In 1848, a partnership was formed between Gabriel Cannon, Joseph Finger, and Henry Kestler to form the Pacolet Manufacturing Company. The original investment was about $5,000. Located at what came to be called Fingerville, the factory produced yarn only and eventually had 396 spindles and 13 workers. Gabriel Cannon also operated a large store at New Prospect near Fingerville. During the Civil War he would serve as president of the Manufacturers' Association of the Confederate States.

The building of new mills by James Bivings and Joseph Finger in the 1840s shows the influence in the Spartanburg District of a broader movement. During this decade Southern agriculture suffered from a severe depression, which was especially felt in South Carolina. As overproduction again caused a decline in the price of cotton, advocates appeared who actually argued that manufacturing was becoming more profitable than agriculture. The result was a popular and fairly widespread movement to "bring the spindles to the cotton."

The strongest of these advocates was William Gregg. Although Gregg was not from Spartanburg District, he influenced manufacturing everywhere in the state by lobbying strenuously to overcome opposition to large-scale manufacturing. His series of articles, published in the Charleston *Courier* and

William Gregg, president of Graniteville Mills and father of the textile industry in South Carolina (Courtesy, Louise Foster)

De Bow's Review, had great influence. One of his arguments held that the absence of capital need not be a hindrance. Mills could be established, he argued, on credit; afterwards the necessary capital would be generated and the loans paid off. Gregg followed his own advice. In 1846, his new mill at Graniteville in Aiken District was the largest in the South. It was Gregg who also built the first mill village in the South. A Spartanburg attorney, James E. Henry, was an investor in the Graniteville mill.

Prior to the Civil War, coarse yarn continued to dominate the product list of the early mills. Looms produced osnaburgs, bleached and unbleached shirting and sheeting. In the 1840s some mills experimented with cotton bagging. James Bivings was one of the promoters of this product. Woolen cloth continued to be produced in homes rather than in mills. However, the wool yarn had to be carded before it could be woven. The larger mills at Fingerville and Bivingsville acquired machinery for carding wool.

THE DISTRIBUTION OF ALL the products of Spartanburg's mills remained primarily local. Goods were sold through local merchants on commission. More distant markets in the western part of South Carolina and North Carolina and in Georgia still had to be reached by the wagon trade. The coming of the first railroad in 1859 made possible shipments of Spartanburg goods to a wider market, including the Northeast. The railroad worked in both directions, however. Textile goods from the northeastern mills could now compete more easily with local products.

In the Lowcountry and central section of South Carolina, the textile mills primarily used slave labor, along with some white operatives. Some of these mills used slaves exclusively, except as superintendents. Some still argued that slaves could never become satisfactory mill operatives. The record shows that slaves, like any workers when given proper training, were efficient operatives. One powerful influence argued in favor of the use of white workers only. That was William Gregg. Advocates of slave labor argued that white workers would not voluntarily submit to the conditions of work in the mills, which proved to be true in many cases. The argument over white versus slave labor was driven by questions of economics and profitability.

In Spartanburg District and the Piedmont as a whole in the 1840s and 1850s, the labor force was almost exclusively white. Much of the district's population lived on small, self-sufficient farms and grew barely enough to feed themselves. In the latter part of the eighteenth century, a traveler through the district was disturbed by the condition of the inhabitants. "The lower class in this gouging, biting, kicking country are the most abject that perhaps ever peopled a Christian land. They live in the woods and deserts [sic] and many of them cultivate no more land than will raise them corn and cabbage, which, with fish and occasionally a piece of pickled pork or bacon, are their constant food. Their habitations are more wretched than can be conceived." Although there may have been an improvement in conditions during the first decades of the nineteenth century, there were enough of these poor farmers remaining locally to supply the limited needs of the mills for operatives.

According to local legend, Fingerville founder Joseph Finger first crossed the North Pacolet River with his wife on his back.

There is no question that white workers came cheaply. In the 1820s children were paid between $1.25 and $2 a week (about $16 and $25 today). Adults were paid as much as $6 a week (about $75 today). Average figures for 1860 show that there was no improvement. Female workers were paid an average wage of $89.05 annually (about $1,603 today); males received $138.96 (about $2,501 today). These very low figures included children, who were at the bottom of the wage scale, along with adult operatives. Wages also varied from mill to mill. At Crawfordsville, in 1880, the average wage was 33+ cents a day. At Valley Falls, not far away, the 15 workers were paid an average of 40 cents a day. The average work week was 12 or 13 hours a day for six days a week. Low wages resulted, in part, from the limited money economy and low profits. In many cases, the mill owners simply could not pay more.

An element of paternalism evolved during the early stages of textile development. One way the owners could hold white workers at low wages was by not only providing employment but free housing, schools, and churches. While not being able to pay higher wages, the mills could afford to provide these amenities. The large mill village at Graniteville in the Aiken District was not typical in the antebellum period. Nonetheless, villages did emerge at the larger mills in the Spartanburg District. The Bivingsville Cotton Factory is the best example.

In the decade prior to the Civil War, South Carolina's struggling textile industry was becoming more organized. William Gregg of Graniteville became president of the newly formed South Carolina Institute in 1849. Throughout the 1850s, the institute's annual fairs promoted the virtues of manufacturing but with limited success. Curiously enough, at the very time the textile industry was becoming organized, the founding of new mills declined. Expansion reached a peak about 1849.

Joseph Finger is buried on a hillside overlooking the Fingerville mill village. His tombstone reads: "Born and died a Methodist."

"Here a mere shadow (physically speaking) with gold frame spectacles over his face all the time..."

—Description of Dr. James Bivings in the Carolina Spartan, 1880

During the 1850s, cotton planters saw their profits rising again. As growing cotton became more profitable, manufacturing cotton products became less attractive. Higher profits for the cotton growers meant higher costs for the cotton manufacturers. By 1852 the movement to "bring the spindles to the cotton" was losing momentum. During the decade prior to the Civil War, only one new textile mill was built in the entire state of South Carolina. In 1857, John Weaver (the only brother who remained in South Carolina) and William D. McMakin built a factory five miles north of Spartanburg at Valley Falls on Lawson's Fork Creek, which employed only 14 workers. This mill was later owned by Henry White and William Finger. It was destroyed by fire in 1891, after being struck by lightning.

The Valley Falls factory was definitely an exception to prevailing trends. During the 1840s and 1850s, eight mills in South Carolina closed. The number of workers employed in South Carolina textile factories as a whole declined 12.5 percent in the 1850s. In addition to the repercussions of agricultural prosperity in the 1850s, South Carolina's fragile textile industry lacked trained managers and technicians. Operating a

Ware Manufacturing, chartered in Massachusetts in 1821, was one of many cotton mills in New England, where the U. S. textile industry was centered in the nineteenth century. (Courtesy, the American Textile History Museum)

textile factory required constant attention. Few owners in the state devoted their full time to industry, and few were willing to devote sufficient effort to acquiring the essential expertise to operate a mill successfully. Upcountry textile factories also faced increasing labor costs caused by the growing scarcity of white labor. All farmers profited from the rising price of cotton in the decade, and this meant fewer potential textile workers. Far from expanding, the industry seemed to be contracting. Although manufacturing in the southern states at one time exceeded that of all of New England, by 1860 output in Virginia, North and South Carolina, and Georgia was only 40 percent of output in Massachusetts alone. Nevertheless, in 1860 Spartanburg was one of the leading manufacturing centers in the entire Piedmont.

While some mills were closing in the 1850s, Bivingsville survived but was experiencing difficulty. After James Bivings sold his interest, controlling interest was purchased by George and Elias C. Leitner. The large debt incurred by the Leitner brothers soon forced them into bankruptcy. The mill sold in 1856 for less than one-third of its original value to John Bomar and Company. Among the other investors were John Conrad Zimmerman, Simpson Bobo, Vardry McBee, S. N. Evins, and Edgar Converse.

Employment in Spartanburg County mills stood at 114 in 1850. That included 52 operatives at Bivingsville, 22 at Cedar Hill (then known as the South Tyger Manufacturing Co.), 15 at Fingerville (then known as the Pacolet Manufacturing Co.), 14 at Hill's Factory, and 11 at Crawfordsville.

In addition to James Bivings, the dominant figure in the textile industry in Spartanburg District prior to 1880 was Dexter Edgar Converse. Edgar Converse, as he called himself, was born in Swanton, Vermont, in 1829. His father, Olin Converse, was a woolen manufacturer. Edgar's education was not extensive. His letters, even in later life, contained misspellings. Although lacking in formal education, Converse grew up with a thorough knowledge of machinery and the management of a mill. After his father died when he was three, and his mother's remarriage, young Edgar was raised by an aunt, Permelia Twichell Brown, who lived in Canada. It was from her husband, Albert G. Brown, that he learned about the manufacture of woolen goods. As a young man, Edgar Converse was given a job at a textile mill in Cohoes, New York, by his uncle Winslow Twichell. While living with his uncle and aunt, young Converse doubtless noticed his pretty cousin, Nellie (Helen Antoinette), who was ten years his junior.

With experience behind him, Edgar Converse migrated south in 1854 to work for the cotton mill in Lincolnton, North Carolina. He remained in North Carolina for only one year, being attracted in 1855 to a job as superintendent at the Bivingsville Cotton Factory.

Edgar Converse arrived at Bivingsville about the time the mill was entering bankruptcy proceedings. When offered the tempting salary of $2 a day to return to Lincolnton, he considered leaving Spartanburg District and returning to North Carolina. He wrote to his uncle, Albert Brown: "People here are the most cautious about doing anything on a large scale of any I ever knew. They are particularly afraid of the manufacturing business." Fortunately, John Bomar, who knew little about the operation of a textile mill, was impressed by young Converse, kept him on as superintendent, and invited him to become a partner in the new company. Toward the $19,500 that was needed to buy the mill in the 1856 bankruptcy sale, young Converse put his life savings of $1,500 (about $27,245 today).

The *History of Spartanburg County*, compiled by the Writers' Program of the Work Projects Administration, describes Converse's impact upon the area as follows: "Not only did he organize a paying enterprise at Bivingsville, but he manifested a civic spirit which made him invaluable in the building of Spartanburg. The part he played in time of war and later of reconstruction was outstanding in the development of city and county. He brought into being in Spartanburg a new stability and perseverance in the textile industry. The example he set inspired others to surmount obstacles."

Under the careful management of Bomar &

Dexter Edgar Converse, founder of Clifton Mills (Courtesy, Louise Foster)

> *"At the Bivingsville factory, owner E. C. Leitner faced a dire situation by the mid-1850s. Listed as 'hopelessly insolvent' with outstanding judgments against him totaling over $50,000, a desperate Leitner apparently committed forgery to obtain the necessary cash from a variety of banks in the state. By January 1855, Leitner had fled to the West with law enforcement 'after him with a sharp stick.'"*
>
> **—Author Bruce Eelman**

Company and young Edgar Converse, the Bivingsville Cotton Factory tripled in value within three years. Having recently married his 17-year-old first cousin, Nellie Twichell, Converse brought from New York her 18-year-old brother, Albert Twichell, and employed him as a bookkeeper and clerk in the mill. The brothers-in-law, who were also first cousins, would be closely associated with each other in the textile business and the civic life of Spartanburg.

Both Converse and Twichell had been in South Carolina for only a few years when the calls for secession, published weekly in the *Carolina Spartan,* put them in an awkward situation. In politics, Converse considered himself a Republican on the national level and a Democrat on the local level. When the Civil War began in 1861, both young Northerners were held in considerable suspicion. They promptly enlisted as privates in Company C of the 13th Regiment (McGowan's Brigade) of the South Carolina Volunteers. Converse served only a few months, when he was discharged to return to Bivingsville to manufacture material for Confederate uniforms. Even the workers of the mill had petitioned the governor for his discharge. They feared losing their livelihood if the mill closed in the absence of its superintendent and chief mechanic. Young Albert Twichell remained in the Confederate Army for the entire war. He was at Appomattox when Lee surrendered, after which he walked all the way back to Bivingsville.

Albert Twichell, treasurer of Glendale Mills (Courtesy, Louise Foster)

At the beginning of the Civil War, the Bivingsville Factory had 1,435 spindles and 26 looms operated by 58 workers. Cloth was not the only article produced for the Confederate Army. Using a cupola furnace for smelting iron ore, the enterprise manufactured Bowie knives and swords as well. Bivingsville was also the only location in the Confederate States where wooden shoe soles were manufactured by machine, rather than by hand, for the troops.

The demand for cloth created by the Civil War provided a temporary stimulus for Spartanburg's textile mills. Unfortunately, during the conflict, many cotton mills in South Carolina were destroyed by the fighting. Spartanburg's mills survived intact but, in the aftermath of the war, faced a depressed economy and the absence of capital for investment.

Bivingsville Mill as it appeared in 1856, when it was renamed John Bomar & Company (Courtesy, Clarence Crocker)

While some mills faced bankruptcy, Edgar Converse fought to keep his mill in operation. Since few had money to purchase textile goods, Converse advertised in the newspaper that the Bivingsville factory would accept barter in exchange for its produce. Wagons laden with cloth and yarns were sent to North Carolina and returned loaded with agricultural products and other essential goods, which were then distributed to the mill operatives in the company store as payment of their wages. Naturally Bivingsville was not the sole survivor or the only mill that prospered. In 1867, Joseph Walker published an almanac in Charleston that listed the existing textile mills in South Carolina. Six Spartanburg factories were cited; some of the information, however, was not entirely up-to-date:

> **1) Lawson's Fork Factory [Bivingsville], five miles east of Spartanburg, with 1,000 spindles, 25 looms, and 60 operatives;**
> **2) Valley Falls Factory on Lawson's Fork, five miles north of Spartanburg, with 500 spindles;**
> **3) the Fingerville Factory, 15 miles north of Spartanburg on the Pacolet River, with 500 spindles and 15 looms;**
> **4) Hill's Factory [closed in 1866], 13 miles south of Spartanburg on the Tyger River, with 500 spindles;**
> **5) Barksdale Factory, 20 miles south of Spartanburg on the Enoree River, with 100 spindles and 50 operatives;**
> **6) Cedar Hill Factory, 18 miles northeast of Spartanburg on the South Tyger River, with 1,000 spindles and 20 looms. [James D. Bivings managed the Cedar Hill Factory.]**

THE DANGER INHERENT IN THE FAILURE to diversify economically was clearly demonstrated for the South by the Civil War. At the end of that conflict the prosperity once found in agriculture had disappeared. Although the economic climate of the postwar era was not favorable to economic expansion, farsighted leaders glimpsed the opportunities that were there and knew that risks had to be taken if their homeland was to recover from disaster. Furthermore, the abolition of slavery meant that it was now possible for the South to move in a new direction. The pre-war preference for agriculture and bias against manufacturing were gone. It was now possible to create a New South.

Changing attitudes were reflected in postwar legislation. In 1865 the South Carolina Legislature passed a law exempting machinery from the state property tax. This perhaps was designed to offset the high cost of machinery immediately after the war. Other problems facing those desirous of reviving cotton manufacturing were the high price of cotton and the scarcity of investment capital. In 1873 the Legislature exempted from taxation, for 10 years, profits from manufacturing.

In 1865 Bomar & Company was reorganized as Bomar, Converse & Zimmerman, indicating the greater role now being played by Edgar Converse. Converse, as manager, continued to show initiative. In 1867 a new building was constructed at Bivingsville. In 1869, at an agricultural fair in Columbia, the Bivingsville Factory won prizes for the best bale of osnaburg, the best shirting, the best sheeting, the best cotton yarn, and the best fabric and denim. Such recognition enhanced the national reputation of the mill and made it possible, in the 1870s, to develop markets for its products in the Northeast and Midwest. These accomplishments are all the greater, considering that many southern mills were declaring bankruptcy during the Reconstruction Era.

Suffering from declining health, John Bomar died in 1868. In 1870, the Bivingsville mill was sold to Edgar Converse and Albert Twichell, with Converse owning the larger share and John Conrad Zimmerman retaining some interest. From this time on, both Converse and Twichell not only played a major role in the expansion of the textile industry in Spartanburg but also would make significant contributions to the civic and cultural life of their community. Edgar Converse provided the model and inspiration for other business leaders who participated in the explosive expansion of the county's industries in the 1880s and 1890s. He was a generous benefactor of cultural organizations and was a liberal contributor to the college for women that came to bear his name. Albert Twichell, an amateur musician himself, was organist at the First Presbyterian Church and one of the organizers of the South Atlantic States Music Festival, which later became the Spartanburg Music Festival.

At the end of the Civil War 8,000 Spartanburg County slaves became free laborers. Few, if any, were allowed to work inside the cotton mills.

When Converse and Twichell assumed control of the Bivingsville mill, the name of the company was changed to D. E. Converse & Company. The mill continued to prosper

Resolution creating John Bomar & Co.
at Bivingsville

We the subscribers have associated ourselves together for the purpose of a Cotton Manufacturing Company at a place known as Bivingsville at a Capital of Thirty Thousand Dollars divided into Three Hundred Shears of one hundred dollars per share and will take the sums annexed to our names. Company to be formed with stock is taken.

Signed,

Simpson Bobo, S. N. Evins, Vardry McBee, John Bomar, D. E. Converse, John C. Zimmerman

March 22, 1856

in the 1870s, partly due to Edgar Converse's belief in re-investing profits. In 1873 a serious national depression resulted from over-speculation and over-expansion. More that 5,000 businesses went bankrupt nationwide. D. E. Converse & Company survived the Panic of 1873 and the rigors of Reconstruction. Converse even purchased new machinery and expanded operations. By 1875 the mill had 5,000 spindles and 120 looms. In a village of 60 cottages, 175 operatives lived with their families. Edgar Converse helped construct some of the cottages himself. Two cotton gins and a sawmill operated in connection with the mill, as well as a wool-carding factory that could handle 12,000 pounds of wool a year. A six-acre meadow provided feed for the animals connected with the mill operations.

> ***"So far as I am concerned I would put my last dollar in cotton manufacture in this state..."***
>
> **—Albert Twichell**

In 1878 the name of the village of Bivingsville was changed to Glendale—a name suggested by Nellie Converse. When first married, the Converses had lived in the company boardinghouse near the mill. Eventually they occupied the imposing, antebellum Bivings mansion overlooking the mill.

The expansion of the Upcountry textile industry that occurred after 1880 had its beginnings in the 1870s. Although South Carolina, in the 1870s, still suffered from wartime damage and Reconstruction policies, other developments aided Edgar Converse in reviving his textile company. Spartanburg County grew more cotton after the Civil War than before the conflict. In 1874, the introduction of chemical fertilizers into the county for the first time increased the crop yield still more. The increased supply produced lower prices for raw cotton and lowered the cost of producing cotton yarn and cloth. Investors, as after 1815, again had incentive to lean toward manufacturing. While the price of cotton was falling, the price of textile machinery was declining from wartime highs. Enterprising owners like Edgar Converse in Spartanburg and Henry P. Hammett at Piedmont Manufacturing Company in Greenville County were able to re-equip existing mills or begin new ones. In the textile industry as a whole there was a new movement to "bring the factory to the field." The mill at Glendale operated literally adjacent to cotton fields.

Some lingering effects of the Civil War were actually beneficial to cotton manufacturing. High tariffs on foreign textile imports had been imposed during the conflict. After 1865, the tariffs were continued and increased. While the American textile industry was enjoying some protection from foreign competition, it also benefited from the opening of new markets for cotton cloth in the Far East.

There were still additional factors that favored a revival of the southern textile industry. From the time that the first small factories were built on the Tyger River, distribution of the product had been

The original company store in Glendale was built around 1850 and stood where the Glendale post office is now located. The building was torn down about 1930. (Courtesy, Clarence Crocker)

Converse College, founded in 1889, was an early example of philanthropy by mill owners. Edgar Converse provided the start-up funds.

a problem and had remained largely local. Although the railroad first came to Spartanburg in 1859, a more important development in transportation occurred in 1873. In that year the rail line between New Orleans and New York was completed through Spartanburg—and very close to Glendale as well. Not only were northern markets more open to Spartanburg products, the rail connection to New York opened the possibility of tapping world markets.

New technology in the 1870s was further reducing the cost of production for mill owners and making possible the greater use of unskilled workers as opposed to higher paid experienced workers. The new ring spindle, introduced in the 1870s, worked automatically for the most part and eliminated the need for skilled spinners. In the next decade, the automatic loom would produce the same results in the weaving room. The Piedmont region and adjacent mountains continued to yield an abundant supply of poor whites. The Civil War had reduced many white farm families to poverty. The post-war era was no kinder. A system of sharecropping and crop liens condemned small farmers to an insecure and marginal existence. It was not necessary for the expanding textile industry to go far into the hills for workers. In the 1870s most of the new operatives came from the immediate area of the mills. Many of these farmers actually preferred the textile mill to hard and uncertain labor in the field and even found, for a while, a welcome security in the paternalism of the mill village. In the next decades, the overproduction of cotton and fall in crop value would increase still more the flow of poor farmers to the mill village.

In the early decades of the nineteenth century, a powerful combination of factors had worked against the development of manufacturing in South Carolina. Most of the early mills failed or achieved only limited success and profitability. By 1880 a combination of factors was finally working in favor of southern manufacturing. The South Carolina Upcountry was about to experience a period of rapid expansion in the textile industry, and Spartanburg County had a head start. In 1880 Spartanburg was the most industrialized county in the southern Piedmont.

Grave of James Bivings, Fairmont Methodist Church (Courtesy, Mark Olencki)

Because daughters were, on balance, a liability in the farm economy but a boon in the mill, farm families with several daughters were more likely to move to the mill than their counterparts with several sons. In 1880, women between the ages of 16 and 24 made up more than 29 percent of Southern mill operatives, and 57 percent of the entire Southern cotton mill workforce was female.

—Author Toby Moore

Martin Meek

Spartanburg architect Martin Meek lives on a historic property in Enoree that was alternately owned by textile pioneers James Nesbitt Jr. (owner of the Barksdale Mill, 1866) and Leonard Hill (owner of Hill's Factory, 1819). His 170-year-old home, called Mountain Shoals Plantation or "the Nesbitt House," was the outpost for business activity in the southern tip of Spartanburg County for much of the nineteenth century. During the past 25 years, Meek, Frank Coleman (1921-1994), and Bill Cooper, Laurens County Librarian (and descendent of the early textile pioneers Clarks), have restored it to its former glory. Shuffling through old documents and books, Meek talks about its history and how he came to live there.

Martin Meek at Mountain Shoals Plantation in Enoree (Courtesy, Mark Olencki)

Exactly why the Clarks, the Sheldons, and the Hills came here out of Rhode Island is to me a big mystery. I think they were looking for two things. They were looking for some good water sources and they were looking for cheap labor…These people were developing the same textile machinery as the early pioneers of the textile industry in New England. They were doing it here in South Carolina at the same time—because the real massive expansion of mill technology in New England didn't occur until 1815 to 1830.

In 1836 Leonard Hill and James Nesbitt buy this property. The next year, Leonard Hill sells his half interest in the 430 acres to James Nesbitt…Nesbitt was the money behind a lot of Spartanburg ventures. Churches, mills, agricultural schemes and finances. I was interested in [researching] how many times did James Nesbitt sue somebody, foreclose. I call him the Snidely Whiplash of Spartanburg County because the list just goes on and on and on. We have strong evidence that the master builder of Mountain Shoals was a man named Thomas Badgett, Bill Cooper's great-grandfather, and we have a document that states that James Nesbitt sued him for a $300 loan.

Fortunately Nesbitt had enough money to survive the Civil War with his estate fairly well intact. He gets on the soldier's relief board. He's somebody with enough money to help the soldiers' families after the Civil War. In the 1860s Census it lists James Nesbitt's net worth. I averaged the net worth for [regular] citizens of Spartanburg County, and the average net worth was about $3,000. James Nesbitt had $74,475 worth of real estate and $54,225 worth of personal estate. He was an incredibly wealthy man in the 1860s. You get into the 1870s and his net worth was incredibly depressed because of the Civil War. That census lists his real estate value at $30,000 and his personal property at $3,000.

In 1866 Nesbitt buys the looms from the Hills Factory and moves them to the Barksdale Factory at Mountain Shoals. He's got shares in railroad companies and other bits and pieces. We know the Barksdale mill was there as late as 1887 when the Enoree mill was built, because it appears on a plat of that date.

Nesbitt dies in 1876 and leaves in his will a parcel of land known as the "Farrow tract" to his second wife, Caroline Burton Nesbitt. But in this section of his will he deducts the "Mountain Shoals tract embracing the store-house block (c. 1860)-Smith Shop, Mills etc. and some 350 acres." He leaves this tract to his youngest daughter, Cassy A. M. Nesbitt.

After James Leonard Hill gets the property, he dies in 1889 and his son, William Allison ends up with the property. When William Allison dies in 1899, for some reason—probably indebtedness—they sell off all of the goods in this house. I have the estate papers and an auction flier for that sale.

The Hills continue to live here until 1924 when Dr. W. H. Irby and his wife come to live here. Mrs. Irby was a bit of a character. She was from Punxsutawney, Pennsylvania. She was an actress, and she staged some mill productions [at Enoree Mills, later called Riverdale], some entertainments for the mill workers. She comes to town, and Dr. Irby is here running a drugstore, they fall in love. He was from a very prominent family in Laurens. The family wasn't real happy that number one, he was marrying a Yankee, and number two, he was marrying an actress, and worst of all, number three, she smoked! She was quite a character, emerald-green nail polish, silk kimonos and live chameleons on gold chains that sat on her shoulder, just to name a few. They only lived together nine years and Dr. Irby died. He died in 1933 and she lived here until 1972 by herself with a couple of Airedales.

When I was a little child I used to dream about this house, but I had never seen this house. I lived in north Texas, in the very harsh and barren plain, and I would get up and tell my mother, "I dreamed about that place again, that place that has rolling hills and beautiful green trees." My father worked with Phillips Petroleum—and later for Phillips Fibers—so we were transferred from Texas to north Georgia. I thought I had died and gone to heaven. It was spectacular, all these rolling hills and beautiful trees I had dreamed of. But I kept having this dream about this house.

So when we moved to South Carolina I asked [friend and local historian] Frank Coleman, "Sometime when you have a Saturday or Sunday free, would you take me around the county and tell me the history?" One Sunday we ended up in the south part of the county, and he said, "You need to see the Nesbitt house." This was about the time when I was finishing college, thinking about graduating, and I wanted to stay in this area. Little did he know that in my request I was actually looking for a specific house.

One afternoon we drove up in the driveway and knocked on the door and nobody was here. Very unlike me, instead of listening to Frank's historic spiel, I drifted off by myself into this historic garden and walked through the gate and down to the end of the boxwood *allee*. Turned around and faced the house. That's when I had a little experience in the yard, and an entity spoke to me and said, "Someday this will be yours." Just as clean a voice—I always thought it was me…my mind wanting this to happen. Anyway I didn't tell anybody, just thought everybody would think you were a freak if you explained to them that you heard voices in the yard.

Several years later, Frank comes to my mom and dad's for dinner, walks in and says, "You know I ran into two of the Nesbitt granddaughters today, and they said Mrs. Irby had died."

I looked at Frank and said, "You want to go partners and try to buy that house?"

He says, "You're crazy. You'll never get it…Mrs. Irby left it to the Enoree Methodist Church."

Well, I grew up a Methodist. I said, "Frank, all we gotta do is find out who the chairman of the board of the Methodist Church is. They can dispose of their own property, and we find out whether it's available or not. That's the key." He grew up down in Woodruff, that area. I said, "Why don't you make some calls and find out who it is?"

So he calls me the next day and said, "The key is not who the chairman of the board is. The key is, who is the chairman of the board's wife?"

I said, "Who?"

He said, "Marguerite Cooper McCarley."

I said, "Who's she?"

Frank said, "Just my best girlfriend in high school."

To make a long story short, the Methodist Church sold us a 5,000-square-foot house and three acres of land for $12,500 in 1975.

—Interview by John Lane and Betsy Wakefield Teter

Early industrialist James Nesbitt Jr. about 1880 (Courtesy, Martin Meek)

The Power of Shoals
Magnets for Cotton Mills

by John Lane

All that's left of the 1819 Hill's Factory on the Tyger River in southern Spartanburg County are the remains of a long millrace filled in with dark earth and a series of imposing (both in size and workmanship) low granite foundations quarried block-by-block from the precipitous adjacent shoals.

The shoals, located 15 miles south of Spartanburg, are still a powerful spot today, but for different reasons than the Hill brothers would have understood. Today someone walking along the river would see the raw beauty of a Piedmont stream as it takes on qualities usually associated with the Blue Ridge Mountains still 40 miles distant—steep banks, dark boulders, swift currents, and the raucous clatter of whitewater. The Hill brothers, early textile pioneers from Rhode Island, saw power in the spot, and measured the benefits in captured horsepower.

Over the next 70 years, shoals such as these drew cotton mills to all corners of Spartanburg County. Joseph Finger found a spot on the North Pacolet (above present-day Lake Bowen) for his factory; John Montgomery picked Trough Shoals in Pacolet for his enterprise; while Edgar Converse built Clifton No. 1 at Hurricane Shoals and No. 2 at Cannon's Shoals.

The Tyger rivers drew mills at Apalache (then known as Cedar Hill), Fairmont (then known as Crawfordsville), and Tucapau. Downstream from these, at a place where an island divides the Tyger River, is the site where the Hill brothers chose to locate their mill, the second textile mill in Spartanburg County. On one side of the island the water flows swiftly over a few ledges, but on the south side, the river drops, stair-stepping down over a dozen feet of exposed rock, enough gradient to once turn a great water wheel, creating 100 horsepower to distribute among the Hills Factory's looms and spindles.

Geologically, these spots on Piedmont rivers are known as "nick points," places where bands of resistant rock slow the headlong erosion of a stream. These nick points can consist of a small ridge of rock barely noticeable in the moving current or drops approaching waterfall status, such as can be seen at the shoals below the old mill in Glendale.

Spartanburg County's river shoals were significant cultural landscapes long before a few Northern industrialists wandered through the backcountry looking for fresh economic territory. Native Americans knew the shoals on Piedmont rivers offered shallow fording sites and easy constrictions for the construction of fish weirs. Villages naturally grew up near water. Transportation on Indian trading paths meant movement through river drainages. Trails headed from the coast to the mountains and crossed the Enoree River near Mountain Shoals and the Broad River at Cherokee Ford.

These shoals on the Tyger River gave birth to Hill's Factory in 1819. (Courtesy, Mark Olencki)

As soon as Europeans began to settle the interior in the 1750s, log dams were constructed and grist mills, iron works, saw mills, and a little later, cotton gins, began to cluster along significant shoals such as Nesbitt's Shoals on the Tyger, Mountain Shoals on the Enoree, Hurricane Shoals on the Pacolet, and Glendale Shoals on the Lawson's Fork. These small operations, located on what industrial entrepreneurs called "mill seats," captured the power of falling water and generated 10 horsepower or so, enough to turn a saw, grind corn, or work a bellows or trip hammer at an ironworks. These gave way to textile manufacturing, and by the end of the 1800s, almost every significant shoals in the county had its red-brick cotton mill.

A visitor to the Upstate in 1895 was impressed by the power of water in Spartanburg County. "The Pacolet River, by reason of its great volume of water, its fall and the character of its banks, lends itself excellently to mill sites; and it turns more cotton spindles and weaving looms than any other mountain stream in South Carolina," wrote visiting Northern industrialist Edward Porritt in an article he called "The Cotton Mills in the South."

Van Patton Shoals above Woodruff eventually became the site of a power generating plant for mills in the south of the county.
(Courtesy, Anthony Tucker)

The age of waterpower lasted until the late nineteenth century when the reliance on falling water was finally replaced with steam and grid electricity. In the early twentieth century mills were not tied to the rivers so directly, though many mills continued to locate near streams as a place to dispose of their effluent. Still, community boosters continued to market these shoals to industry. A special industrial edition of the *Spartanburg Journal* in September 1906 devoted two pages to the undeveloped shoals of Spartanburg, including such remote places as Ott's Shoals on the North Tyger, Linder's Shoals on the Pacolet, and Wofford's Shoals on the Enoree. Editor Gibson Catlett wrote of the farmers who owned the remaining shoals: "They have lived by these roaring cataracts for years, have witnessed the great volume of water pouring down and over them since the earliest recollections of childhood."

Today, as the mills are salvaged and leveled, their materials—bricks and pine beams—now more valuable than their location, it's possible to once again see these "nick points" as early farmers in the county saw them. The place that gave birth to Hill's Factory is now inaccessible by road, and the Tyger River falling over the shoals is the only sound in the deep woods. Cotton has given way to synthetic fibers made from petroleum, and the mills have moved to other countries, just as they once moved into this raw frontier.

Textiles in the Civil War
Shortages of Labor and Cotton

by Bruce W. Eelman

Although the Civil War brought hardship and privation to many upcountry residents, Spartanburg's manufacturers experienced a wartime boom as a result of demands for uniforms, weaponry, and ammunition.

The Coopersville Iron Works, which straddled both Spartanburg and Union districts, was under contract to supply iron to the Confederate arsenal at Charleston. Spartanburg's South Carolina Manufacturing Company also manufactured iron war materials for the Confederacy. Textile factories in the district produced essential clothing for the military while they continued to provide for the local community. By January 1863, the demand for products from the Bivingsville textile factory were reportedly "so extended that neither the machinery nor operative force [were] adequate to supply it." James Hill's textile operation and James Bivings'

Crawfordsville factory each increased production and helped make both men rich by the war's end.

Military demands and civilian shortages also resulted in a greater diversification of products made at the mills. In 1864, John Bomar's Bivingsville factory was reportedly turning out 600 wooden shoe soles per day. When the need for military clothing peaked, the mills ran short in production for local consumers, driving up prices for yarn and cloth. "The yarn fever is quite as high now as ever," farm wife Elisabeth Lipscomb, who lived along Big Thicketty Creek, reported to her sister in October 1863. Local farmer David Golightly Harris found factory yarn selling for $1.50 to $2 per bundle "& not much to be had."

Wartime production demands also drove up the value of skilled and unskilled factory labor. When a Confederate lieutenant took workers from the Coopersville iron foundry to a "camp of instruction" in October 1862, factory superintendent A. M. Latham convinced Governor Pickens to return the men "to avoid future interruption in work so important to the Government." Even those looking to make their own cloth were dependent upon skilled mechanics to provide looms. In February 1863, for example, 19 residents from a community in neighboring Greenville District petitioned the Confederate Army for the release of J. B. McDowel "for the purpos[e] of supplying this Beat with looms as he is a good mechanic an[d] says he will furnish the great demand for looms at a uniform Price of ten dollars a peace [sic]." Fourteen of the 19 petitioners were women who stressed the need to "make clothes for our children and Friends in the army."

The wartime success of the textile and iron factories suggested to both local and Confederate leaders the necessity of industry and wage labor in the South as an effective means to adjust to downturns in the agricultural economy and to prevent future dependence on the North. Confederate president Jefferson Davis had long supported a stronger industrial capacity for the South and through the Civil War viewed internal improvements as "an integral part of his program of southern resistance." Robert Barnwell Rhett, Jr., an early advocate of secession, agreed with Davis that industrial self-sufficiency was necessary "to rid ourselves of Yankee domination."

For most Spartanburg leaders, the war confirmed their antebellum views that white community progress was only possible with economic diversification. A contributor to the *Carolina Spartan* found in the Bivingsville factory one of the few successes of the war. "In these days, cheerless and gloomy," the commentator noted on January 23, 1863, "it is a matter of gratification that the improved water power of our district is doing so much to alleviate the evils of war." Besides providing for all the "temporal wants" of the district and surrounding states, the factory also provided good employment for over 60 operatives, "all of whom seem to be cheerful, and exceedingly happy."

Although generally commended for their operations, factory managers were not ones to ignore market advantages resulting from war emergencies. With Union General William Tecumseh Sherman's army operating in the state in early April 1865, Bivingsville's John Bomar and Edgar Converse urged their agents to buy as much cotton as possible since desperate planters were unloading their crops at bargain prices. "Cotton is the first thing to [be] burn[ed] by the Yankees when on a Raid," Bomar and Converse informed one of their agents, "& this is becoming a matter of alarm to the planter and it will induce many to sell to get it out of the way." Yet these calculations of profit were overshadowed by a general view of manufacturing as a contributor to the war effort and employer of needy families.

Because Sherman's march through South Carolina never reached as far west as Spartanburg County, the region's mills were able to function throughout the war without substantial interruption.

The English Manufacturing Company
Early Interest from Abroad
by Bruce W. Eelman

Following the Civil War, Spartanburg's industrial opportunities attracted both northern and foreign investors. When Edgar Converse organized the Clifton Manufacturing Company in 1880, for instance, Boston businessmen purchased stock totaling $41,000. But postwar Spartanburg's largest non-native industrial endeavor came not from the North, but from across the Atlantic.

In 1875, a group of British investors formed the English Manufacturing Company and purchased 3,000 acres along the Pacolet River from Simpson Bobo's defunct iron works. The company had the authority to raise $300,000 capital for its planned factory complex, which was to include a cotton mill of 10,000 spindles, a shoe factory, and a hosiery mill. All capital subscribed to the company was tax exempt until 1885. Part of the plan for raising funds involved subdividing the land into six-acre plots that would be sold to settlers who took stock in the operation. Company representative Alfred Peete toured both England and New England soliciting subscriptions and recruiting mechanics for the planned operation. Early in 1876, Peete secured $16,000 in subscriptions from a group in Holyoke, Massachusetts, under condition that the shoe factory would be staffed entirely by Massachusetts workers.

Locally, Peete lectured to Spartanburg's mercantile and professional community on the important link between prosperity and the division of labor through industrial diversification. Farmers were encouraged to support manufacturing because it created both internal and external markets for agricultural products. "[B]y encouraging the building of factories," *Carolina Spartan* editor H. L. Farley was convinced, "we make them the handmaid of agriculture."

Local interest in manufacturing was not restricted to wealthy merchants and professionals. Some people who lost everything of their old agrarian lives saw the potential for ground-floor opportunities in the new industrial developments. Lalla Pelot, a Newberry resident who had lost both her husband and her financial stability by the mid-1870s, reported that she was "literally living day to day not knowing where a meal is coming from tomorrow." Prior to the English Manufacturing Company's failure, Pelot placed great hopes for future stability in selling her home for $4,000 to $5,000 and investing the money in the English Manufacturing Company. In addition to the investment, Pelot wanted her family to live on the company's land where life would be "comforting." She also urged her son to find employment at the cotton factory which, when completed, would be a "great work."

Through 1876, creditors expressed great confidence that the English Manufacturing Company was moving along with "every indication of success." Yet less than half of the capital stock had been paid in by the fall of 1877 and work on the proposed operation ceased completely in December. A host of financial difficulties may explain the company's failure, but it seems clear that local investment for the endeavor was lacking. Spartanburg's growing mercantile community appeared much more inclined to invest in established firms or to start their own operations that they could control.

Three years later, just south of the English Manufacturing site, Edgar Converse purchased 250 acres and began building what the British had only dreamed of. His Clifton Manufacturing Company was an instant success and finally brought cotton manufacturing to the lower Pacolet.

Paving the Way
Iron: The State's First Industry
by Terry Ferguson

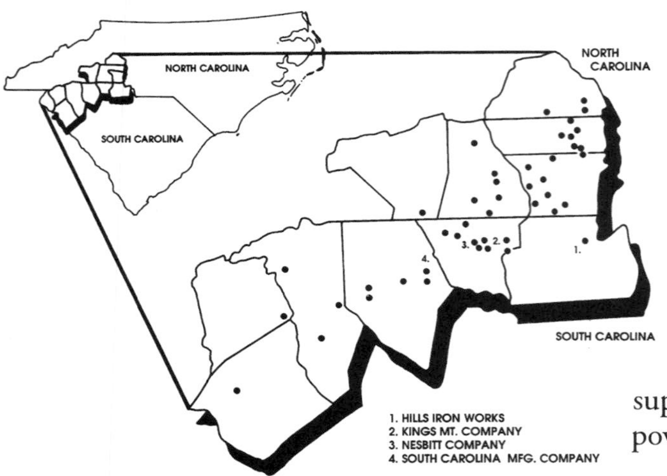

Documented iron-manufacturing sites in the Piedmont of the Carolinas (Courtesy, Ferguson and Cowan 1997)

Many people think the textile industry was the earliest manufacturing in South Carolina. But iron manufacturing, with its beginnings a couple of decades after the onset of widespread settlement of the Piedmont in the mid-1700s, marked the coming of the Industrial Revolution to South Carolina. In fact, many iron manufacturing locations later became sites of textile mills.

The iron industry developed in the Piedmont of the Carolinas due to the presence of five key resources: iron ore, marble (needed as a purifying agent), crystalline rock (for building material for furnaces and forges); abundant hardwood trees (to produce the charcoal fuel supply), and waterfalls and rapids, commonly known as mill seats, for power. These resources were only available in the Piedmont.

The earliest documented iron manufacturing in South Carolina began in the 1770s, almost half a century before the earliest known textile manufacturing operation. Between the late eighteenth and mid-nineteenth centuries, numerous ironworks developed across the Carolinas. Though some of these have been documented, the locations of others have been lost over the decades. The earliest well-documented ironworks were those of William Hill in York County and William Wofford in Spartanburg County in the 1770s.

Development of many other ironworks continued across the Carolina Piedmont into the nineteenth century, with the peak period being the late 1820s through the 1840s. Three major operations—the Nesbitt Iron Manufacturing Company, the King's Mountain Iron Company, and the South Carolina Manufacturing Company—define this period. These three companies consolidated earlier operations and controlled more than 50,000 acres in Spartanburg and Cherokee counties. Their locations contained hardwoods, ore bodies, furnaces, forges, rolling mills, cutting mills, and other associated buildings, as well as agricultural fields. Unlike the better-known agrarian-based plantations of the Lowcountry, these were plantation industries focused on iron production. Farming was done for subsistence and to supplement principal income from the iron. Slave labor, as well as wage and contract labor, was used to run these operations. In contrast to the newly developing textile industry, slaves held key skilled labor positions in many cases.

By the 1850s the iron industry in South Carolina was in decline. This was due in large part to the depletion of resources, particularly hardwood trees needed for charcoal. Another factor was competition from other regions of the country with better quality ore and coal-based manufacturing, which could produce iron three to five times more cheaply. These factors, along with the collapse of the plantation and slave-based labor systems at the end of the Civil War, ensured that no major charcoal-based iron manufacturing operation survived into Reconstruction. The collapse of the iron industry in the Carolinas left an economic void, which would be filled by the rise of the textile industry.

View of Ironworks at Cooperville, Nesbitt Iron Manufacturing Company, from east bank of Broad River (Courtesy, Lossing 1859)

With this collapse, mill seats became a valuable untapped resource, and textile companies stepped in to purchase them. Notable examples include Cherokee Falls in Cherokee County, Clifton, Glendale, and Tucapau (Startex) in Spartanburg County, and High Shoals in Gaston County, North Carolina. Proximity to water also was important when textile operations later turned to steam for power.

The first 100 years of industry in the South Carolina Piedmont was dominated by iron production and manufacturing. Textiles dominated the next 100 years. With the decline of textiles in the late twentieth century, one can but wonder what will be the defining industry for the next century.

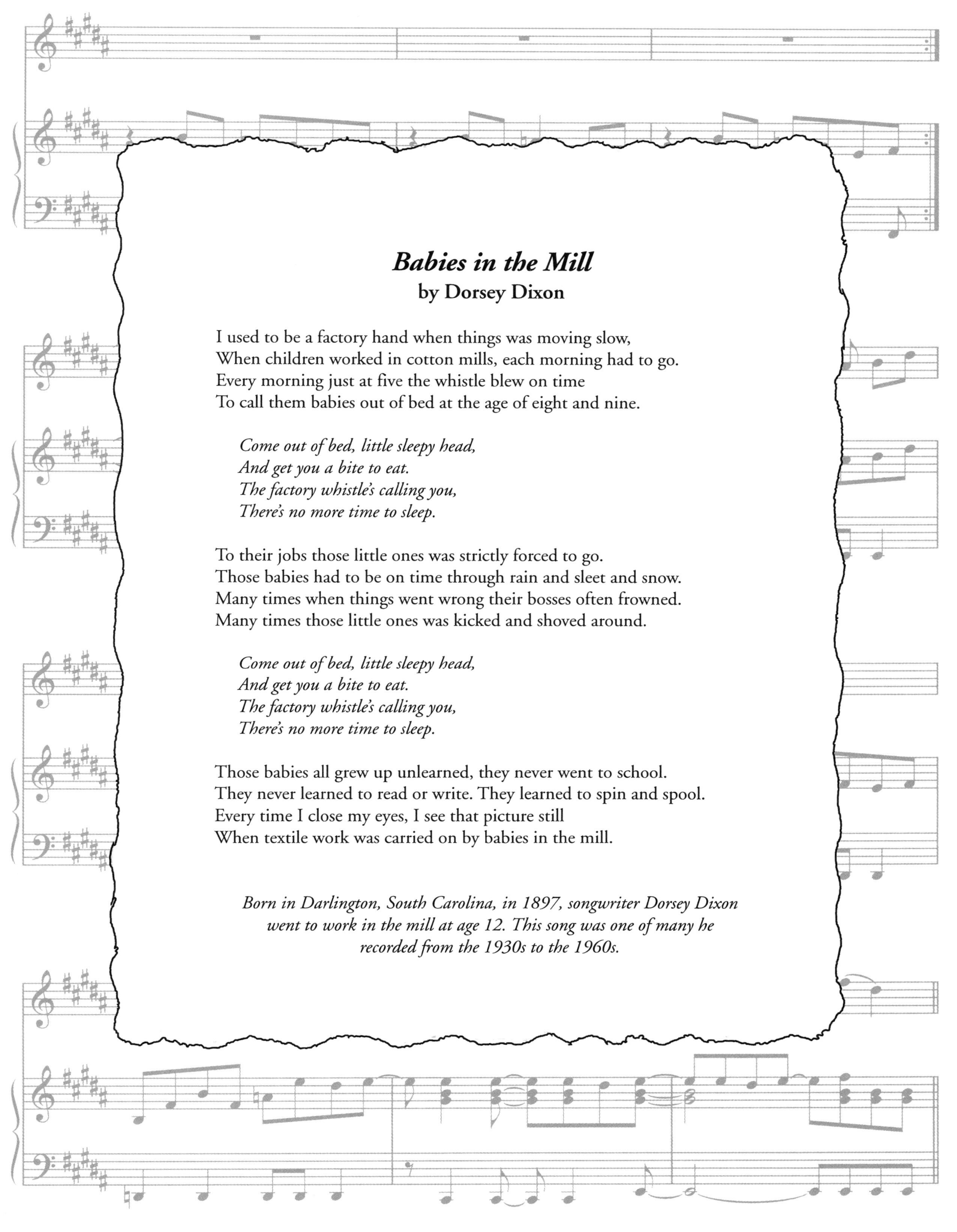

Babies in the Mill
by Dorsey Dixon

I used to be a factory hand when things was moving slow,
When children worked in cotton mills, each morning had to go.
Every morning just at five the whistle blew on time
To call them babies out of bed at the age of eight and nine.

Come out of bed, little sleepy head,
And get you a bite to eat.
The factory whistle's calling you,
There's no more time to sleep.

To their jobs those little ones was strictly forced to go.
Those babies had to be on time through rain and sleet and snow.
Many times when things went wrong their bosses often frowned.
Many times those little ones was kicked and shoved around.

Come out of bed, little sleepy head,
And get you a bite to eat.
The factory whistle's calling you,
There's no more time to sleep.

Those babies all grew up unlearned, they never went to school.
They never learned to read or write. They learned to spin and spool.
Every time I close my eyes, I see that picture still
When textile work was carried on by babies in the mill.

Born in Darlington, South Carolina, in 1897, songwriter Dorsey Dixon went to work in the mill at age 12. This song was one of many he recorded from the 1930s to the 1960s.

Almost all the spinners and spoolers at Enoree Manufacturing in 1896 were children. (Courtesy, Inman Mills)

2 Boom Time in Textile Town 1880 to 1909

by Philip Racine

During the 1870s Spartanburg investors built some cotton mills; that interest became an obsession in the next three decades. Investing in agriculture held out no promise for Spartanburg's businessmen. In the last quarter of the nineteenth century, Spartanburg shared with the rest of the South an agricultural system that discouraged innovation and investment. Landowners and farmers were caught up in a vicious circle of hope, renewed expectation, and inevitable disappointment. Before the Civil War slaves had worked the county's largest farms, but with the destruction of the slave system those farms had become too large to be farmed by one family. The owners subdivided the land into 35-50 acre plots and rented them to white and African-American tenants or sharecropper families. During the 1880s, Spartanburg County farmers grew more cotton than had been grown before the Civil War, but the price they received for that cotton hardly repaid their investment.

Tenants and sharecroppers were barely able to get by; they had little to no money of their own to buy tools, seed, food, and clothing to hold them over until harvest time. So farmers had to borrow at high interest from landowners or neighborhood stores. Even in good years, after tenants and sharecroppers had harvested the crop (mostly cotton) and repaid their creditors, purchased a few necessities and then, perhaps, a few extras such as a little candy, a new dress, a hat, or overalls, there was little to nothing left. And in the last quarter of the nineteenth century, there were few good years.

For people in the county with money to invest, agriculture seemed a dead end, and they turned to promoting industrialization. Investors recognized that Spartanburg County had a ready supply of its own cotton, abundant labor that resulted from depressed agriculture, and plentiful waterpower to drive the machinery of cotton mills, all of which combined into a mighty incentive to create a cotton mill boom.

There was a nationwide depression during the early 1880s, and Spartanburg's potential investors took advantage of the economic slump by building more cotton mills right where cotton grew. It seemed obvious that money could be saved by not having to ship the cotton 1,000 miles to be processed in New England mills. Cotton could be

An 1899 brochure used to promote Spartanburg in New England as a textile center (Courtesy, Spartanburg Regional Museum)

bought locally and greige cloth sold to customers through the large New York brokerage houses. Northern money, which eventually was invested in Spartanburg's mills, did not come from Northern mills seeking to expand but rather from Northern firms that engineered or manufactured mill machinery (Lockwood Greene) and from New York cotton brokers (Seth Milliken).

The profits of the mills of the 1870s further fueled interest in expanding the industry. Edgar Converse's brother-in-law, Albert Twichell, reflected the feelings of his fellow investors when he said: "So far as I am concerned I would put my last dollar in cotton manufacture in this state." Despite the inflation of the Civil War and the troubles of Reconstruction, there was considerable local money to spend; that money combined with interested investors in Charleston and in the North meant the ability to build mills that could easily compete with Northern establishments. Spartanburg had much to offer potential investors—in addition to the availability of money for investment, there were relatively low labor costs (in 1890 the average weekly wage of a cotton mill worker in Massachusetts was $8.05 compared to the $5.17 earned in South Carolina), and a potential workforce of entire families used to working hard all day long, caught in an unrewarding agricultural system. Also the South had little interest in labor organization, lower construction costs than anywhere else in the nation, and much potential waterpower.

> ***"We felt no danger from the South until 1880. In that year, I called attention of my stockholders to the position in the South. The cloud was no bigger than a man's hand; but it was there and it was threatening us."***
>
> **—New England textile industrialist Edward Porritt**

> Mr. Editor:
>
> Do the people of Spartanburg realize that within the last two years there has grown up in their midst a town of more than one thousand inhabitants? Do they realize the practical and financial advantages of such a town to any county?...Sir, do the editors of the Spartan and the Herald realize that there is such a place and if they would take the trouble to visit the place they might secure a few subscriptions to their county papers?
>
> *—Letter appearing in the Carolina Spartan in 1882 describing the emergence of Clifton as a town*

The local newspaper, the *Carolina Spartan*, a fervent supporter of industrialization, published news stories and opinion pieces predicting a happy, prosperous future built on cotton mills. In 1880, Edgar Converse, owner of Glendale Mill and an enthusiastic promoter of the building of even more cotton mills, made a trip north to buy machinery for his new mill, Clifton No. 1, which he wanted to build on the Pacolet River. In covering that story the *Spartan* wrote: "Instead of bleak, barren hills, a thrifty population with comfortable houses and productive farms, will be seen...a magnificent factory will add to the wealth and prosperity of the County." Such were the images created by local developers to persuade Spartanburg's residents that mills would improve the condition of the county. Before the Civil War, many leaders of the South had worked to prejudice their fellow Southerners against industrialized economies. The leadership class had argued that industry violated Southern values by destroying and polluting the land and by creating "wage slavery" that took advantage of workers and then threw them away once their usefulness was gone (in contrast to Southern slavery that allegedly "took care" of slaves from birth to death). In these ways, Southern leaders had argued, industrialism was contrary to the image of the planter, the small farmer, and the system of slavery that was the backbone of the Southern "way of life"

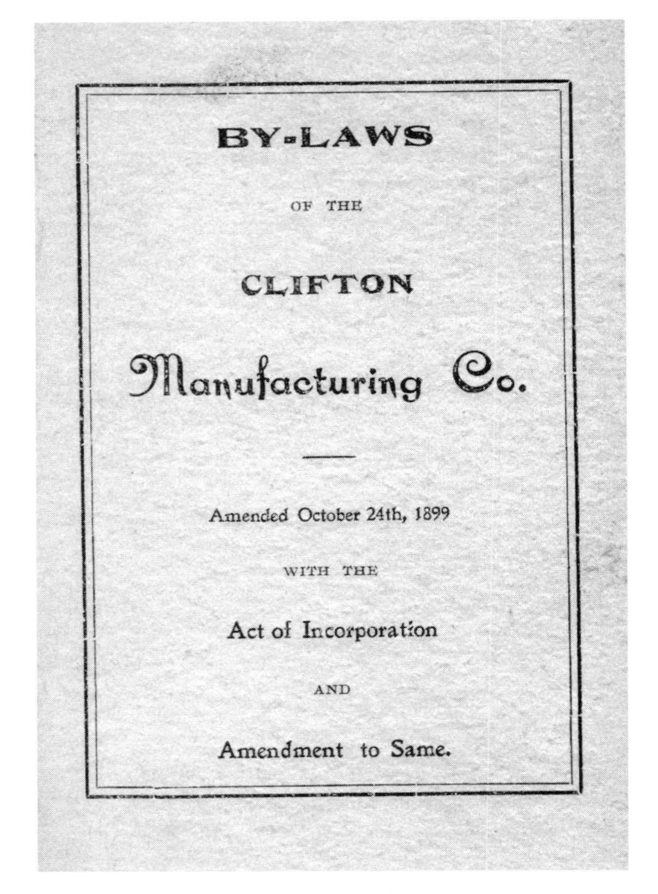
BY-LAWS

OF THE

CLIFTON

Manufacturing Co.

Amended October 24th, 1899

WITH THE

Act of Incorporation

AND

Amendment to Same.

By-laws of Clifton Manufacturing, 1899
(Courtesy, Converse College)

"Mr. Converse does not stand around with his hands in his pocket and wish for Northern men and Northern capital to come down this way and build up our waste places, but he goes to work..."

—Carolina Spartan, January 7, 1880

that so many Spartanburg white men had defended during the Civil War. After the war, Spartanburg's leaders reversed themselves, boosting industrialism as the cure for the county's economic problems.

Persuading the general population that cotton mills under local Southern supervision would be beneficial was only part of what had to be done. To make the cotton mill enterprise workable, it was essential to improve transportation, and Spartanburg's business community worked methodically to bring the railroads to town. Having brought the Spartanburg & Union Railroad to the town in 1859 and the Atlanta & Charlotte Air Line (Southern Railroad) in 1873, Spartanburg's business community added the Spartanburg & Asheville (opening South Carolina to the rich Ohio River Valley) in 1880 and the Carolina, Clinchfield, & Ohio in 1909. By 1910, Spartanburg could receive goods from and export them to all four points of the compass, a significant accomplishment for the cotton mill industry. Since most mills were not located directly on the main railroad lines, some owners built connecting lines from their mills. For instance, Edgar Converse built such a connecting spur from his Clifton Mills along the Pacolet to the Southern Railroad. Other mill owners considered it only a minor inconvenience to transport their finished goods to the nearest station.

The construction of Spartan Mills in the early 1890s was a typical example of the process of creating a new cotton mill. In preparing for the new mill, Captain John H. Montgomery, who had made money in Spartanburg selling fertilizers to his fellow farmers, raised considerable capital from local investors and then counted on his ties with Northern businessmen for much of the rest. It was normal practice for people who were planning on building a mill to advertise a subscription to raise capital in the local newspaper. Nearly a decade earlier, Captain Montgomery had secured for his Pacolet Mill an investment of $10,000 from Seth Milliken of New York City, a selling agent for several Southern mills (at that time Milliken had said that it would either be the best or the worst $10,000 he had ever spent), and in 1890, for his Spartan mill project, Montgomery once again went North for part of his capital. Also, since there were no Southern manufacturers of cotton mill machines, he bought his equipment from Lockwood Greene, a Northern engineering firm.

Northerners were a significant part of the Spartan Mill enterprise—four of the nine members of the board of directors were from the North. When asked what prompted their interest in Southern mill development, these Northern investors spoke most often of cheaper labor and construction costs. Throughout the twentieth century, when investors commented on the advantages of putting money into Spartanburg, they would mention "cheaper labor" over and over again. Spartan Mills offered to build within the city limits—in the 1890s there were no mills within the city—if the city exempted the mill from

During the construction of Spartan Mills, between four and five million bricks were manufactured under the direction of Thomas M. Bomar, an African American who was also a stockholder in Spartan Mills and Pacolet Mills.

Spartan Mill as seen across the mill pond, circa 1900. (Courtesy, Walter and Betty Montgomery)

The following is newspaper editor Charles Petty's account of the completion of the Spartan Mills smokestack, then the tallest of its kind in the South.

The smokestack of the Spartan Mills, which has been rising slowly for weeks, was completed last Saturday. Being on low ground, one cannot appreciate its height until he gets near the base. This is by far the tallest structure in the city. The diameter of the foundation is 40 feet...the diameter of the top 12 feet, 8 inches height, not including foundation 182 feet. The building of this stack was a very particular job and was considered somewhat dangerous. Thomas M. Bomar was in charge of the work. Some of the masons, who began the structure with him, abandoned it when they began to get so high that men on the ground looked like little boys...Captain Montgomery is to be congratulated on the progress he has made in this work. Mrs. Montgomery prepared an elegant dinner, consisting of turkey, ice cream cake and many other good things, which was served on the top of the stack last Saturday. The editor of the Spartan was invited by President Montgomery to partake of this "high dinner" but pressing engagements prevented his acceptance. Of course he was not afraid to go to the top of that structure. He would go there just as quick as President Montgomery would go.

—April 9, 1890

taxation for 20 years. The city agreed to what would become a typical incentive to investment in Spartanburg, city or county. Because of sharp dealing and the advantages granted to investment groups by the city of Spartanburg, investors in the area succeeded by 1895 in building mills to house 30 percent of all the spindles in South Carolina. By 1901 the county contained more mills than any other county in the state.

From the 1880s to the turn of the century, investors earned significant returns on their investments in cotton mills. In the 1880s, annual dividends ran from 22 percent to 75 percent. Even in hard times, such as the period around 1900, profits hovered around 8 percent. And the mills kept building. Clifton No. 1 kicked off the building boom in 1881, coinciding with the construction of Cherokee Falls Manufacturing Company (which was part of the Spartanburg District at that time). From there,

1883: Pacolet Mill
1888: Pacolet No. 2, Enoree Manufacturing Company, Fingerville Manufacturing Company (rebuilt), and Whitney Manufacturing Company
1889: Cowpens Manufacturing Company, Clifton No. 2
1890: Spartan Mills, Beaumont Mills, and Fairmont Yarn Mill
1891: Pacolet No. 3
1892: Gaffney Manufacturing Company
1895: Victor Mills, Pelham Mill, Island Creek Mill, Gaffney Mill No. 2, Arlington Mill (an expansion of Cedar Hill Factory)
1896: Arkwright Mills, Tucapau Mills, Converse Mill, and Spartan Mill No. 2

Spartan Mills with cotton fields and railroad tracks in the foreground. (Courtesy, Walter and Betty Montgomery)

Directors of Spartan Mills, 1880

Seth M. Milliken, *New York*
John W. Danielson, *Providence, Rhode Island*
J. L. H. Cobb, *Lewiston, Maine*
Stephen Greene, *Newburyport, Massachusetts*
Albert H. Twichell, *Spartanburg*
D. R. Duncan, *Spartanburg*
John B. Cleveland, *Spartanburg*
W. E. Burnett, *Spartanburg*
John H. Montgomery, *Spartanburg*

Capt. John H. Montgomery, president of Pacolet and Spartan Mills (Courtesy, Walter and Betty Montgomery)

1899: Woodruff Cotton Mill, Valley Falls (rebuilt)
1901: Inman Mills
1902: Saxon Mill, Drayton Mills
1903: Arcadia Mills
1906: Jackson Mills
1907: W. S. Gray Cotton Mill in Woodruff
1910: Chesnee Cotton Mills

Of the 21 million spindles in the United States in 1901, 6 million of them were in the South, 1.8 million in South Carolina, and 555,000 in Spartanburg County. The New England states had about 13 million spindles and dominated the industry, but the South's cotton mills grew steadily. In Spartanburg County alone, the number of spindles had grown from about 8,000 in 1880 to half a million by the turn of the century. Although the South did not surpass New England in the total number of spindles until 1925, the rapid growth before 1910 accurately foreshadowed the future domination of the industry by the Southern states, with Spartanburg County playing a central role in that domination.

Indeed, investors built at least 37 mills in Spartanburg County between 1880 and 1910. In 1901, Gibson Catlett, editor of the *Spartanburg Herald*, wrote about this remarkable building spree in a publication for the South Carolina Interstate and West Indian Exposition held in Charleston:

The mill hill of Pacolet (Courtesy, Pacolet Elementary School)

"Beginning in a small way a few years ago, these mills have paid regular dividends and doubled up from time to time out of accrued profits until the dollars first invested are now worth thousands, and the little factories have become industrial giants." Such reports excited continued building, yet so many mills appeared so quickly that by 1906 there was a severe labor shortage. After hunting for laborers all over the Upstate, recruiters tried importing workers from Europe. After all, at that time immigrants were flocking to the Northern and labor unions were a significant factor in their politics. Immigrants seeking industrial work would find none of these conditions in Spartanburg.

So labor recruiters went to the mountain regions of western North Carolina and especially eastern Tennessee where they found many mountain people eager for any steady employment. The meager wages of the cotton mill industry seemed to them a blessing. It was steady work—it included housing, and the family wage was more than could be earned on a farm. As a result, the folks from

Early Spartanburg County mill mechanics about 1905 (Courtesy, Martin Meek)

cities of the United States. But the workers recruited for Spartanburg—especially from Austria and Belgium—did not work out. Spartanburg did not seem an inviting place to Europeans. There were few immigrants in the region (slavery had discouraged immigrants before the Civil War). The area had no history of large-scale immigration except for English, Scotch-Irish, and Germans, each group having come to the region in significant numbers more than a hundred years previously. Most importantly, wages were low by Northern standards. In addition, many European countries had significant social welfare legislation (wage and hours laws, child labor laws, and old-age benefits),

The total number of spindles in Spartanburg County increased from 8,000 to more than 400,000 between 1880 and 1900.

—Author William W. Thompson Jr.

western North Carolina and eastern Tennessee flocked by wagon and by train to upstate South Carolina. Typically, workers recruited from the mountains signed a contract with their new employer, such as this one signed January 5, 1882, by Robert Atkins, hometown unknown.

This Witnesseth, That I, Robert Atkins, do this day contract with Clifton Manufacturing Company to move my family to Clifton, S.C., and occupy one of the tenements at that place, and for the entire year of 1882 furnish my faithful services when wanted, and also that of my children as operatives in their Factory, and do hereby bargain to so continue in their employ until all advances made to me by them are paid in full. I also agree to submit to and support all Rules and Regulations existing in the factory and on the place customary in such cases.

Although this contract dates from the 1880s, the nature of such contracts remained the same over time. Mr. Atkins noted on the agreement that he was the father of four girls, ages 16, 15, 14, and 13.

Sometime after these intense efforts by recruiters to attract laborers from the mountains, a survey of Spartan Mill village indicated "75 percent of its population came from Polk County, N.C." Any concerns the workers from the mountains may have had about competition from African Americans or the possibility of labor trouble were laid to rest by advertisements like the following: "Labor of purest Anglo-Saxon stock; strikes unknown." The same message appealed to investors and these potential workers alike.

The mill village was the heart of the cotton mill industry. Since Southerners were mostly farmers, they had lived separated from one another, and their towns and villages had been small and few. There were no concentrations of population to constitute a workforce for a mill. Before steam power, mill builders needed water to power their machinery so they had to build their mills on rivers and construct nearby villages to house their workers. Such a simple idea created a wholly new social relationship within the white community. The industrial model of employer-employee, boss-worker, upper class-lower class came to Spartanburg County. Although similar relationships had long existed between farm owner and farm worker and between large farmer and small farmer, industrialism changed the degree of dependence and the distribution of power that made up the relationship. Mill work was different, but it was not just the nature of the work that changed. The impact and full extent of that difference would become increasingly obvious only with time.

The villages proved useful for more than just keeping a workforce together. Because of the village, workers did not have to spend time and money getting to work; mill owners could offer inexpensive and flimsy houses as part of the wage, and the mill store could provide for almost all the needs of the mill villagers, including credit on the owner's terms. In addition, having workers live in mill housing gave owners and superintendents extra leverage to impose their own ideas of morality and behavior on workers and their families. Finally, having workers live in mill housing provided a ready threat in case of trouble, for instance when laborers tried to organize. Even Spartan, Beaumont, and Saxon Mills, all located in or near the city of Spartanburg, built villages.

The first of these city mill villages was Montgomeryville, the original name of Spartan Mill Village built concurrently with the mill in the early 1890s. In his 1890 report to his stockholders, Captain John H. Montgomery wrote: "No one business enterprise has ever entered the town limits

WANTED

500 Operatives to Work in a Cotton Mill

The Pacolet Manufacturing Company of Pacolet, S.C., can furnish steady employment for over 300 days in the year for boys and girls over age 12 years old, men and women at average wages as follows:

- Experienced 12 to 16-year-old boys and girls from 50 cents to $1.25.
- Experienced boys and girls over 16, and men and women, 75 cents to $1.50.
- Old men, 60 to 70 years old, 75 cents to $1.

Beginners can make enough for a good living, and, as they become experienced, will increase their wages. In a short time, they will become experienced enough to draw regular wages. Some beginners have come here, and after three days, were making $1 a day.

We furnish you good, comfortable houses at 50 cents a room per month. We furnish you wood, coal, and provisions laid at your door at market prices. Pacolet Mills houses are located on a hill and the place is noted for its health and is free from all malarial diseases. Only a short distance from the North Carolina mountains. We have good water, a splendid system of free schools, churches of different denominations; in fact everything that appeals to one who wishes to improve the condition of his family. Our mill is heated with steam and is warm and comfortable in the winter. We make coarse cloth and can have our windows open in the summer to give nice cool air through the mills. If you are a poor man there is no better location for you to select than Pacolet. It behooves every man to either educate his children or place them in positions to learn good trades. If you are only able to give your son or daughter a common school education, and then they teach school, they can make thirty-five dollars per month for four months in the year. If he or she could learn to be a good weaver, $40 a month year round could be their wages and have a good comfortable home, and no exposure to the rain, snow or cold. We will advance you your transportation, and if you remain with us six months, the same will be given you.

We want families with at least three workers for the mill in each family.

If you are interested write us and tell us how many you have in your family; how many whole tickets and how many half tickets you want, where you will take the train and name of your depot agent and on what date you will start, giving us plenty of time to send your depot agent transportation for you.

PACOLET MANUFACTURING COMPANY

This poster, which circulated in the mountains of North Carolina and Tennessee, was reprinted by author August Kohn in his book, The Cotton Mills of South Carolina, *published in 1907.*

of Spartanburg bringing with it such an influx of population, increase of business, and expenditure of money as that of Spartan Mills. You should be proud of it... and help to encourage and elevate the people who have thus come among you, recognize the dignity of the labor, and help make those who live by it feel they are citizens of the town, and with

> *"We have got the best people as operatives that can be found in any factory town in this country. They are industrious, intelligent, kind-hearted people, good citizens and with every principle pregnant with splendid ideas of good citizenship. A better class of people does not exist in this country."*
>
> **—John H. Montgomery, commenting on Pacolet workforce**

The shop men from Tucapau Mills, 1909. From left to right: George Hill, C. E. Howe, J. O. Miller, L. J. Davis, J. E. West, Tom Davis, and Adam Bright. (Courtesy, Junior West)

good behavior will be respected accordingly."

Although these were fine sentiments, things did not actually work out that way. That Captain John bothered to make his point so forcefully indicated that he did not expect townspeople to give his workers respect so easily. Montgomeryville had 152 cottages for workers and superintendents, making it a substantial suburb on the edge of town. Former residents of Montgomeryville recalled its early days in an article published 50 years later in the *Spartanburg Herald*:

> **When the first mill was constructed, a wooden office and a number of "boxcar" houses were built. These houses were lit by lamplight and there were deep gullies along the alleys that made travel at night unsafe. The town itself had few amusements and according to J. N. Dye, master mechanic who has been with Spartan Mills for 45 years, there was nothing for the workers to do "but walk to town and back." On Saturday nights, Mr. Dye reminisced that the crowd stayed in the barbershop talking until 11.**
>
> **When the mill first started, the working time was from 7 a.m. to 6 p.m. with a half-hour for dinner and until 4:25 on Saturdays. Houses were rent free. Weavers were paid 65 cents a day. Many of the spinners were 10 years old and were paid 8 cents to run one side. One employee who began spinning when she was 10 recollects that she was paid 64 cents a day to run 8 sides.**
>
> **One other diversion that Mr. Dye recollected was buggy riding. He recollects that the young men saved their money so that every 2 or 3 weeks they could rent a horse and buggy for $1.50 "to ride our girls around the country."**

In 1907, the village was wracked by catastrophe. A fire broke out that destroyed an entire block of houses and left nothing but a row of chimneys standing like the remnants of a burned forest. Miraculously the fire did not touch the Duncan Memorial Methodist Church, which sat right in the middle of the devastation. When Walter S. Montgomery, the owner of Spartan Mills and the first generation of three men to bear that name, toured the burned-out area, he paused at the church and said: "This must indeed be holy ground." Mr. Walter carried on his father's indomitable spirit.

Typically, mill villages had several rows of identical unpainted houses on dirt roads, each with an outhouse in the back and access to a well. In the last decades of the nineteenth century, most of the people who came to live in these villages had been farmers who came by necessity. Most had been unable to continue farming because they simply could not survive on the depressed land, and now they had to learn a new way of life. They

> *"Spartanburg is the cotton mill centre of the South. Here the industry was born in the South and has held a constant lead."*
>
> **—The Exposition, April 1901**

Many early village houses, like this one in Glendale, were one-room "shotgun" houses (so-called because you could fire a gun from the front door to the back). Some also called them "boxcar" houses. (Courtesy, Clarence Crocker)

had been used to working alone, able to set their own routine according to their knowledge of the land, weather, and crops. They had worked at their own pace from dawn to dusk in some measure of love/hate relationship with nature—the land and the weather could be bountiful, but they could also be cruel. They had worked as a family, each with her or his assigned part of the chores that changed as the seasons changed. The work had varied according to the state of the crops, sometimes slow and routine but at other times anxious and frenzied, such as at harvest or hog-killing time. The

SOUTH CAROLINA.

No 1641 — 50 Shares

PACOLET MANUFACTURING COMPANY.

PACOLET

CAPITAL STOCK $300,000.

3,000 SHARES $100 EACH.

This is to certify, that John H. Montgomery is entitled to Fifty Shares of the Capital Stock of the Pacolet Manufacturing Company, transferable only on the books of the Company in person or by Attorney duly authorized on surrender of this Certificate

PACOLET, S.C. May 17th 1884

Treasurer — John H Montgomery — President & Treas

A stock certificate issued by Pacolet Mills founder John H. Montgomery to himself, 1884. (Courtesy, Bill Lynch)

move to the mill village changed much of that life.

Living in the mill village meant abandoning self-sufficiency. Families became less dependent on homemade clothes, on food from the garden, on the father's firewood. There was a company store for much of that; in any case the mill provided it. There were next-door neighbors with whom each family shared good and bad times as well as an inflexible routine. Mill and village drastically reduced nature's role in family life. The mill whistle marked the time of waking and the beginning and ending of work. The continuous racket of the mill, especially the weaving room with its sharp clanging of racing shuttles, replaced the relative quiet of farm labor. A reporter from the *Carolina Spartan* reported about Clifton No. 1 in 1885: "With 542 looms in constant motion, slamming and banging, clicking and kicking, working and jerking, unwinding and rebinding, unfolding and rerolling, the busy noisy machines keep up such an everlasting clatter that one is induced to wonder if the bedlam of the infernal regions can out do a weaving room." The cotton dust, which at times made operatives disappear to their fellow workers down the line (this was especially true of the carding room), replaced the clean air of outdoor work; the almost constant attention of supervisors replaced the solitude of farm labor.

One thing that did not change, one habit that transferred directly from the farm to the mill floor, was a dedication to work. Working on a farm was constant and hard. It had to be, for it meant the difference between being fed and being hungry, though it did not ever mean being anything but poor. In that agricultural life a work ethic had developed that went hand in hand with the ethos of evangelical Protestantism. People were put on this earth to work, and they would reap what they had sown. Moving from the farm to the mill village changed none of that. If the

"When I came from the farm, I had nothing. You could have carried my things on your back. Now I would not be ashamed for any man in the state to come to my house, or to sit down at my table."

—Anonymous quote in the *Carolina Spartan*, 1885

CLIFTON MANUFACTURING COMPANY.
$5.00 CLIFTON, S. C., Mch 31st 1897
RECEIVED OF A. H. TWICHELL, TREASURER,
Five DOLLARS,
D E Converse on a/c Paid to the Clifton Orchestra
D. E. Converse

CLIFTON MANUFACTURING COMPANY.
$18.15 CLIFTON, S. C. 10/8 1907
RECEIVED OF A. H. TWICHELL, TREASURER
Eighteen + 15/100 DOLLARS
Transportation on A B Murrell's family

These receipts are from Clifton Manufacturing. The bottom receipt for $18.15 was for "transportation on A.B. Murrell's family" to Clifton Mills in 1907. (Courtesy, Mike Hembree)

On March 26, 1907, fire roared through the mill village of Spartan Mills, leaving 300 homeless. The following morning, the Spartanburg Herald *ran this statement from mill owner W. S. Montgomery at the top of its front page:*

"Fifty-three dwellings houses were destroyed containing 372 rooms at an average cost of $100 per room. The Fitting school buildings were of brick and cost considerably more—I should say twice as much as one of the dwellings.

"I have always been against my mill people carrying petitions to the city for help but this disaster is so broad and the cause for which the petition that will be circulated today is so worthy of support of all people that I gave my consent for them to ask aid of the city. In fact, I think it is so worthy a cause that I gave personally $200. Mr. Britton, the superintendent, gave $100 and there are a number of other large gifts.

"The board of all the sufferers who work in the mill will be paid by the company until their new homes are ready."

In 1907 an entire block of houses burned in the Spartan Mill Village, leaving only chimneys. Vernon Foster tells the story that when Walter S. Montgomery visited the site after the fire and stood on the church grounds, he was heard to say: "This must indeed be holy ground." (Courtesy, Herald-Journal Willis Collection, Spartanburg County (SC) Public Libraries)

Mill hill residents in Enoree shared a common well, shown at center in this 1895 photograph. (Courtesy, Inman Mills)

knowledge of industrial work was substituted for knowledge of the land, crops, and weather, it did not alter the need, the determination, and the insistence on working hard and long. The Lord expected, and the mill owner welcomed, no less.

Initially, workers chose to live in the mill village because of the lure of a steady income. In 1910, 87 percent of the South's mill workers lived in mill villages. Once in the village, however, workers actually had little choice. The low or non-existent rent was necessary because of the low wages. If the mill owner charged rent and the mill temporarily had to shut down, he often stopped the rent in order to keep his workforce intact. Owners often required that everyone who lived in mill housing work in the mill, at times even children. The living conditions in the villages varied considerably from village to village. Some, such as Pacolet village, eventually became showplaces with recreation halls, grass and flowers, and organized activities, and one suspects they were often constructed to show to "outsiders" who came south to "investigate" mill conditions. But much of that activity came after 1910; before that time most villages were drab places. Of course, living on a farm would not have been physically all that different. But these conditions remained unchanged long after improvements were made in nearby towns and in the city. The difference between mill villages and the city of Spartanburg became obvious early on. In the 1880s many mansions were built on East Main, South Pine, and Church streets. Owners of mills and their villages typically lived in the city, not in the village. Even Edgar Converse and Albert Twichell, who lived in Glendale for 20 years, eventually moved to town. James Alfred Chapman drove a horse and buggy to the mill in Inman from his Spartanburg home during the early years of the mill. Later he bought one of the first Cadillacs in

Within a year of its opening, Clifton No. 1 employed four times as many people as Glendale, which had held the top spot for the previous 40 years.

1899 Work Rules at Spartan Mills
written by an anonymous overseer

Don't Let the Spinners throw Whit Cotton and Clean Wast on Flors
Don't Let Spinners get in Back ally and talk in bunches.
Don't Let Spinners work unless got a Big Pocket on apron.
Don't Let Doffers Nock off Filling on nothing But Lether or Ruber.
Don't Let Eney one fuss in Side of the Mill or in Side of the Fence of the Mill.
Don't Let Sweepers Leve Flore Durty at Stoping time.
Don't Let the Sexian Hands Fale to Do his Duties a tale.

—From a weekly time book provided to author William W. Thompson Jr. by Henry Calvert

Almost half of exports of U. S. cotton mills in 1899 went to China. Many Southern mills, including Spartan, Whitney, Clifton, and Pacolet, were deeply dependent on Chinese business.

the area and was immediately recognizable on his commute. By the end of the nineteenth century, many parts of the city of Spartanburg enjoyed city water, paved roads, electric trolleys, and electricity. The difference between the physical conditions in the mill villages and those in the city of Spartanburg did much to form the prejudices against the mill workers felt by Spartanburg's city residents.

There were other factors besides the physical look of the villages that contributed to their quality of life. For instance, a considerable number of mill workers were sickly. In 1897, a smallpox epidemic swept through Spartanburg's mill villages. By 1900, many suffered from hookworm (estimates for Spartanburg County mill villages set that number at over 30 percent), usually infested through bare feet walking around the outhouses behind the village homes. Hookworm caused diarrhea, listlessness, and general fatigue. As late as 1916, a study of seven Spartanburg mill villages showed that only two had sanitary waste disposal systems. The other major health problem was pellagra, which began to rear its head by 1905. Workers with the mysterious disease suffered from diarrhea, a reddening of the skin on the arms, hands, feet, and face that crusted over and peeled away, leaving telltale butterfly lesions on the face. In severe cases, it caused insanity and death. Ten years would pass before scientists began to unravel the connection between mill village diets and pellagra.

Both hookworm and pellagra existed on farms, but country people were relatively isolated; especially in the case of pellagra, those who suffered need not go to town. Mill villages plagued by hookworm and pellagra had concentrated populations that made the health conditions all too public, and so the mill village easily became associated with them. The "laziness" that city people attributed to folks who lived in mill villages typically resulted from these two health conditions.

"A" Street in Inman, October 1901. The mill had not yet opened. (Courtesy, Inman Mills)

MILL OWNERS ALSO RECOGNIZED the need to provide for the spiritual well being of their workers. Mill owners provided churches for the village, usually Baptist or Methodist. In most of these churches the mill owners paid the ministers, and the ministers preached a message similar to that once preached

A streetcar line was built from downtown Spartanburg to Glendale, Clifton No. 2, and Converse mills in 1900. The streetcar ran until 1935 when it was replaced by buses. The bus line ran to Glendale until 1992.

—Author Mike Hembree

by white ministers to antebellum slaves: "This is a good life, or the best you can get; look forward to the afterlife as your reward for obeying and rendering deference to your bosses, and respect authority." As an example of the tone of Sunday meetings in these "official" churches, one Spartanburg minister was quoted in the *Carolina Spartan* as saying: "If there are such things as wrongs existing under the present labor system, I know of no men more capable or more sure of righting their wrongs than

An artist's rendering of Drayton Mill, organized in 1902. (Courtesy, Bill Lynch)

Spartanburg, S.C. June 28th 1894

With great pain the Directors of Tucapaw Mills announce to you the death of our late President, Dr. C. E. Fleming, which event occurred on Saturday, June 23rd, 1894. Suddenly and without warning, while the President was on his way to the work at Tucapaw.

At a meeting of the Directors held June 25th, Mr. Thomas E. Moore, of Wellford, S.C., was elected President of the Company, and Mr. H.E. Ravenel, of Spartanburg, was made Treasurer.

Mr. Moore is in the very prime of life, in robust health, and is a gentleman of high character, excellent judgment, good executive ability, and unlimited energy. He has the confidence of this community. He was Vice-President of the Company and had been giving valuable assistance to Dr. Fleming in the construction. We feel sure that under Mr. Moore the Mill will be well and cheaply built, and he has youth and adaptability enough to fit himself for future duties by the time the machinery is in the mill.

The work is well under way, the foundation of the Mill dug out, a great deal of lumber delivered, 400,000 brick ready for burning, and a large amount of preliminary work of all sorts is done, such as cleaning up, cutting wood etc., etc.

During a recent visit North, Dr. Fleming had received so much encouragement in the way of contract for machinery that the proposition is before the Stockholders to increase the Capital Stock from $150,000 to $300,000.

The subscriptions now are more than the amount first proposed to be raised, and it is hoped that each stockholder will interest himself in procuring the increase.

D. E. Converse,
Jno B. Cleveland,
H. E. Ravenel,
Committee of Directors of Tucapaw Mills

Children dominated the spinning room at Saxon Mills in 1908. (Courtesy, Sue White)

the noble mill presidents. I, for one, am willing to confide it to their wisdom and goodness. Let us leave it to them." Recently, historians have emphasized the existence of Holiness or Pentecostal churches on the periphery of many mill villages. These churches did not receive funds from the mill owners, and in times of labor trouble their preachers were much more likely to oppose the owners in favor of the workers. It appears as if many workers who were on the church rolls of Baptist or Methodist congregations actually attended these Pentecostal churches, or often, the services of both.

The Northrop loom was introduced in 1894. It reduced weaving labor costs by half, allowed the weaver to tend more looms, had an automatic thread changing device, and reduced cotton dust. This loom hastened the industry's move to the South.

Education, meanwhile, was not a highly prized commodity in the mill village in the early days. In the "model" mill village of Pacolet there were four teachers for 400 children. Parents needed their children to work in the mills as soon as they possibly could, for the "family wage" depended on it. In 1880, about 25 percent of cotton mill workers in Spartanburg were children under the age 16; some 30 years later, that figure had only dropped to 19 percent. A father's mill wage was not enough to support a family so the wife and the children had to work, and the total amount they brought in as a family was more than they could make on a farm. In the 1880s, a typical farm family in the South made about 54 cents a day; a male mill worker made about 53 cents. However, a wife made about 30 cents, and children earned from 12+ to 25 cents a day. So a family of five could make as much as $1.58 a day, which was three times what a farmer could make.

Not only was it necessary for wives to work to make up the total family wage, but also for some jobs the cotton manufacturers preferred female to male workers. Ever since the beginning of cotton mill manufacturing in the county, mill owners had tried to hire female workers. The owners and their supervisors and superintendents believed that women were more reliable, more industrious, worked more quickly, were neater, and possessed greater manual dexterity, and thus were less

Several of the mills provided schools for village children at the turn of the century, though the schools quickly became overcrowded, with student-teacher ratios of 1:100 or higher. According to a 1900 report by the State Superintendent of Education, Pacolet had 369 students and three teachers; Clifton had 716 students and six teachers; Glendale had 125 students and one teacher; and Whitney had 99 students and one teacher.

They not only worked for less money, they also worked more hours. In the 1880s, Spartanburg's mill laborers worked a 16+-hour day—from 4:30 a.m. to 9:00 p.m. By 1900, that day had dropped to 11 hours, six days a week. Many

likely to break thread. Most important, however, mill owners knew that females could be paid less than males and believed that they were less likely to strike. Superintendents also liked having children around. In addition to getting valuable work for very little money, small children came in handy when something had to be done to a machine where space was limited. Children could be expected to work, for the father and mother were around to keep the children "in line." Family discipline was maintained. When the state Legislature passed child protective legislation at the turn of the century, children continued working, being then known as "helpers" and therefore not having to be entered on the company's books. Everyone worked: father, mother, children—all in all, the "family wage system" is what brought people to work in the cotton mills.

In 1910, Spartanburg mill workers had a more dependable and greater income than they could make on a farm, but their incomes were about 46 percent less than comparable workers in other parts of the nation. If you factor in that it cost 8 percent less to live in the South than in other places in the nation, Spartanburg's cotton mill workers were making 38 percent less than workers elsewhere.

The Graded School in Pacolet Mills was built in 1896. Charleston author August Kohn wrote, in 1902, "Certainly there is no better school than in any of the graded school systems in this State."

Watson School, a one-room school near the village of Enoree, in 1904 (Courtesy, Anthony Tucker)

Tucapau Mills, 1895 (Courtesy, Junior West)

workers were aware that they were being exploited, but given the alternative of farm work, what could they do?

Some cotton mill workers, believing these wages and hours unfair, explored the possibility of forming labor unions. As early as March 1886, Edgar Converse, the president of Glendale and Clifton Mills, received a letter warning him of union organizers coming to Spartanburg. In August, the Rev. J. Simmons Meynardie, who had organized a Knights of Labor strike in Augusta, Georgia, showed up in Spartanburg to start organizing. The Knights of Labor was an early labor organization somewhat like a combination of a labor union and the agricultural grange. The *Carolina Spartan*, a supporter of the mill owners, denounced Meynardie and his efforts. Several workers at Converse's Clifton Mill joined the Knights and had a meeting with the mill president at which they asked him to recognize their organization. Converse refused and locked these workers out of the mill and evicted them from their mill village homes. The workers hired a young Spartanburg lawyer, Stanyarne Wilson, to contest the eviction. The workers lost the case, and by December they left the state to seek "an area more hospitable to organized labor." Converse's actions broke the back of the Knights of Labor in Spartanburg.

Disappointed at the outcome of his efforts, Stanyarne Wilson ran for the South Carolina House of Representatives, and the workers supported him in victory. While in the House, Wilson supported a working hours limits bill that was subsequently

"Everything at all mills is moving along satisfactorily except a little advance in wages at mills around Greenville. It has not affected us, and was a very foolish thing to do."

—John H. Montgomery in a letter to Seth Milliken
March 25, 1900

Oct. 15, 1887
A.H. Twichell Esq.
Clifton, S.C

Dear Sir,

Yours of the 14th is in hand with contents as stated. I have watched with great interest your handling of the strike at Clifton and think it was handled very judicially and fair. No doubt it taught a lesson which I hope will be heeded. I hope you will be fully supplied with hands again and have no further trouble. We are hard at work at our new mill and are laying bricks as fast as possible.

Yours truly,

Ellison Smyth

Capt. Ellison Smyth, president of Pelzer Mills, is commenting on a short-lived strike at Clifton organized by the Knights of Labor, a union organization from the Northeast.

Stanyarne Wilson
Photo from 1900 Spartanburg Herald.
(Courtesy, Spartanburg Regional Museum)

defeated in the South Carolina Senate. So in 1892, he decided to run for the state Senate, setting off a storm of controversy when he was quoted in the local newspaper saying, "Go to the mines of Siberia and you will find more humanity shown to convicts than in the factories of Spartanburg County. It is grinding human flesh into money." One hundred workers from Pacolet Mill endorsed the statement, but it created a furor elsewhere. Eighty employees of Spartan Mill signed a letter denying mistreatment: "We consider ourselves not only better than convicts but our condition is better in every way than if we were employed on farms." Regardless, Wilson won, and the legislature passed the hours bill (lowering working hours from 16+ to 11) in 1893. Mill owners were determined to resist such legislation, and as soon as they heard that it had passed, owners who had never charged rent on their mill houses began to do so. Almost all the mills shifted to piece work, which effectively reduced wages. When workers threatened to strike because of these actions, mill supervisors tightened discipline and increased penalties for absenteeism. The workers, unable to defend themselves, suffered when they were supposed to benefit.

The degree to which many mill owners resisted a shortened workweek was illustrated in a letter to a fellow mill owner by Captain John H. Montgomery written in April 1900. Here he chastises a colleague for considering shorter hours:

> **I am utterly astonished at the position you take—voluntary concessions would be a fatal move on our part. To talk about 66 hours per week being too long is the merest bosh. I have ploughed 13 hours per day or more, week in and week out, and did just as much Friday and Saturday as I did Monday and Tuesday. I have not time here to discuss the matter but I hope you will "hold up" as stated in a former letter. I fear your association in Washington has corrupted you.**
>
> **Do you believe the adult operatives, women and men, would by a majority vote favor shorter working time at correspondingly reduced wages? I do not think one tenth of them would so vote and if they do not—they do not want shorter working time themselves.**

Well-dressed picnickers visit the shoals at Enoree at the turn of the century.
(Courtesy, Anthony Tucker)

> *"New England has been cursed and is today seriously handicapped by labor unions which interfere in the management of the business and stir up strife between the employers and the operatives ... We have never been cursed with these organizations as experience shows that they have never thriven except in communities composed largely of foreigners and it is to be hoped that they will continue to give us a wide birth."*
>
> **—Col. James L. Cox, as quoted in The Exposition (March 1901), produced for the South Carolina Interstate and West Indian Exposition in Charleston**

Two operatives converse in Glendale Mills. (Courtesy, Charles Hammett)

"At Whitney, there is also an interesting little lad, named Ike Messner. He did not know his age, but I judged he is about 6 years old. He helps his older brother and, although he did not know how old he was, he seemed to be able to chew tobacco; and unfortunately, this habit is very common among the children of the cotton mills."

—Author August Kohn, 1907

Workers' efforts to organize finally stopped. Part of the reason was an experiment in the Lowcountry that took place around 1900. John Montgomery and Seth Milliken (whose interest in Spartanburg mills was growing beyond merely acting as a selling agent for their products) opened a cotton mill in Charleston manned exclusively by African-American operatives. At the time there existed an unwritten agreement among workers and owners that African Americans, who made up about one-third of Spartanburg County's population, could work only outside the mills, mostly in the yards unloading and loading trains and the like. As a result, almost all worked in agriculture. Opening a mill with an all-black workforce sent a powerful message to Upstate white workers—mill owners would be willing to break a long-held social taboo if unrest among white workers interfered with production. Historians disagree about the reasons for this experiment and as to why the Lowcountry mill soon closed, but the message that came on the heels of efforts to organize labor seems clear—mess with unions or make any effort to organize locally and lose your jobs to African-American workers. Afterwards, several decades passed before there were further efforts to organize labor in Spartanburg County.

The picture drawn here has been grim, and surely the hard times and circumstances were mixed with joy, love, friendship, and happy experiences. Were these mill village people happy? Was there anything good about mill village life between 1880 and 1910? There was certainly a sense of belonging, what is now referred to as "a sense of community." There was the joy of the church, the social activities as well as the religious ones, the joy of love and, especially in the first generation of mill village people, a comforting sense of having found a steady income and of being relieved of the anxiety of rural life. The joys and wonders of family life, of neighborhood, and yes, even of work, were all there to be experienced as surely they were. The baseball teams and social clubs that marked mill village life later in the century were not yet widespread, but village people had the church and they had each other. Like other human beings, they

Clyde Crocker's birthday party, Saxon Mill village, about 1910 (Courtesy, Caroliniana Library)

loved, married, had children, grew old, and died. Through all of it, they lived the joys and sorrows of the age-old rituals of being human.

Along with all of the age-old and new experiences that were part of the emergence of two new social classes, Spartanburg's industrialists and Spartanburg's industrial working class, there also emerged a new social attitude. The creation of an industrial workforce in Spartanburg County had far-reaching social consequences. According to historian David Carlton, the relationship of mill villages to the city of Spartanburg affected the area's history for most of the twentieth century. Initially, the professional and business classes of the city of Spartanburg were enthusiastic about

After John H. Montgomery left $10,000 in his will to Cooper-Limestone Institute, the school tried to rename itself Montgomery College, but finally decided on Limestone College in deference to Montgomery's wishes that this honor not be bestowed upon him.

In the late 1880s and early 1890s Trough Shoals was surrounded by "drug stores" which were thinly disguised saloons. Tompkins Clay, a reporter for the Carolina Spartan, wrote in December 1892: "All that was necessary to start one was a little calomel, castor oil, paregoric, and a few barrels of whiskey. Back behind the whiskey a doctor sat, pen in hand, ready to sign a prescription for anyone who stated that he did not feel well, and that he thought a little whiskey would do him good."

—Author Ruth Trowell Watson

industrialization, but by 1900 those same early boosters had come to believe that deep differences existed between the people who lived in the mill villages and themselves. Many people in the city—the city's middle class—came to view the mill villages as "alien," unruly, and even dangerous. Mill employees living in the mill villages were separated from the city; they had their own stores, churches, and schools. It became an acceptable view in the city, so acceptable that it appeared in one of the city's daily newspapers, that mill village people were sullen and "necessarily vicious in their tendencies and, hence, undesirable citizens." When mill village people came to town, they felt unwanted and strange; they could tell that people on the streets shied away from them; they were familiar with epithets such as "lintheads." So they kept to themselves, conducted their business, and left the city as soon as they could.

THE TURN OF THE CENTURY BROUGHT many reform efforts across the country, commonly called the "Progressive Movement." Progressivism took various forms, but among these was a desire to clean up corrupt government, improve education,

Construction of Inman Mills, September 1901. (Courtesy, Inman Mills)

and generally open up the promise of American life—unlimited opportunity—to all social classes. All of this activity, whatever form it took, was ultimately aimed at creating better citizens, which, for instance, in the case of recent immigrants, meant making Americans out of them. In that vein, the middle class of the city of Spartanburg attempted to "better" the people who lived in the mill villages by improving their character. Yet, changing adults was much more difficult, reasoned the reformers, than changing children. The city people believed that if they could reform the children of the current mill workers, a better class of operative, a worker who thought and behaved more as the city's better classes did, would emerge. So the city reformers set out to support child labor reform, school attendance laws, and legislation limiting working hours.

Many cotton mill workers saw middle class efforts as attempts to change their lives, habits, and points of view. The implication of the reforms was that the mill workers were not as "good a people" as they ought to be, and understandably, the workers resented that attitude. When mill workers went to town, they could feel the hostility; one worker told a student of mill village life that the merchants "were glad to get your paycheck, glad for you to come up there and leave it with them, but to uptown people you were still cotton mill trash." Another worker remembered what it was like to be from a mill village in the city limits and to attend a city school: "The other children would kind-of look down on you. You'd go to school and they'd call you a linthead and all that stuff. You was kind-of from the wrong side of the tracks."

The term "linthead" was the most common derogatory name city people used; the cotton dust was so common within all the mills that it stuck in workers' clothing and, especially, in their hair. It was very difficult to get all of it out. A mill worker who went into town on a quick errand, most likely by trolley car, was sure to have some lint on her clothes or in her hair, and people on the street were sure either to stare at her or quickly turn away. As such behavior became more common, people born in mill villages grew up expecting it, and the bad feeling between them and the city people was always present whether there was any obvious bad behavior or not. In addition, many mill workers read the comments made about them in the city papers, especially around 1900 when the issue of their "strangeness" was a matter of public comment and later, after every election, when their voting preferences drew such negative reactions from the city's middle class. Mill workers were likely to read comments like the following written in 1901 by *Herald* editor Gibson Catlett: "It is true (the employees) are making but a little more than half as much as the New England cotton

In the early 1900s, a columnist called "The Idler" wrote periodic anonymous articles for The Free Lance, *a Spartanburg weekly newspaper. In 1902 he busied himself with visiting the new cotton mills springing up around the county:*

Whitney
August 8, 1902

A few years ago this mill site was nothing but an ordinary shoal on Lawson's Fork with a dilapidated grist mill on the north side of the creek and…was on a poor, dangerous road. Men had to ford to pass the place and nothing but jack rabbits, coons, minks, and weasels perambulated in the water, but what is in evidence now? It stands second to no mill site in the state, with good roads centering there from all directions, with a number one bridge and a village of neat and comfortable dwellings and the occupants apparently as contented as can be…All the operatives in the mill were neatly dressed, especially the girls. The carding, spinning and weaving is very light work, in which, if I were engaged, I would have to go out and chop wood to rest myself. It has only one drawback. Like the fellow in jail said, it's too constant.

Beaumont
August 1, 1902

I was driving yesterday afternoon and happened to pass near Beaumont mill and the electric power house. I'll declare it was hard for me to realize that I was down once more by Chinquapin creek where I used to fish so often for minnows and catch so few and so short! Gentry's pasture is now cut in two and Vass's meadow can scarcely be recognized—all this as "industrial growth of the city"…What Spartanburg boy of ten, twenty or thirty years ago has not fished and bathed in Chinquapin to his heart's content? How many knives have been lost in its waters as we bent over them peering into the seine? How many snipe have we shot at along its banks wading after the birds in the mire?…Of course all of us believe in business push, and progressive spirit, but it does take all the poetry out of life and I'm not sure that sentiment is utterly an evil.

Inman
August 8, 1902

They are engaged in putting in the machinery and a big force is engaged now in excavating and digging for the foundations for a duplicate factory so that by the time No. 1 starts off in the ides of September, No. 2 will be well on the way. The mill has a beautiful location. Comparatively level, it is more so than any factory I have seen heretofore. They have a railroad track running right up to the door of the factory. All the surrounding country is level, full of farms. These Inmanites are certainly energetic, highminded and withal pleasant and sociable people, and some day will compete with the most prosperous country town or city. They have great natural advantages, new houses are going up and they are now having a new road made from Boiling Springs, which will connect them with another prosperous farming section.

Apalachee
July 8, 1902

With characteristic nerve the Greenville people are already claiming the Apalachee as their own product. It may be just as well to say right now that the mill is to be located at Arlington, which is several miles on this side of the Greenville County line. The Journal was the first newspaper in the state to print the news about the mill and said that it would be in Spartanburg County. But since the mill is to be built near the town of Greers [sic], part of which is in Greenville, and since its president is to be L.W. Parker of Greenville, the Greenville folks find it easy to confuse the minds of the public. I don't know why Mr. Parker prefers to build the mill in Spartanburg County. It is just possible however that as a shrewd businessman he recognizes the fact that Spartanburg mills can always be counted on to pay good dividends.

Two weavers at Saxon Mills pose for a picture in the early 1900s. (Courtesy, Sue White)

mill operatives and there is no restrictive legislation as to age or hours, but there is no such thing as labor organization and people are so far advanced above the conditions from which they have come that instead of a single democracy, there is a semblance of a social aristocracy among the people in many of the mill settlements." Reading such comments reminded mill workers more of the "social aristocracy" that existed between themselves and the townspeople.

City people criticized the mill workers' opposition to proposed legislation in the South Carolina state legislature on child labor, compulsory schooling, and working hours. That legislation, Spartanburg's middle class kept insisting, was for the workers' own good. The city's middle class would not try to understand the situation from the workers' point of view. For instance, workers opposed child labor legislation because it decreased their already meager income, interfered with a worker's control over his family, and attempted to impose middle-class values on them. Mill workers saw these reforms as an effort to "better" them when they did not need "bettering." Mill employees were hard-working people who had to depend on the extra wages their children could earn to make a living in the mills. Had they been paid more, they might have been more willing to do without that extra income. Mill workers' attitudes played into the hands of those mill owners who resisted child labor reform for their own selfish reasons, as illustrated in a letter by Captain John H. Montgomery written to Ellison Smyth in May 1900: "I understand you and others are to make a minority report from the Industrial Commission. I hope it will be strong and such as all of us will approve. No foolishness such as 10 hours child labor law etc." On the same day Montgomery wrote another mill owner: "I hope you will *pile it on* Smyth every opportunity you have about 10 hours child labor

and everything that looks so much like he is working for himself at the expense of others." Reformers never attacked the owners for the low wages they paid; instead, these reformers attacked the workers as selfish parents who took advantage of their own children.

Workers also opposed compulsory schooling because it would place the village children under the teachers from the city; they thought these teachers would teach mill children to despise their parents. Also, parents believed that extra schooling would not make a significant difference in the lives of their children. Compulsory schooling would not open up new opportunities that could take a mill child out of the village in which he was born; the schooling available was not worth what it would cost the family in income, they thought. The child was better off beginning to earn money to contribute to the "family wage" as soon as possible.

Workers also remembered what had happened to them the last time hours legislation had passed. When "reform legislation" had passed, mill supervisors often had made working conditions worse, and in some cases the cost of living actually increased for village residents. After visiting Pacolet Mills in 1895, New England textile industrialist Edward Porritt wrote an article to describe the impact on workers of the 1893 South Carolina law that established an 11-hour maximum work day: "When the 11-hours a day law was adopted in South Carolina, the mill corporations in the county districts did not reduce wages; but instead of doing so, they adopted a system of charging rents for their houses so that the shortening of the work-week from 72 hours to 66 hours has since 1893 been costing each operative 25 cents a week."

In order to protect themselves, workers supported a politician named Cole Blease for governor, a declared enemy of the Progressive reformers. Blease, in addition to being a racist, opposed everything that smacked of reform and change. Spartanburg city's middle class could never understand why mill workers supported a politician who opposed all the legislation that would seem to benefit them. But the workers were not opposed to the reforms themselves; the workers simply did not trust the people who were sponsoring those reforms, and they would eventually end up supporting those same changes when they were sponsored by someone workers could trust.

The period from 1880 to 1910 witnessed the creation of Spartanburg's cotton mill culture. It was the heyday of the building of mills and mill villages, spurring the creation of a ruling industrial class, an industrial workforce, and the development of an industrial culture and a set of attitudes that proved almost impossible to change. But change they would.

Weave room at the Enoree Manufacturing Co. (Courtesy, Anthony Tucker)

The mill village in Glendale counted 160 children in 1906. Of those, 154 were enrolled in school, though average daily attendance was only 65.

—Author August Kohn

Converse Mill before the flood of 1903. After the flood, the mill was rebuilt higher on the hill. (Courtesy, Converse College)

Voices from the Village

James Chapman

James A. Chapman, son of the founder of Inman Mills, presented his memories of the earliest days of the mill village in this address to "Ladies Night" at the Spartanburg Rotary Club in May 1959. Here, he recounts visiting the mill as it was being constructed in 1901 and making the often treacherous journey back and forth from Spartanburg to Inman.

James A. Chapman, known as "Mr. Jim," pictured among mill workers in the 1920s. He is in the second row with the open shirt collar. (Courtesy, Inman Mills)

I think back over the early days of Inman and my first contact with it, which was when my father brought me up here one day while the mill was being built. Someone who was working on the mill brought him a bag of cherries, and I got in them and started eating them and ate so many that I had the stomachache. At that time the office was where No. 1 C Street is and the store was No. 3, and as the roads were so bad, the only way of getting here was to drive a horse and buggy or we had to come up on the train and go back on the train or spend the night here.

Another thing I remember in the early days was coming up here with him the day I got my first pair of long pants and going blackberry picking up around where the upper pond is today. I remember how those briars plucked at my new pants. Of course I had to go on and wear them. Then in 1909 when we were enlarging the mill, [brother] Bob and I came up here and worked during the summer, working in the card room for a while and then the spinning room erecting machinery. At that time Mrs. Canaday, whose husband was running our store, ran the boarding house and we stayed there. I think we paid $12 a month.

Of course, there were no conveniences. No lights except a lamp and no water except what was brought in. There was a mighty good well there, though. The same thing applied to the town of Inman, which, of course, had grown some from the two stores. I continued working here through the two summers through 1913 and came back permanently in the fall of 1916, at which time Mr. Cobb was superintendent. There was one thing that I remember in coming up here with Father as a boy, at which time Bob and I were along, and he had an old one-cylinder Oldsmobile. Whenever he would get up here and get back home, he would always say, "Well, I made it again." It would take about an hour to make the run and many is the time I have seen him drive off to the side of the road and go and lead a horse which someone was driving past so that the horse would not get scared of the automobile. He would have to stop the engine when he did this, and that sometimes meant a good long period of cranking to get it started, and cranking was by twisting with your hand.

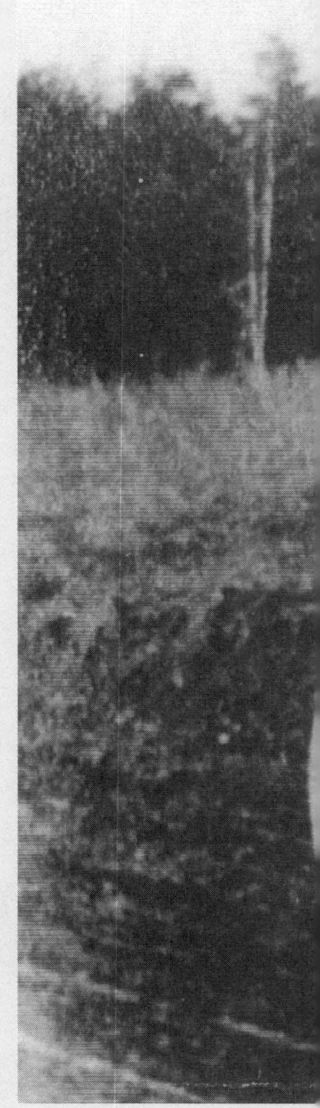

One day Bob and I came up here with him, and when we started home at the usual time in the afternoon, we began to have trouble. We got down almost to the city limits, and we had to stay there and work awhile and finally we got it to running and got as far as Spartan Mills office. We had to then call Blowers' Dray to send a horse and wagon to pull us in. Bob and I were pretty small boys at that time, and Bob was ashamed to be pulled in so he got down in the wagon and stretched out so nobody would see him. I never did understand why he was worrying because it was after ten o'clock at night when we got home…

Of course, I can remember when there were no lights in either the town of Inman or down at the village. When we started running the mill on a 24-hour basis, I did have lights in the superintendent's house for five nights out of the week. I remember also when there was no water or sewerage in either one of the communities [Inman and Inman Mills] and when they were both put in, when there were no paved streets, no paved sidewalks, and all of that has not been that long ago.

The first James Chapman drove to his plant in Inman in this car. (Courtesy, Dorothy Josey)

VOICES FROM THE VILLAGE

David Camak

David English Camak, founder of Textile Industrial Institute (now Spartanburg Methodist College), published his memoirs, Southern Gold from Human Hills, *in 1960. In this passage of the book, he describes a visit to the Spartan Mill village in about 1902.*

David English Camak in his eighties (Reprinted from Human Gold from Southern Hills)

This clinging to the soil was pathetic and everywhere evident. Here and there were relics of the family farm, usually guarded tenderly by the head of the family. One had an old buggy, another a saddle or a set of harness, while the most fortunate possessed horse buggy and plow. The buggy was used on Saturday afternoons and Sundays, while the plow—followed by a lonely town-caged man who once called himself a farmer—did service by the hour in neighbors' gardens, or perhaps in a nearby rented plot, where the erstwhile tiller tried to make himself think that he was still essential to the support of his family; whereas in truth the children were now the bread-winners, and he was a mere fossil of a bygone era, a sad forlorn failure.

The streets ran straight with tragic disregard for topography. However graceful the slope and curve of a hill or the dip of a valley there was no deviation from the checkerboard pattern. As most of the villages were located on hillsides, there were many houses built unavoidably very high from the ground on one side, supported by tall, stilt-like pillars giving full underscope alike to the eye of the observer and to biting winter winds, which forced themselves through every crack in the floor. As a rule, there was no subfloor.

It seemed it never occurred to anybody to enclose the vast gaping space thus left beneath the hillside homes. Some of the rooms were heated with small grates burning soft coal, while others had nothing. Furnaces were unknown. Water for cooking, scouring floors, washing clothes and bathing was toted by the women and small children from the wells in the middle of the street, each well serving about twenty families. There was no community center—no gymnasium, no swimming pool, no playground, no park, no central building where clubs and societies might meet, and no clubs and societies except an ancient fraternal order or two that usually used a hall over at the community store…

The public school…was located on the extreme edge of the village as far as possible from the center of population, and in the opposite direction from the uptown section of the adjacent city. It was a square red-brick building with a square red tin roof on a red-clay lot without a sprig of grass or a single tree to break the color scheme. Moreover, it was adjacent to the most depressing tenements—the beyond-the-rent-line type. There was no occasion to pass the school in going anywhere, and there was little inducement to visit it. What few children there were who did not work in the mills were, after a fashion, taught here by three or four girl teachers. The teachers boarded in the adjacent city and took no interest in the social life of the mill people. The curriculum was old-fashioned and did not extend beyond the fourth grade. While the mill corporations paid the major portion of county taxes, there seemed to be no objection to their being spent on the uptown schools to the neglect of the mill children whose toil produced the revenue.

The churches came nearest to being community centers, but even they had no social program, even of the crudest sort. They were built to preach in, which, in the thinking of the average minister who served the mill communities, consisted of a sort of ecclesiastical spanking of the people for sins the vast majority of those who attended church never committed. It was a negative, fearful gospel; no social conscience, no positive expression of Christianity except the usual collections of small coins "to send the gospel to the heathen"…

As for old the folk, they did not need any social life, or happily did not know they needed it. They would "jist set" and smoke and gossip and dip snuff—just that and nothing more! As a whole, it was from week's end to year's end one dead level monotony of work, eat, sleep, work, eat, sleep. They did not live; it was as it they camped a while on a strenuous job with the expectation of breaking camp at any time and returning to their homes. And no doubt the vast majority did expect to do that very thing "someday."

I visited beyond the rent line and on the best streets. I saw places of human habitation so filthy I could with difficulty remain even a few minutes in them; and I saw homes so neat and spotless it seems a sacrilege to enter. I saw people of culture and refinement whom reverses had reduced to common toilers. I saw ill-kept premises and beautiful flower gardens; women who had husbands, and grass widows who had none, and actual widows struggling to keep the family together and the wolf from the door while sending their children to school year about, and the family making a desperate fight for education and respectability.

Visiting in the homes one saw mainly afflicted people, old folk, small children, some mothers—and occasionally a Negro servant. Everybody who could reach a spinning frame on tiptoe or handle a broom was at work early and late. Much of my pastoral visiting had to be done in the mill, where the workers were delighted to see their ministers. I soon learned that by throwing my voice against the roof of my mouth to carry on simple conversations in the mill in spite of the terrific noise. My mechanical turn served me well in learning something of the various machines, so that I could talk intelligently to the workers about the one thing that engrossed their lives and so gain ready access to their friendship.

Thus I spent the first months of my pastorate getting acquainted with the great mass of people who, as a distinct social group, had gripped me with such strange fascination.

Voices from the Village

Carolyn Law

Carolyn Leonard Law was born on Hampton Avenue, the daughter of a prominent local physician, on August 27, 1909. When she was 23, she married into another prominent family when she wed John Law Jr. Here she relates her memories of her father-in-law John A. Law Sr., who built Saxon Mill and Chesnee Mill, and her own move to the mill village.

I'm over 90 years old. I'm a walking monument. My father-in-law had Saxon and Chesnee Mills. I thought he was a very kind person. He was interested in the people who worked for him. If any of them were in difficulties, he was interested in them. I think it was at Christmas that they used to have a celebration and have fruit they would distribute to everybody. Had tables set up in that front yard. It's just different than now. He was handsome. He was about 6'2" and he was a native South Carolinian. His mother was an Adger. He had Central Bank, too. A lot of the banks closed [in the Depression] and Central paid off every penny to its depositors.

He was one of the kindest people, very devoted to his mother and father. His "sainted mother," he always referred to her. They lived on Spring Street. Dr. Law, he was our first full-time Presbyterian minister.

Carolyn Law at the former home of John A. Law Sr. in Saxon (Courtesy, Mark Olencki)

I met my husband—I imagine it was at Sunday School, Christian Endeavor, I'm sure. That's where all these young couples met, Christian Endeavor! That was at First Pres. I went to Converse. My husband went to Wofford. I was about 23 when I got married.

My husband and I lived in one of those houses out there. I lived on Saxon Heights. It had crepe myrtles down the street. It was a four-room house. I wanted the floors painted because they were so ugly. My mother was very concerned because the fireplaces were about this big [holds hands up, indicating a narrow opening]. But we had an oil stove, and then you could get it by the gallon and have a tank outside. Mother would call and say, "How are those poor little children?"

I would say, "Mother, what poor little children?"

"Your poor little children. I know they are freezing." I was very happy there. Loved my neighbors. At one time they had a pond somewhere, but people weren't swimming in there when I was there. I don't even remember where it was.

The street I lived on, we had flowerbeds. We were interested in the yard. We had neighborhoods in those days, and now you have streets. Mrs. Law's house had a right big yard to it, and she was crazy about irises. Flag Lilies we called them. She was such a good person. She was from Augusta. She was a Sibley. They had a big old-fashioned house, had sleeping porches. Mr. Law would sleep there. The snow wouldn't come in, but it would be cold, and he'd have an oilcloth on top of all his cover because he loved outdoors. I don't think he ever took a big salary. As far as money is concerned, I don't associate him with being spend-thift at all, very simple wants.

The people in the mill were just hard-working, good people. I remember somebody was sick and Mr. Law called his selling agent in New York. They needed mosquito netting for him. They needed it to be sent down. He had a big heart. I don't think there was a big division between the real owners of the mills and the workers, it seemed to me. There was a friendship there. I think there was a personal relationship between the bosses and the workers.

I had two girls. See the picture there on the wall? They had two dresses. You know how children have hand-me-downs. They were at a birthday party one day and somebody said, "Louise, I like your dress." And she said, "No, my dress is on Sibley. I'm wearing someone else's dress, my first cousin." If you liked them, you just handed them down. It had glamour to it.

We lived out there a goodly while. My daughters loved to go [see their grandparents] on Hampton Avenue. You see, we didn't have paved sidewalks, streets, in Saxon. They said they liked the "pave-MENT."

I remember the strike [in 1935]. I had some towels out on the clothesline and they disappeared. I suppose there were hard feelings, but I don't remember any antagonism. I was friendly with everybody.

—Interview by Betsy Wakefield Teter and Don Bramblett

Saxon owner John Law with grandchildren, Bill Child and Polly Hill, 1932 (Courtesy, Susan Woodham)

The Birth of Clifton

Boom Town on the Pacolet

By Mike Hembree

As with so many other communities across the Southeast (and, indeed, in countless other landscapes around the globe), the Spartanburg County textile village of Clifton was born because a river ran through it. Cotton mill entrepreneur Edgar Converse, a Vermont native who saw the potential of the hilly riverland in upstate South Carolina, formed the Clifton Manufacturing Company in 1880 with plans to develop a chain of textile plants along the Pacolet River.

Its waters rolling south toward the fall line and on to the Atlantic, the Pacolet was the perfect vehicle to fuel Converse's dream. It would create the hydropower needed to turn the wheels of his cotton mills, three of which would be built along the Pacolet's banks in the Clifton area (formerly known as Hurricane Shoals) from 1880 to 1896.

The genesis of a company and community that ultimately would crowd the once quiet Pacolet River valley with thousands of mill workers occurred January 19, 1880. The Clifton Manufacturing Company was formed in a meeting in Spartanburg. Ten directors—Converse and nine other prominent Spartanburg businessmen—were elected. The minutes of the meeting include the decision to name the community Clifton, a choice apparently made because of the cliffs overlooking the river as it flows through the area.

Construction on the first of the three mills began in February 1880. Sixteen months later, Clifton No. 1—as it would be known by generations of workers (and as the area around the mill remains known today)—was operational with 7,000 spindles turning.

The mills needed employees (or operatives, as they were known then), of course, and the workers needed places to live. Thus did Converse create what became commonplace along many of the larger rivers of the Carolinas in the closing years of the nineteenth century: a self-sufficient village that attracted workers from nearby farms and the mountains to the north and west.

As Converse's huge brick mills went up within mere feet of the river, so did hundreds of houses, saltbox-style structures built in neat rows along narrow dirt roads cut into the hillsides surrounding the plants. So, too, were schools, churches, and community buildings constructed. By July 1882, the village had 92 houses—all built by the company—and a 10-room boarding house. Three years later, the Clifton population had grown to 1,400, and the "free" library operated by the mill company contained 250 volumes. Only 25 years earlier, the Hurricane Shoals population had been counted at 160.

A construction crew works on the top floor of Clifton Mill No. 2 sometime in 1887-88. Note the horse and wagon at top right. (Courtesy, Mike Hembree)

With the first mill quickly becoming a profit base, the company authorized a second mill at a location a short distance downriver in August 1887. Construction began on "Clifton No. 2" in 1888, and the mill started operations in 1889 with 21,512 spindles. In May 1895, Clifton Manufacturing decided to build a third plant, ultimately choosing a location north of the No. 1 plant. The mill opened in 1896 and, with more than 1,000 looms, was considered one of the finest textile facilities in the country.

Typical of those who traveled to Clifton to work was Robert Walker, a farmer who lived in McDowell County in North Carolina. Responding to a newspaper advertisement seeking mill workers, Walker, impressed by a visit to the area in 1904, moved his family—his wife, Hattie, and six children—by train to Clifton. They settled in a company house at Clifton No. 1. Two days later, Walker and five of the children started work in the mill.

Similar stories are told by families throughout the Carolinas. Farms and furniture were left behind in the Tennessee and North Carolina mountains as families became mill workers—in this case, Cliftonites. Sons and daughters followed fathers and mothers into the mill workforce—often as children themselves. The cotton mill culture became a way of life.

By 1920, the population of the Cliftons had reached 3,000. Although nearby Spartanburg held its attractions, life remained largely insular for most of the mill villagers. That would change gradually in the 1940s and 1950s, particularly when families reached a financial level that allowed for the purchase of their first automobiles. The post-World War II years, with dozens of soldiers returning to Clifton to start families and returning to jobs in the Clifton Manufacturing Company operations, perhaps were the glory days of the operation Edgar Converse had started on the Pacolet riverbanks more than a half-century earlier.

The 1960s brought years of decline, eventually leading to the closing of the three mill operations and radical social change within what had been a close-knit society. Many left the hillside communities, either for newer, better housing or for jobs at newer, better factories outside the circle of the Cliftons.

John H. Montgomery
Father of Textiles

By Karen L. Nutt

A career change at age 50 sent Captain John H. Montgomery into the textile industry, where he made an indelible mark with the founding of Pacolet Manufacturing Company in 1881 and the formation of Spartan Mills in 1889.

Born December 8, 1833, at Hobbysville, a small farming community 14 miles southwest of Spartanburg, Montgomery was the oldest of 12 children born to Benjamin F. and Harriet Moss Montgomery. His ancestors, originally from Scotland and Ireland, first settled in Pennsylvania and in the 1780s came to South Carolina.

Montgomery attended the neighborhood school near the Montgomery farm, and historians describe his education as "better than average." One of his instructors was Richard Golightly, who, according to historian J. B. O. Landrum, was well trained in English, Latin, and higher mathematics.

When he was 19, Montgomery would put his math knowledge to use as he worked as a clerk in an Enoree country store owned by James Nesbitt (who also eventually became a textile pioneer). Earning $5 a month plus room and board, Montgomery worked for Nesbitt about a year with his tasks reaching beyond those of a typical clerk. For example, Montgomery worked around Nesbitt's house and barn and drove a four-horse team loaded with flour to the iron works at Clifton, about 30 miles away.

Four generations of Montgomeries pose in 1895. Seated: Benjamin Franklin Montgomery and his son, Capt. John H. Montgomery, founder of Pacolet and Spartan Mills. Standing: Victor M. Montgomery, president of Pacolet Manufacturing, and his son, John, who died in his early 20s. (Courtesy, Walter and Betty Montgomery)

Montgomery then left the Upstate for the state capital, working for four months for Robert Brice, who also operated a store. Returning to Hobbysville, Montgomery entered a partnership with his brother-in-law, E. R. W. McCrary and continued working in the general merchandising business. The store was just a few miles from that of his first employer, Nesbitt.

The partnership was short-lived, however, because McCrary, along with several other members of Montgomery family, left South Carolina for Texas. Thus, Montgomery was left with the mercantile business, which he continued to operate for the next three years or so, although his limited resources made the operations difficult. In 1857, Montgomery married Susan A. Holcombe, whose father, David Holcombe, was a native of Union County who later settled in

Spartanburg. The following year, Montgomery moved the stock in his Hobbysville store to his new father-in-law's business a few miles away. The new location included a small tannery, which Montgomery helped to operate until the beginning of the Civil War.

In December 1861, Montgomery became Private Montgomery in Company E, 18th Regiment, South Carolina Volunteers. Named commissary of the regiment, Montgomery then became Captain Montgomery, a title that would be used to refer to him long after the war ended.

With the end of the war came the beginning of a new life for Montgomery, who had little money or property when he returned home. Basically, he had a small stock of leather in the tannery and a small farm. Initially, it seemed his career would lead him into the agriculture industry as, in 1866, he began the use of commercial fertilizers on his farm, demonstrating to his neighbors their importance to stimulate plant growth. Then he began to sell fertilizers to local residents, and his business started to grow. At the same time, he resumed operations at the tannery and continued his merchandise business.

Montgomery's venture into textiles, along with the activity at Edgar Converse's mill in Glendale, was a turning point for the industry in Spartanburg. Abandoning farm work for a career in business, Montgomery moved to Spartanburg in 1874 to work in the mercantile firm of Walker and Fleming. Legend has it that he had "too much sense" to plow for the remainder of his life. The firm eventually became the largest cotton buyer in the county, in addition to selling guano, farm equipment, and insurance.

Prompted by Converse's success, Walker, Fleming, and Montgomery began to consider building a cotton factory of 10,000 spindles and 300 looms. With a group of friends, including Converse, they went to Trough Shoals (Pacolet) to examine a potential site for their mill. According to legend, Converse told Montgomery that day that it was not necessary for the head of a mill to have been brought up in textiles and that the main qualification was that he be a good businessman.

Montgomery, with little knowledge of the industry, first consulted fertilizer dealer John Merriman, who in turn connected him with friends in New York, including Seth Milliken. Milliken had ties as a selling agent for several New England mills. Although Milliken had no official connections with Southern manufacturers, he loaned Montgomery $10,000 for the Pacolet venture.

Montgomery was close to 50 years old when he became president of the Pacolet Manufacturing Company in 1881. In 1887, the Pacolet Manufacturing Company had 26,224 spindles and 840 looms; it expanded in 1894 to 57,000 spindles and 2,200 looms. Montgomery also founded and served as president and treasurer of Spartan Mills, which opened with 35,000 spindles in 1889 and later expanded to 85,000 spindles with 1,100 broad and 1,490 narrow looms. He shepherded Spartan Mills from earnings of $16,854 in its first year to $360,368 by 1900. He also participated in the founding of Whitney Mills, Drayton Mills, and Lockhart Mills in Union County, and was a stockholder in Clifton Manufacturing.

Montgomery died while working in the industry that made him a legend. Nearing the age of 69, he fell from a scaffold at a textile mill he was building for Pacolet Manufacturing Company in Gainesville, Georgia, and was killed. Active in the First Baptist Church and as a trustee of Limestone and Converse colleges, Montgomery was a well-respected member of the community. At his funeral in November 1902, Dr. L. D. Lodge of Limestone spoke of Montgomery: "Captain Montgomery was a man without affectation in his manner, without cant in his speech, and without guile in his heart. I speak of him as a knightly man whom the manhood of South Carolina will be glad to honor."

Women and Children
Backbone of the Early Mills

By Alice Hatcher Henderson

Women and children who worked in early Spartanburg County cotton mills played a significant role in making the new textile business profitable. While historians have usually attributed the post-Reconstruction economic miracle primarily to the men who founded the mills, equal credit must be given to the families who loaded their belongings into wagons or trains and made the switch from digging in the dirt to spinning in the mills. A small army of women and children, comprising over half of all Spartanburg textile workers before 1914, were willing to stand 72 hours a week by their machines, earning five to ten cents an hour.

The South won the cotton textile mill battle over the North because Southern women and children tacitly agreed to spend two-thirds of their waking hours in the mill. This enabled the Spartanburg mill owner to pay all workers, male and female, less than textile workers in New England. There, fewer women continued working after marriage, and male workers' wages had to be higher. And, if each home in a Southern mill village contained at least three workers, housing construction and maintenance costs could be diminished. Later on, when Southern mills ran two shifts a day and parents alternated childcare responsibilities, they had an even higher return on machinery investment than their one-shift Northern counterparts.

For some women and children, mill village life was preferable to the life they left behind on their Piedmont or Blue Ridge farm. Gradually, the mill homes added electricity, running water, and indoor plumbing, none of which had been available in their farm homes. Stores carried a wider variety of food and clothing items than those to which were accustomed. Cash income that came in regularly every week, rather than at the end of the crop-growing season, seemed like a luxury. Opportunities for social life increased tremendously, with church circles, baseball games, Sunday picnics, and sometimes even a mill village band to add variety to the week.

Children who liked school had an opportunity to go regularly eight months a year, a longer school year than the rural schools provided. The librarian at Kennedy Public Library reported in 1925 that 20 percent of its readers were from local mill villages. Widows or women with disabled husbands could continue to earn money in the mills, supporting their families, an important consideration in the days before Social Security. Even the work within the mill was psychologically satisfying for some who took pride in their skillful weaving or careful inspecting. Finally, some women liked the mill owner's rules for behavior, which kept their husbands' drinking, gambling, fighting, and womanizing under control.

The number of children in the mills was declining when this photo was taken of the spinning room at Saxon in 1915. Overseer Ben Wofford stands in the third row. (Courtesy, Sue White)

For others, however, the long hours on their feet in the mill, followed by a busy evening at home catching up on cooking, cleaning, clothes washing, and childcare, were almost unbearable. Since many were pregnant or recuperating from a recent childbirth, constant fatigue became a fact of life. Others complained of pressure to speed production or tend more machines, with no increase in pay. Some had supervisors who showed too much favoritism or too little compassion. Nutritional deficiency diseases, particularly the prevalent pellagra, robbed many women and children of health and vitality. Tuberculosis also found easy targets where people worked in close proximity to each other and passed the disease to their fellow workers. It is thus not surprising that women in textile communities became active in unions a decade or two earlier than women in the rest of the country.

Pictures of Spartanburg mill workers shock us today with the sight of young, skinny children with pinched faces lined up solemnly along with the adults. In 1902 a bill passed the South Carolina General Assembly making age 10 the minimum age for employment and age 12 the minimum age for working the night shift. However, children of indigent parents were exempt from the provisions. After 1903, children who were not yet 12 could only work if they had attended school at least four months during the year and could read and write. Again, an exception was made if a parent testified that the child's employment was an economic necessity. In 1907 the Clifton Manufacturing Company reported that it still had 72 boys and 69 girls under the age of 12 working, one of the highest numbers in the state. In 1915, after a heated debate in the South Carolina Legislature, a law was passed establishing a minimum age of 14.

The 1916 Federal Child Labor Law limited children between 14 and 16 to a 48-hour workweek, but the U. S. Supreme Court found this law unconstitutional. Finally, in 1938, the minimum age was raised to 16 years. Before that, according to historian G. C. Waldrep, children were sometimes forced to quit school as soon as they reached the minimum work age, or else their family could be evicted from the mill village house for not providing the optimal number of workers. It was also not unusual for 10-year-olds to be told to stay home from school to take care of younger children in the family while both parents and older siblings worked.

By the Depression of the 1930s, the years of high profits in Southern mills had temporarily come to an end. What remained was a city very different than the struggling town of the 1870s. Few of the women and children who had contributed to this prosperity received recognition for what they had done. They had worked longer hours for less pay than white women and children in the rest of the country and had helped Spartanburg recover from the consequences of a lost war. Very few had been given the opportunity to rise to a supervisory position or a well-paid loom fixer's job, according to a 1921 report by Mary Lane of the Women's Bureau of the U. S. Department of Labor. After visiting the Saxon Mill, she wrote, "All the supervisory positions are held by men, Why Not Some Women?" However, aside from their exclusion from supervisory positions, women in most jobs in the mill received equal or almost equal pay to the men by the 1930s.

Woman tending loom at Spartan Mills. (Courtesy, Walter and Betty Montgomery)

There were occasional examples of women serving in decision-making roles. Marjorie Potwin at Saxon Mill combined some of the modern functions of a director of human resources with the self-appointed role of "dormitory matron," for the whole village, in the 1920s and early 1930s. Molly Spears of the small Shamrock Mills in Landrum became the first female mill president in South Carolina upon the death of her husband. Many women also took leadership roles in writing letters to officials in Washington, D. C., to complain about working conditions in the 1930s.

In some ways, the late-nineteenth and early-twentieth century textile mill families pre-dated by a hundred years the busy two-wage earning family of our generation. Few mill village preachers told them that "a woman's place was in the home" because the preachers, the storekeepers, the mill owners, and the bankers all knew that these women and children in the mills were necessary to keep money flowing into the city. Like the African-American women who took charge of cooking, cleaning, and child tending in the homes of the mill owners, they were not accustomed to praise for the double tasks they had assumed. Stone and brass tablets listed the names of their sons, brothers, and fathers whose lives had been lost in World War I, but no memorials named the women and children whose lives were shortened or narrowed by their extraordinarily lengthy confinement inside the mills. The new histories of the South are beginning to finally give them the attention they have long deserved but seldom received.

Seth Milliken
The Money Behind the Mills

By Karen L. Nutt

The $10,000 Seth Milliken invested in Captain John Montgomery's budding textile business on the Pacolet River was no small potatoes.

Milliken, a former potato farmer from Maine, realized putting such a large sum of money in textiles during the South's Reconstruction period would be "either the best or the worst investment he would ever make," according to historians. But such an investment was certainly within

Milliken's character. Earlier, he had been among a host of farmers who had grown a bumper crop of Maine's most-harvested product, potatoes. Investing about all he had, Milliken planned to sell the potatoes in the seaport of Portland, Maine, but the influx of farmers doing the same thing prompted him to seek an outlet for his crop elsewhere. So he chartered a ship, loaded it with potatoes, and headed for Boston. But even there, farmers and their potatoes flooded the market. He had still another idea: to sail to New York City, where he found plenty of buyers in the wholesale market and sold his potatoes at a profit.

Milliken first ventured into business through a general merchandise store when he was 20 years old. Five years later in 1861, with $8,500 in hand, Milliken and his brother-in-law, Dan True, went into business together, opening their own mercantile store in Portland, about 30 miles from his hometown of Minot, Maine.

In 1865, Milliken's association with William Deering established a name that would endure for more than a century and become prominent in the American business community—Deering Milliken. Deering and Milliken operated a general store and became "jobbers" who bought dry goods and woolens in quantity, then sold them to dealers. Its first textile client was Farnsworth Mill of Lisbon, Maine.

Shortly after the potato incident, Milliken, Deering, and their families moved from Portland to New York City. The company's headquarters, logically, also moved to New York, specifically to lower Manhattan. Although Deering resigned from the company not long after the company's relocation, the company bearing his name remained intact. Deering moved to Chicago and established a company that manufactured a new grain-harvesting machine. This company later became part of International Harvester Company.

Soon after the move, Deering Milliken was selling products from five woolen mills in addition to the Farnsworth Mill. By the end of the century, the number of mills grew to 16. Then Milliken turned his attention to cotton fabrics as a result of his association with Montgomery. He became the exclusive selling agent for several mills in Spartanburg County, including Spartan Mills, Pacolet Mills, Drayton Mills, and Whitney Mills. Milliken took a typical commission of 4 percent of everything sold.

Seth Milliken
(Courtesy, Louise Foster)

The joint venture of Deering Milliken and Pacolet Manufacturing was beneficial to both companies. Pacolet added two additional mills near the original site on the Pacolet River, as well as a new mill at New Holland, located near Gainesville, Georgia. After Montgomery's death, nature dealt a devastating blow to Pacolet in 1903 when floodwaters destroyed the first two mill plants. Only the third, and newest, plant remained. Later that year, a tornado roared through Pacolet's fourth mill at New Holland, transforming it into a pile of wreckage. Milliken was instrumental in restoring the New Holland mill and built Pacolet Mill No. 5 where the first two had been located.

In the meantime, Deering Milliken grew as a selling agent as other Southern cotton goods manufacturers hired them to sell their products. With business relationships all over the Southeast, Milliken continued to work until age 80 and trained his son Gerrish to take over the business. Four years after retirement in 1916, Seth Milliken died.

But the foundation he laid led to the creation of a powerhouse American manufacturing company. During the Great Depression, Deering Milliken was able to save many Southern cotton mills from bankruptcy and wound up owning dozens of them, including Drayton Mills of Spartanburg, Lockhart Mills in Union, and Gaffney Manufacturing. Over the course of several decades, those mills were brought under the umbrella of Milliken & Company, which stands today as the largest textile company in the United States.

Engineering the Mills

Lockwood and Greene Come South

By Glenn Bridges

Two New England engineers were key figures in the development of the early Spartanburg County cotton mill industry, and the company that still bears their names is now recognized worldwide as a global leader in industrial design and construction.

Founded in 1832, Lockwood Greene Engineering, Inc. is America's oldest engineering group operating under a continuing name and is regarded as America's pioneer of mill engineering. Even before Amos Lockwood and Stephen Greene were surveying the mill hills of Spartanburg, company founder David Whitman, nicknamed "The Mill Doctor," provided consulting services to textile mills throughout the New England states.

Top, Amos Lockwood
Bottom, Stephen Greene
(Courtesy, Lockwood Greene)

The company's connection to Spartanburg began in 1882 when Seth Milliken advised his new Southern business partner, Captain John Montgomery, to secure the engineering services of Amos Lockwood for the design of the mill in Pacolet they planned to build. Lockwood, who had inherited the engineering staff after Whitman's death in 1858, initiated the first of five plants Lockwood Greene would design in Pacolet.

Lockwood also brought a new partner into the fold to capitalize on a fast-growing Southern industry—a highly regarded architect by the name of Stephen Greene. Now under a new name, Lockwood Greene & Co., the young partner aggressively toured the South for additional projects, which led to a strong relationship between Greene and the Montgomery family. By 1888, Greene was busy implementing designs for the first of several Spartan Mills operations (Spartanburg Manufacturing at the time), as well as drawing up plans for mills in Clifton and Whitney for later that year. In 1889, the partnership between Greene and Spartan Mills solidified when a New England mill where Greene served as treasurer announced it was closing. Greene persuaded the owners to sell the machinery to Spartan Mills for $200,000 in cash and $250,000 in Spartan Mills stock.

Shortly after, Greene was named one of the first board members for Spartan, setting the pace for a number of investments by Greene and his company within the Spartanburg community, including the purchase of Tucapau Mills, which it had designed in 1894, and a heavily vested interest in Lyman's Pacific Mills (1924).

By 1923, the engineering firm was designing so many Spartanburg projects that it decided to establish its Southern headquarters in the city, opening offices in the brand new Montgomery Building downtown. A big influence was its proximity to a major hub, the Southern Railway, and a widely accepted opinion that Spartanburg was the geographical center of the entire Southeastern textile industry.

In addition to Pacific Mills, Saxon Mills (1900), and Fairforest Finishing (1929), numerous non-textile related projects took place within the next two decades, including the Memorial Auditorium, Mary Black Clinic, Spartanburg General Hospital, Cleveland Junior High, First Presbyterian Church, and the Montgomery Building, which sits on the original site of the home place of Captain John Montgomery. Expansion work for Jackson Mills and The Draper Corp. were eventually added to Lockwood Greene's storied portfolio, as were hundreds of others in the Upstate.

As the first of its kind, the firm literally revolutionized the integration of building construction, power equipment, plant layout, and other intricate details of day-to-day mill operations. As the textile industry evolved, Lockwood Greene was at the leading edge—taking it from the mid-1800s when multi-storied mills depended on adjacent rivers as a source of power, to its modern-day version featuring windowless, one-story air-conditioned facilities that crank up at the touch of a button.

The capacity of the cotton mill industry more than quadrupled between 1890 and 1901, with Lockwood Greene accounting for 40 percent of the design work. Thirty-nine projects were in South Carolina alone. In 1893, the company set in motion a major change for the cotton mill industry when it designed the first prototype for textile machinery driven by electricity—the catalyst for freeing future mills from the cumbersome dependence on water. A year later, Lockwood died suddenly, but Greene kept his partner's name and continued securing new textile work, as well as diversifying into other industries, such as schools, publishing houses, and manufacturing facilities.

In the early 1920s, the textile industry began to shift south as a severe depression hit the New England states and companies eyed the South's lower wages and less regulation. When Lockwood Greene realized the South was where the textile industry could prosper, the company purchased two profitable South Carolina mills of its own—Tucapau and Pelzer Manufacturing (Anderson County). But by 1928, those mills were deep in debt and were being run by a committee of creditors that had taken over all assets and were liquidating the company. The only truly valuable asset remaining was the engineering division, which was operated separately and was still earning a profit.

During liquidation proceedings, creditors sold the engineering business to the men who had been running it in exchange for preferred stock to be held by a group of banks representing the creditors. Twenty-one men became the owners under a new corporate name, Lockwood Greene Engineers Inc. Even so, the company almost didn't survive the 1929 stock market crash. At the end of World War II, the industrial regulations, restrictions, and controls that were necessary in wartime were lifted, and the company regained its former glory.

As Lockwood Greene's former board president, Samuel B. Lincoln, best describes it in his historical account of the years 1832-1958, the firm's roster of major projects "reads like a 'who's who' of American industrial progress." Shortly after moving its corporate accounting firm to Spartanburg in the 1970s, Lockwood Greene experienced one of its biggest growth surges. To establish a broader geographical base, the Boston office was closed and account responsibility was transferred to Spartanburg, the new corporate headquarters.

The company also opened a new office in Dallas, Texas, and plunged into international waters with an office in Athens, Greece. Among its major design clients were Campbell Soup Co. (New Montana, Canada), the Atlantic City Auditorium (New Jersey), Xerox Inc. (Oklahoma), McGraw Hill Publishing Co. and *The Wall Street Journal* (New York), and the Yu Yuen Cotton Mill in Tientsin, China.

In 1966, having outgrown its office space in downtown Spartanburg, the firm moved to a new facility on Interstate 85. Then on Easter weekend 1983, the Spartanburg office celebrated another milestone in its prosperous history when it opened a new 123,000-square-foot corporate headquarters on Interstate 26 that—for the first time in 25 years—placed all Spartanburg operations under one roof. As the twenty-first century dawned, Lockwood Greene employed 400 people there and 2,000 more at strategic offices in Atlanta, Dallas, New York, and Oak Ridge, Tennessee.

A War of Words
Controversy Over Working Conditions

By Betsy Wakefield Teter

In late 1887, a representative of the Knights of Labor, a Northeastern-based labor union, visited Spartanburg County mills and wrote a report in the *Philadelphia Press* that set off a storm of controversy among local mill owners. The union representative, Thomas B. Barry, wrote of low wages, difficult working conditions, and widespread child labor.

At Clifton, he reported: "The working people have no stoves in their houses, but warmth is obtained and cooking is done only by means of open fireplaces." The employees at Clifton were paid "by checks, not cash, and many of these people never get a dollar in their hands, no matter how long they may have worked at a place." Those checks, he claimed, were only for use in the company store.

Barry reported seeing children as young as five or six working in the mill: "These little folks I found to be pinched and depressed in health, wilting in their young days like faded and drooping flowers. Many of them are, in fact, more stooped and debilitated than aged persons, owing to over physical exertion and close confinement." As to wages, Barry wrote, they ranged from 15 cents per day for children to $1 a day for adults.

When mill owners in Spartanburg County learned of the article in the *Philadelphia Press*, they shot back with editorials printed in the *Carolina Spartan* on January 18, 1888. The *Spartan* ran these on the front page under the headline, "Wholesale Slander and Refutation."

Edgar Converse, owner of Clifton, questioned whether Barry had even been to his mills and refuted the charges, point by point. As to children working in his mills, few were of the age mentioned by Barry, he said. "Their work is light and admits of moving around and taking much exercise. During the day, they have several recesses, ten to twenty minutes, when the boys go out on the sand and play and the girls romp in the halls. The fact is, the children beg to be able to work in the factory."

John H. Montgomery, owner of Pacolet Mills, also defended the local mill system: "As to wages, I would like to say that while the prices paid here seem low when compared to New England mills, we are not competing with those mills. Our competition is here at home, and to show you that hands can make more at the mills and live easier and more comfortably than on the farms, we have several families that have rented out their own lands and moved to the mills."

Both men pointed to the facilities available for their workers' educational and moral needs. "The sober and virtuous habits of our factory population should be emphasized," Converse stated. "I have visited them in their everyday dress and with their Sunday clothes on, and they are modest, polite and courteous." Montgomery also spoke of the workers' moral character. He asserted that the "morals of the people are…much better than in towns and cities where disreputable women congregate. Such characters are not permitted in our mills."

Later, the *Carolina Spartan* followed up its stories about mill conditions, reprinting a letter from the *Springfield* (Massachusetts) *Republican* written by L. T. Oatman, an inspector with an insurance company that insured southern mills. Oatman, who also toured the South in 1887, seemed to support the mill owners. "From a comparison of the Northern mill operatives with those of the Southern, or Eastern mills, the advantage, it seems to me is in favor of the South," he wrote.

Mill Architecture
The Art of Building a Village

By Allen Buie

"I call architecture frozen music."
—Johann von Goethe, 1829

The songs of Spartanburg County's textile culture crystallized on local riverbanks more than a century ago and still today sing to the landscape. Perhaps nothing speaks as strongly of a culture than the physical places where people once lived and worked. Spartanburg's history as a textile manufacturing center is most tangibly defined by the remains of its textile mill buildings, most familiar to all as hollow brick shells whose noble arches still watch over and reflect upon the waters that gave rise to them. The mill buildings of the late 1800s showcased the most modern theories of factory design and the oldest skills of the brick mason.

The typical southern mill village was built almost completely from the ground up, and its unique requirements as a manufacturing center dictated a layout almost opposite that of the traditional southern town. Instead, the founders of Spartanburg's mill villages used the model established in New England as their template for these new communities.

In the 1800s, the shorelines of South Carolina were replete with the homes of the well-to-do, who pushed close to the water to catch cool breezes as the town centers pulled back safely toward higher ground. The typical textile mill village eschewed this traditional layout to place its most important structure, the mill itself, on the river to draw power in a time prior to electric generators.

The remainder of the mill village focused upon and radiated outward from the mill, establishing both physically and symbolically its central position in the village hierarchy. Those with bell towers announced the mill both audibly and visibly, the ringing of bells a timekeeper for the workers below.

Whereas in a typical town the public structures—the town hall, the church, school, and store—were at the center of town and of daily life, in the mill village they were subordinate to the mill building. Occupying the inner perimeter of the village, they were dependent on and acquiesced to the textile mill proper. The outer perimeter of higher ground typically hosted mill worker housing.

The mill buildings displayed high quality design and detailing, built almost wholly of red brick often fired from clay dug at the site of the mill. These utilitarian structures were often finely detailed with corbelled brick cornices and copper-trimmed roofs atop the towers. The long rows of windows, though expensive, were specified to maximize light and ventilation for the workers. Heavy timbers, often cut from the surrounding forest, and cast-iron columns helped carry the enormous loads of looms and spinning machines. Though the masonry walls were typically up to two feet thick, the structures were surprisingly delicate; their facades dissolved into glass panes between brick piers. In many cases, the mill workers themselves built the mill structures. Though textiles have a relatively short life, the buildings are a more permanent testament to the skill and craftsmanship that defined an industry.

Many mills boasted a tower, sometimes two, and these towers were often unique to each building. Because the mills were laid out efficiently with only one plan, the only opportunity for architectural whimsy lay with the towers. Within his villages, Edgar Converse erected Romanesque and Beaux Arts towers as both landmarks and sentinels of the young communities.

Traditionally a town's public buildings are the most lavishly detailed, but in the mill village the public structures were usually simple and stark, often in contrast with the finely detailed mill structure. These buildings—churches, stores, and schools—were built simply of wood and were detailed in the vernacular Piedmont style of architecture, once common throughout the rural areas of Upstate South Carolina. These public buildings were the most uniquely southern and were more consistent with the context of Spartanburg County than the mills and houses inspired by those in New England.

Though town halls were often used as school buildings, separate schoolhouses were often built. An interesting case study in Spartanburg County was the school that served the Glendale community throughout much of the 1900s. Unlike the other public buildings, this was a handsome brick structure that reflected Edgar Converse's belief in educational opportunity.

The typical mill house in Spartanburg County was the simplest of all building types in the village. Symbolic of those who built and inhabited them, mill houses were simple but efficient; their architectural style, however, indicates the larger influence of the mill owner. Just as the mill was modeled on those in New England, so too were the worker houses. Built in the saltbox style, they contrast with typical Piedmont houses, though they did make concessions to the warm southern weather with lower-pitched roofs and large front porches.

Often villages had a boarding house or hotel for temporary accommodations, though houses were the primary residences for workers. In some villages, such as Clifton, houses were almost entirely duplexes, which afforded a stronger sense of individuality than boarding houses but were less expensive to build than single-family dwellings.

Archaeology and anthropology are founded on the principle that the culture of times past is accurately reflected in the places it leaves behind. The built environment is a window into the beliefs of past communities, telling the story of how the inhabitants moved, lived, worked, and worshiped. The form of a community is directly related to the way of life, modes of transportation, and the means of survival of its people. Mirrored in the waters of the Pacolet River are not only textile mills, but also the songs of a culture and the way of life of a people.

Clifton Mill No. 2. (Courtesy, Converse College)

Newspaperman Charles Petty
A Textile Cheerleader

By Allen Stokes

When the cotton mill campaign began in the 1880s, its greatest public booster was a newspaperman whose editorials rallied Spartanburg citizens and businessmen to support a major transition in the local economy.

Charles Petty, born January 15, 1836, near Gaffney, was the son of James Petty and Ruth Cannon Petty. He grew up on his parents' farm and was first educated at nearby "old field schools." He later attended D. D. Rosa's school at Limestone and St. John's College in Spartanburg. After graduating from Wofford College in 1857, Petty began teaching at Spartanburg Female College. He married Julia Davis of the Wilkinsville community in Union District in 1859.

He continued teaching until the summer of 1861. In August he volunteered in Co. C, 13th Regiment, South Carolina Volunteers. Petty was elected lieutenant and served with the unit at Gettysburg and in many battles in Virginia. Returning from Appomattox, Captain Petty began farming at Wilkinsville and was elected to represent Union District in the 47th General Assembly. Petty and Thomas Bomar purchased the Limestone Springs property in the early 1870s and reopened the school for girls, which Petty headed as principal in 1874. He remained in this position until 1878. During this time he was elected to represent Spartanburg County in the 52nd General Assembly, known as "the Wallace House," and he also traveled as a textbook salesman for Ginn, Heath & Co.

Petty began a new career in 1879 when he purchased the *Carolina Spartan*, Spartanburg's weekly newspaper, from F. M. Trimmier. He assumed the position of editor at about the time that Edgar Converse purchased Hurricane Shoals (Clifton) and Walker, Fleming & Co. purchased Trough Shoals (Pacolet) on the Pacolet River. In his role as editor of the *Carolina Spartan*, Petty was the town's booster and promoter of industry, commerce, education, and civic improvements. In countless editorials Petty sang the praises of industrialists who were investing their capital in ways that would benefit the larger community by providing employment for poor whites, a market for local cotton, and a stimulus to local merchants.

The factories organized in the early 1880s were located in the county, but in 1888, campaigns were begun to organize two factories, Whitney and Spartan, near the railroads in Spartanburg. Petty enthusiastically welcomed the prospect of mills closer to the town: "What our town needs now is an increased interest in factories of some sort. If a cotton mill should be put in successful operation here during the next year, it would not be long before we would have a foundry and machine shop, a shoe factory and other similar enterprises."

He also played a role in the naming of Spartan Mills, originally conceived as the Spartanburg Manufacturing Co. On May 29, 1889, he complained that the original name was too long: "It would wear out pens and lives, and waste quarts of good ink in writing and printing this…let the mill have a short euphonious name that one can speak trippingly on the tongue and write with ease and rapidly. Spartan Mills as a name has been suggested by some, and that would be very suitable as that was the original name of our county."

THE
Carolina Spartan.
ISSUED EVERY WEDNESDAY,
At Spartanburg, S. C.,
TERMS--$1 50 Per Annum, in Advance.

PRETTY IS AS PRETTY DOES.

BY ALICE CARY.

The spider wears a plain brown dress,
And she is a steady spinner;
To see her, quiet as a mouse,

Petty edited the *Carolina Spartan* until the paper merged with the *Spartanburg Journal* in 1913 and was published as the *Journal* and *Carolina Spartan*. He continued to write for the paper until shortly before his death in February 1915. Petty was a member of Central Methodist Church, a trustee of the graded school system that began in 1884, an active participant in the United Confederate Veterans, and a member of the local committee to celebrate the centennial of the Battle of Cowpens. He was buried in Oakwood Cemetery and was survived by his wife of 56 years and five of the eight children that were born to the couple: Mrs. Mary P. Calvert; Paul Petty; Mrs. C. J. Shearn, of New York; Mrs. A. F. Pringle, of Charleston; and Mrs. Charles H. Henry.

The Flood of 1903

Terror Along the Pacolet River

By William M. Branham

At 5 a.m. on Saturday, June 6, 1903, Hicks Stribling, storekeeper at Clifton Mill No. 2, heard water gurgling below his second-story room in the company store. As water rose up to meet him, he scrambled to the roof and then to a nearby tree where he spent the next 11 hours. He was naked. As dawn rose and revealed his condition to a woman in a nearby tree, she offered her apron to restore his dignity.

Ben Johnson, a merchant of the settlement of Santuck, just below the No. 2 mill, drifted nine miles with his wife and two children, down the raging Pacolet River on the roof of his house, only to see his family disappear over the Pacolet Mills dam in a swirling eddy.

These are two of the harrowing stories of the "June Freshet of 1903" on the Pacolet River in upper Spartanburg County. When the waters subsided and the catastrophe was totaled up, there were more than $300 million (today's dollars) in damages, 600 people left homeless, 4,000 without jobs, 70 homes swept away, and 65 people killed or missing, some of whose bodies were never found.

In 1903, the 10 miles of the Pacolet River valley between Converse (where present day U. S. 29 crosses the river east of Spartanburg) and the village of Pacolet Mills boasted a bustling community of 10,000 people. Seven major cotton spinning and weaving mills ranged from the giant 50,000-spindle Clifton No. 3 to smaller mills at Clifton, Glendale, and Pacolet. Five dams along the river supplied the necessary power to run cards, spinning frames, and looms to produce a variety of cotton goods upon which the economy of Spartanburg County depended heavily.

Life in those mills meant hard work and long hours. But in off times, the pleasant Pacolet River provided swimming, fishing, and Sunday picnics along its banks. The electric railway offered easy access to downtown Spartanburg and the shopping joys there. The Aug.W. Smith Co. advertised first-quality men's suits at $15, the John A. Walker Co. offered a splendid shipment of ladies' shoes and oxfords at $1.50 a pair.

And so it was that in busy, bustling Pacolet Valley in June 1903, the spring crops up the river toward Campobello and Fingerville had been planted and the rows of corn and cotton were well on their way. Early in June the welcome spring rains began. From its headwaters near Tryon, North Carolina, the Pacolet River meanders from a trickle, crossing the South Carolina line southeasterly to Landrum, growing larger through Clifton, Pacolet, and on to the Broad and to the sea.

Pacolet Mills residents stand by as rising flood waters approach Mill #3. (Courtesy, Town of Pacolet)

The settlers along its banks were accustomed to rising waters, and most were fully prepared. Every year cattle were moved from bottomlands, machinery and tools put away. At the mills, no extra precautions were necessary as almost 20 years' experience brought knowledge of expected river behavior.

But late in the afternoon of June 5, the third day of heavy rains, the North Pacolet River began to swell on its way to Fingerville.

Around Landrum, the Pacolet took the waters from North Carolina and sped them on toward the sea. From Spivey's, Motlow, and Obed creeks, the runoff raised its level to 10 feet over the normal bed of the river. Waters rose to eight feet in the cotton mill at Fingerville, and the machinery was destroyed. Just below Fingerville where the North Pacolet and Pacolet join, the water was higher. Further down at Buck Creek, 15 feet above normal, William Harden's grist mill was demolished and its timbers added to the tons of debris smashing down the channel. Lawson's Fork Creek, as it rushed through Spartanburg, carried with it the major Southern Railway trestle serving the city, dumping it in the river above Converse.

But the *coup de grace* was to come at about 3 a.m. in the Campobello area some 30 miles upstream. Natives called it a "water spout," but it was evidently a small tornado accompanied by a fierce cloudburst. There are no records of how much rain fell in an extremely short period, but it must have been horrendous. Sweeping everything in its path, a wall of water gained strength as it tumbled down the Pacolet Valley. Its first major obstruction was Clifton Mill No. 3 (now Converse) at the bottom of a gorge alongside the river just north of the Southern Railway trestle. Mrs. C. W. Linder describes the scene: "The five-story, 50,000-spindle mill trembled for a while, and then gave way. Like a huge matchbox, it was carried under the trestle and on down the river." The dam gave way at the same time, and the water rose some 40 feet in a matter of minutes. The current there was estimated at 40 miles per hour. Mrs. Linder's house and others followed the wreck of the mill downstream. Three-quarters of a mile further down, Clifton Mill No. 1 awaited the onslaught. Warning had been given but there was little time to prepare. The wall of water thundered down the valley sweeping away everything in its path. All of No. 1 village within a hundred feet of the former riverbank was destroyed. The torrent took away a third of the mill, inundating the lower floors.

Flood damage in Clifton (Courtesy, Bill Lynch)

The upper end of the mill was demolished, exposing crooked and bent machinery and broken timbers. Long dirty streamers of yarn goods hung from the second and third story windows, the ends still attached to the wrecked looms. Portions of the dam and the turbines were gone. Three persons who had taken refuge on top of the mill were lost when a portion of the building collapsed.

The flood's greatest toll was yet to come. At Clifton Mill No. 2, Hicks Stribling was scrambling to his tree. Most of the operatives had been warned and had, in the few short minutes that the water took to flow from No. 1 to No. 2, managed to flee to higher ground. But some veterans of the Pacolet just didn't seem to believe the calamity to come and stayed in their homes.

The fury struck No. 2. It took away half the four-story mill. Normally 100 feet wide at this point, the river had spread to more than 500 feet. Terrified men, women, and children took to the trees, climbing higher and higher as the waters rose. Keeping them company in the branches were snakes, raccoons, and all manner of wildlife.

Just below No. 2 lay the settlement of Santuck, about 16 crackerbox mill homes in a low-lying bend. Here is where the disaster took its greatest toll of life. John Merchant, a second hand in the card room at No. 2, saw his sister, her husband, and three children swept away. The brother-in-law managed to catch a limb and be rescued. One of his children, a boy, was seen at Pacolet, nine miles down the river, fruitlessly crying for help as he was forced over the dam in the eddy below and lost. Mr. and Mrs. J. B. Findley drifted some five miles down the river with the debris. Mrs. Findley was lost. The rescue efforts at Santuck and the Clifton Mills were heroic. George Willis, 17, a weave room attendant at No. 2, and a friend spotted a woman clutching sticks of cordwood as she struggled to escape drowning. They commandeered a well rope, threw it to her, brought her close enough to shore, and carried her packsaddle the hundred yards to higher ground. Hicks Stribling and his apron were finally rescued from the tree. Someone enlisted the help of a star Converse baseball pitcher who tied a string to a ball and threw it to the hapless refugee. A rope was carried over, and Stribling finally made it to safety and clothing about 4 p.m. on Saturday.

Makeshift rafts were made from some of the thousands of cotton bales that were dumped from the warehouse by the force of the water. By tying ropes to the rafts and floating them out to the stranded survivors, most were saved, but some who couldn't hang on for 11 hours were lost.

One black man had been offered a dollar for each person he rescued with his

VOL. III.

BY RAGIN

Cotton Mills, Wa
Goods W

APPALLING CALA

Deluge Fell
Last

BRIDGES S

Most Appalling L
the History o
Over a Mil
Total

Spartanburg a
visited between m
this morning by pr
serious calamity i
the form of a rainfa
resemble a cloudbu
eral hours There
and no damage by
only water. This fe
veritable torrents
hilly character of th
vented the surface o
ing a lake of water t
As it was the rive
their banks to height
known, the creeks bec
and the small branche
became roaring torre
away everything in
Railroad bridges we

cotton bale raft. When he reached 99, he said he'd try for 100. On this last try, his bale overturned and he was lost to the river.

The greatest loss of life was at Clifton No. 2 and Santuck, but the greatest destruction of property was yet to come. Nine miles down the river, Pacolet Mills No. 1, No. 2, and No. 3 awaited the flood. By this time evacuation along the danger zones was complete and it was daylight. Spectators lined the banks.

First to go was Pacolet Mill No. 1, a four-story, 30,000-spindle mill that went down with a crash about 9 a.m. The crowds watching on the bank turned their attention to the adjoining mill, No. 2, half-submerged in the muddy waters. After about an hour, it shuddered and went under with a deafening roar. Pacolet No. 3 partially survived.

As the spectators watched, the body of a woman "cold in death" was carried downstream. "The horrified crowd watched as a small boy, clinging to wreckage near the woman, called for help as he passed before the eyes of the gazing but helpless crowd. An attempt was made to reach the child, but no one was able to brave the fury of the waters, and the little fellow was carried downstream calling pitifully for help."

By late in the afternoon, the waters along the river subsided and the work of picking up the pieces began. There were more than 600 homeless, without food, clothing, money, or shelter. The regular supply routes along the Electric Railway were gone; Glendale was as far as it could go. There was no electricity. Most survivors were just too tired and dazed to do anything.

On Sunday the search for the dead began. Many bodies were never found, washed down the river or buried under tons of sand and wreckage. One woman was found only because her knee projected above the sand. As the clay silt dried out, flies attracted to the cracks betrayed the location of many dead below the surface. The victims totaled 65 with some families completely wiped out.

President Victor Montgomery of Pacolet Mills and his brothers, Walter and Ben, together with a crew of laborers, struggled to save cotton and undamaged goods from the partially destroyed

PARTANBURG JOURNAL.

SPARTANBURG, S. C., SATURDAY EVENING, JUNE 6, 1903

PRICE: $5 A YEAR

time the structure seemed to be in sound condition. Today nothing remains where the trestle stood but two stone abutments, one in mid-stream and the other near the north side of the rushing torrent.

Lawson's Fork, which is ordinarily a small stream about 50 feet wide, is today an angry torrent two hundred feet wide rushing madly down to augment the already swollen waters of Pacolet river. All day today hundreds of people from the city were going and coming to and from the scene of the catastrophe. The sight was one to inspire awe in the breast of all who beheld the scene. Far up the river the angry waters could be seen rushing madly down stream almost like a thing of life. At the bend in the river several hundred feet above the trestle the scene was grand.

POWER HOUSE UNDER WATER.

The electric power and gas plant located beside the trickling waters of Chinquepin creek was flooded before 3 o'clock, the water being four and a half feet deep in the furnace room and completely submerging the fires under the boilers. This put the electric light dynamos out of business and left the city in darkness. It was not until about 10 o'clock that the machinery could be run and no cars were operated until that hour. Manufacturing and other plants using electric power were necessarily left idle. The damage to the fire boxes and pumps was not great, but the retorts were much damaged. However, there is

100 LIVES LOST?

Reported that Clifton No. 3 Was Wrecked This Morning,

AND MANY COTTAGES DESTROYED.

Communication Almost Impossible all Day.

WIRES ARE ALL DOWN.

It Was Cried Across the River to Persons in an Electric Car that No. 3 Was Gone and 100 Drowned.

It is reported this afternoon that Mill No. 3 at Clifton succumbed to the force of the flood this morning and that 100 lives were lost. The report came from parties who appeared on north side of Lawson's Fork, this morning where the railway trestle went down. The men shouted the news across the waters and although it was difficult to hear their voices above the roar of the water, the purport of what they said as heard on this side was to the above effect.

Clifton No. 3, is a four story structure located above the high trestle of the Southern Railway about 200 feet distant. The mill is in a deep valley surrounded on both

when he looked back and saw the bridge going down. He turned on a full head of steam and so close was the call that the last cars were pulled up an incline from off the bridge as it started down.

DAMAGE AT ARKWRIGHT.

At the Arkwright Mills two stores and three bridges were washed away. The stores belonged to John Burk and B. O. Pennington. The water reached the warehouse of the mills but did little damage.

OTHER MILLS REPORTED LOST.

It was reported here that the mill at Lolo, a 5,000 spindle plant, had been completely washed away. This mill is owned by S. W. Scruggs principally and runs on yarns entirely. There was also a rumor to the effect that the Fairmont Mills had been washed away.

THE DAMAGE AT INMAN.

News from Inman today states that the dam of the Inman Cotton Mills was washed away completely. The cotton mill, however, escaped injury. A grist mill belonging to Mr. Wingo was also washed away. The county bridge near Inman was carried away by the force of the water.

Wires All Down.

All wires, both telephone and telegraph, are down and it is impossible to get a mesage through in any direction. One or two wires were working this morning, but even these are gone and no communication can be had with any point. People outside of Spartanburg must think the whole place is swept

3 PACOLETS GONE

The Entire Manufacturing Plant Swept Away by the Waters,

DEAD BODIES FLOATED BY IN STREAM.

Presbyterian Church Among the Destroyed.

THOUSANDS IDLE THERE.

Magnificent Industry Which Ha Been a Most Succesful One Washed Down Torrent of Pacolet River.

The worst damage so far reported was at Pacolet Mills. The water there had been rising for several days but the heavy downpour last night caused to be emptied into the river and its tributaries great volumes of water which rushed madly down stream on their mission of destruction. The big three story mill, No. 1, at Pacolet went down about 9 o'clock this morning with an awful crash. Then the crowds of people about the spot and lining the high cliffs which overlooks the mills, turned their attention to the large four-story mill, No. 2, which stands adjoining No. 1, and which was even then half submerged in the angry waters. The crowd watched for an hour as the fate of

CHOATE'S SON WEDS.

Imposing Ceremony at Albany Joins Happy Hearts.

ALBANY, N. Y., June 6.—Fashionable society people from New York, Washington and other cities thronged St. Peter's Church today at the wedding of Miss Cora Oliver, daughter of General and Mrs. Robert Shaw Oliver, and Joseph H. Choate, Jr., son of the American ambassador at the court of St. James. The bride wore heavy white tulle, edge all around with point lace, which was fastened to her hair with a diamond ornament. Her attendants included Miss Mabel Choate, a sister of the bridegroom; Miss Mary Bowditch and the Misses Elizabeth and Marion Oliver. Following the ceremony at the church the bridal party and the guests including Ambassador Choate and Mrs. Choate, repaired to the home of the bride's parents, where an elaborate reception was held.

The bride of today is the daughter of General Robert Shaw Oliver, one of Albany's richest and most influential man, widely traveled and a famous wit. She is widely known for her beauty and a devotee of outdoor athletics. It was on the golf links that she first met Mr. Choate. There the courtship was pursued until an engagement resulted.

Mr. Choate was an honor man at Harvard, is a celebrated golfer and is a fine stalwart type of American manhood. He has been in service as third secretary of the American embassy in London and is on an indefinite leave. Mr. and Mrs. Choate will probably spend the summer in this country and later take up their residence in England.

SOUTH CAROLINA NOTES:

Deaths in the Pacolet Flood

Clifton No. 2
Julius Biggerstaff
Augustus Calvert, his wife
and two children, Felix and Lou
Bud Emory
Mrs. J.R. Finley
Joel H. Hall, his wife, his mother,
and Ella, Jimmie and Lola, his children,
and five other children
Mrs. Henderson
Mrs. B.S. Johnson and her five children
Oliver Johnson
Roscoe Johnson
The Louin family of eleven
Mrs. Massey and four children
Ed Robbs, Mrs. Robbs and two children
Genoble Sims
Novie D. Sims
Landrum Waddell
Martha Waddell
Dock Williams
Mrs. Jane Williams' baby

Clifton No. 3
Mrs. Fleetia Gosa
Mr. Grier
Mrs. Henson
Miss Maggie Kirby
Mrs. William Kirby
Garland Long and his wife
Mrs. John Owens and child
Roy Owens
Samuel Swearingen and his bride
William Wood

Pacolet
Quay Worthy

—from *The History of Pacolet, Volume II*,
by Willie Fleming

Mill No. 3. But there was little to be salvaged. Most of value was on its way to Columbia and the sea on the waters of the Broad River.

By Sunday the sightseers had begun to gather. Though the electric railway was out, the Spartanburg livery stables were completely rented out of the flashy carriages and buggies so popular. June Carr, a photographer from Gaffney, loaded his bulky 8 x 10 camera and the fragile glass plates and journeyed to Clifton to record the devastation. Curious onlookers, dressed in their Sunday best, roved through the destruction carrying off spindles and bits of cloth as souvenirs.

Reporters from as far away as Atlanta descended on the scene. One issued a dispatch that made the national news concerning the destitution and deprivation of the operatives at the mills in Clifton. And this is where the egg hit the fan: Though both Pacolet and Clifton Mills had made special efforts to pay off employees so they could purchase food and clothing, and a relief committee had been formed that ultimately collected more than $15,000 from as far away as Atlanta, Charleston, and Philadelphia, a Judge Williams had reported on Tuesday the 10th that many of the operatives were without shelter or clothing, went to bed supperless, and had been without bread part of the time since the disaster. He returned, he said, and simply confirmed the report that many of the operatives were on the verge of starvation and without shelter. He reported that he personally canvassed his friends and the neighborhood and gathered what food he could find, and he and his friends did what they could. "Nothing has been done for these people and conditions are getting very bad. Why the committee in Spartanburg doesn't loosen up is more than I can understand. These people need food and money, and they are not being treated right at all. Something should be done and done at once or we are going to have a lot of starving people on our hands."

This report brought forth cries of outrage both from R. H. Chapman, chairman of the relief committee, and from the newspapers that were not noted for their criticism of mill owners. Chapman said he had visited at both Pacolet and Clifton and had talked with the relief committee there and if such conditions existed, he didn't know of them: a head-in-the-sand condition common to many mill owners of that time. Ultimately the relief committee reported that all the money needed had been obtained and asked that further contributions be withheld. South Carolina Governor Heyward donated $50. Thus a grand total of $15,000 was available for 600 homeless and 4,000 unemployed residents of the valley.

No organized relief was provided for the hundreds of small farmers along the path of the flood who lost homes, stock, tools, and crops. Federal aid to the victims amounted to one carload of food and clothing.

The Post Office Department came into its share of trouble. A big flap arose about the mail service. As mail came by train, and bridges were out to the north, west, and east, nothing was moving. It was Tuesday afternoon, after much vocal citizen and newspaper complaining, that mail finally arrived from Augusta.

Among the other rumors prevalent at the time was that the large dam at Lake Toxaway had broken, causing the surge of water. This brought heated denial from the management of Toxaway Company who wished to let it be know that everything "they did was in a first class manner and their dam had not broken…only a small pond had gone over its banks."

A crew digs through the rubble of one of the Clifton mills (Courtesy, Converse College)

By Thursday things seemed to have settled down. The Southern Railway had repaired most of the trestles coming into town. Ferries were operating in a number of places along the Pacolet. The task of burying the dead was underway. One funeral was conducted via a special car on the Electric Railway from Glendale to the cemetery in Spartanburg. Many of the dead were interred in ground on a hill above Clifton Mill No. 2.

Today along the Pacolet Valley in the path of the flood, it's relatively quiet and peaceful. None of the rebuilt mills are operating any longer. Santuck is brush and forest. No one rebuilt there. Most of the stones in the graveyard are tumbled over and covered by weeds. Clifton village is a cluster of duplex mill homes, now converted to single-family use, populated with retired mill workers and a sprinkling of younger families. Modern Spartanburg is creeping steadily closer. Only a mile from Clifton is Broome High School, and $100,000 homes are being built nearby. But like Ole Man River, the Pacolet goes on. At night or on a quiet Sunday morning, the only noise is the rush of the water over the abandoned dams.

—Reprinted from Sandlapper magazine.

Vesta Mills
An Experiment with Black Labor

By Allen Stokes

By the end of the nineteenth century, the concentration of mills in such manufacturing centers as Spartanburg and Greenville placed a heavy strain upon the supply of local cotton and labor and prompted owners to look to other areas for the location of factories. The partnership of John H. Montgomery and Seth Milliken ventured far afield in 1899 with a controversial purchase of a mill in Charleston staffed primarily by African Americans. Their experiment with black labor sent shock waves through the mills of the Upstate and eventually caused a period of cooler relations between the two partners.

The purchase of Charleston Cotton Mills by Montgomery and Milliken came as the National Union of Textile Workers had launched an organizational campaign in South Carolina in 1898. The campaign experienced limited success in the Upstate, but viable locals were organized in the Horse Creek Valley region of Aiken County and in Columbia. The Charleston mill purchase also occurred during a period when mill owners faced challenges posed by critics who began to discuss the "cotton mill problem." Criticism of conditions in southern factories, which surfaced initially in the 1880s, escalated in the 1890s. Manufacturers and industrial boosters were disturbed by the closer scrutiny and placed on the defensive.

Against that backdrop, the two textile pioneers looked to the Lowcountry. Charleston Cotton Mills had failed in 1898 after a tumultuous period of labor unrest between former white operatives and African-American operatives who replaced the white employees when O. H. Sampson, a Boston businessman, acquired the company. Montgomery and Milliken acquired the property at auction for $100,000 in 1899 and renamed it Vesta Mills. The complex story of the unsuccessful Vesta partnership is contained in a book of letters from John H. Montgomery to Seth Milliken and others, now part of the Special Collections at Clemson University.

Montgomery's correspondence does not reveal the extent of his financial investment in Vesta; but in response to inquiries from Charlestonians about acquiring Vesta stock, Montgomery noted that the "back bone [Milliken]" of the recent purchase considered it a bargain and was reluctant to "give it up" when "to own it all would be easy for him." A year later, however, with problems compounding at Vesta, Montgomery advised Milliken that one factory in the Piedmont was worth two anywhere in the state south of Newberry.

By employing black labor in what Montgomery called the "coon mill" in Charleston, Montgomery and Milliken may have sought to demonstrate to their white labor force that there was an alternative source of labor available in the event that they chose to align with organized labor. Montgomery's decision "to try colored help" surprised J. H. Williams, superintendent of the Lockhart Mills: "I have said that I believe the negro can be educated to do mill work and I predict a success for the Capt.…but I do hope he will fail on purpos[e] and forever settle this question if He will do that He will do the greatest piece of work He Has ever done in His life."

The prospect of resuming production at the Charleston factory was applauded by the Charleston *News & Courier* where it was noted that with local management the operation of the factory would be beneficial to African Americans in areas of the South where there was a shortage of white workers. A newspaper in Charlotte, North Carolina, sounded a similar theme as it commended the employment of African Americans and asserted that Vesta's success "would doubtless be 'viewed with alarm' by our manufacturing friends in New England." The editor of the *Yorkville Enquirer* introduced a note of caution when he suggested that the successful employment of African-American labor at Vesta could be a source of endless trouble for the industry in the South.

Montgomery sought to avoid any undue public attention when production began at Vesta Mills, and only brief accounts of the mill's operations appeared in the local newspaper. There does not appear to have been a recurrence of the racial confrontations that plagued the operation of the Charleston Cotton Mills. The approximately 40 white employees worked largely in the weave room. After Vesta Mills had been in production for about six months, Montgomery was pleased with the results in the carding and spinning departments. It is not clear how much, if any, interaction occurred between the races, but Montgomery remained convinced that blacks eventually would be employed throughout the mill: "It is slow but I believe we will 'get there' some time." In fact, Montgomery was more concerned with managerial problems than with the labor. The shortsightedness of the mill's supervisors disturbed him. He cautioned the manager in February 1900: "Not looking ahead has been one of the lame places most of the time…Looking ahead, and closely after details, is what it takes to run a mill successfully." In July 1900 he notified the manager that "the cost of your goods is a weight upon my mind." In Montgomery's estimation, low production and high fixed costs burdened the company.

In the fall of the year Milliken attended a stockholders meeting. The mill's modest earnings apparently led to some discussion about closing the mill. Montgomery strongly opposed this action, which he considered would reflect adversely on his reputation as an owner. "I would rather lose fifteen thousand dollars invested some where else than to have that mill fail," he wrote to Milliken. Montgomery outlined plans for putting the mill on a sound footing. These plans included placing orders for Draper looms, picker-room machinery, and modifications to the spinning frames.

In spite of these plans, Vesta's owners decided to cease production in January 1901 and to move the machinery to Gainesville, Georgia, where Pacolet Manufacturing was constructing a new mill. A period of cooler relations between Montgomery and Milliken followed this decision. Montgomery aligned himself with mill engineer Stephen Greene and advised that a debt owed

Milliken could be erased by a sale of Vesta's inventory of cotton and yarn. Frank E. Taylor, a Charleston fertilizer manufacturer who had sought to acquire stock in Vesta for himself and others, cast the lone dissenting vote among Pacolet's board of directors against expanding to Gainesville.

The closing of Vesta Mills brought forth a flood of comment in the press. The prevailing sentiment attributed the mill's failure to African-American labor. One newspaper argued that the mill's failure removed a threat that had caused apprehension among white employees: "They saw in it either social degradation or starvation wages, and they were not slow to express themselves against the people who were thus forcing them as a class against the wall, that they might enrich themselves with the gains thus forced from the pockets of the poor." In his public pronouncements on Vesta's failure Montgomery cited locational disadvantages. His letters also mention "distractions" in Charleston that lured the operatives from the mill—urban distractions not encountered in the Upcountry. As for the quality of the African-American workers, he thought that the problems encountered at Vesta did not differ from those encountered at Upcountry mills where only half of a "green white labor force" would become capable employees. Montgomery might have agreed with Charleston businessman William Bird, a Vesta stockholder, who attributed the mill's failure to Milliken "[who] took every opportunity to show the colored labor unprofitable. Those negro women could tie a knot as well as white women could."

In 1915 South Carolina passed an act that forbade whites and blacks from working together in the same rooms. The only employees to whom the act did not apply were firemen employed as subordinates in boiler rooms, floor scrubbers, persons responsible for cleaning lavatories and toilets, and carpenters and other employees responsible for the maintenance of buildings. Fifty years later legislation by the federal government and economic diversification that lured the sons and daughters of a later generation of mill workers away from the cotton factories gave African Americans access to employment that had long been denied them.

Migration from the Mountains

Building a Local Labor Force

By Diane Vecchio

Early mill workers in Spartanburg County were mostly native South Carolinians and their descendents, coming from farms where they had been small landholders or tenant farmers who had fallen on hard times. Some had failed at farming; others, particularly tenants and sharecroppers, could no longer survive a system where the landowner received one-half the crop.

But by the early 1900s, the supply of local labor had been nearly exhausted. Between 1905 and 1907, this scarcity of labor needed for the growing number of mills in Spartanburg County led mill owners to recruit new workers from nearby states. Labor scouts penetrated the Blue Ridge Mountains and recruited millhands from the areas around Asheville, Hendersonville, and Waynesville, North Carolina, and further west to Tennessee. Mountain people, confronted by a powerful push off the land and an increasingly attractive pull into the cotton mills, were native-born, many of them descendents from English, Scots, and Scots-Irish settlers.

Fiercely independent and ambitious, labor migrants from the Blue Ridge Mountains turned to mill employment for a variety of reasons. A father, who decided that his daughters could better contribute to the family economy by earning mill wages than by working the fields; a widow, suddenly finding herself the head of a large family of young children; a small farm owner, lacking the resources to compete successfully in a market-based economy. These individuals are just a few representing the migration

A family moving from the mountains of Tennessee and North Carolina to Spartanburg County cotton mills signed a contract such as this. This 1882 contract was for work in Clifton Mills. (Courtesy, Bill Lynch)

State of South Carolina, } Feb 2d 1882
SPARTANBURG COUNTY.

This Witnesseth, That I, Phillip F Baker do this day contract with the Clifton Manufacturing Company to move my family to Clifton, S. C., and occupy one of their tenements at that place, and for the entire year of 1882 furnish my faithful services when wanted, and also that of my children as operatives in their Factory, and do hereby bargain to so continue in their employ until all advances made to me by them are paid in full. I also agree to submit to and support all Rules and Regulations existing in the Factory and on the place customary in such cases.

My family consists of:
2 *Boys, aged*
1 *Girls, aged*
Witness:
H M Chapman

Phillip F Baker

of men, women, and children, seeking steady employment and a better standard of living.

Over 3,500 people left the rugged life of the mountains to become mill workers in South Carolina's Upstate during the first two decades of the twentieth century. Those promised a free ticket to Spartanburg County were often met at the railroad station by a labor scout from one of the mills. Other families packed their goods in a mule or horse-drawn prairie schooner and made their way down the mountains on their own.

Labor migrants from the mountains were accustomed to isolated and independent lives and often had difficulty fitting into mill villages. "Mountaineers" were portrayed as poor, uncultured, and in some cases, "half wild." Yet, in some ways, they closely resembled the local mill workers employed in South Carolina textile manufacturing. They came in family groups and were white native-born offspring of northern European immigrants.

Perhaps the most striking feature of the labor system in South Carolina, as well as in all Southern textile mills, was the exclusion of black millhands. In some parts of the state, slaves had worked in spinning and weaving rooms in antebellum textile mills. When slave prices rose during the cotton boom of the 1850s, Southern manufacturers turned to poor white farmers for cheaper labor. After the Civil War, mill work was for whites only. A number of factors limited black employment in the textile industry, including the promotion of the mills as the salvation of poor whites, the desire to tie blacks to agricultural labor, and the deepening of segregation in the post-reconstruction South. In South Carolina, occupational segregation became law with the Segregation Act of 1915, which made it illegal for anyone "engaged in the business of cotton textile manufacturing...to allow...operatives...of different races to labor and work together within the same room." In some mills, however, black men were hired for the heaviest work, in the yards, for example, moving bales of cotton and loading boxcars and wagons with finished goods. Black masons put up the walls of Spartanburg's Saxon Mills, and black laborers under white supervision built the streets, but when Saxon Mills opened, only whites were hired as operatives.

The lives of local workers and mountain migrants came together in company-owned mill villages that provided workers with single-family dwellings, one or more churches, a modest schoolhouse, and a company store. These amenities were essential to securing a labor force, yet manufacturers also saw in them the means of exercising control over their employees. In a government bulletin published in 1908, federal investigators reported that "the company owns everything and to a large extent controls everybody in the mill village." Wilt Browning, one of the last generation to be born and raised in Easley's mill village, writes that "the mill provided shelter and, in a sense, clothing and food as well. And people took comfort in that. Nobody called it 'Big Brother' at the time, but that's what it was."

Browning's colloquial reference to "Big Brother" was nineteenth-century paternalism. Paternalistic measures were used in Southern mills, as well as in Northern industries, to attract workers, to socialize them to industrial work, to instill loyalty to the company, to curb labor unrest, and to prevent unionization. In the South, paternalism was also seen as a way to protect the mill workers, and thus society, from the demoralization that many South Carolinians saw implicit in industrial life. Mill workers' lives were molded—behaviorally and morally—by the social control implicit in mill village life.

The lives of women, children, and men who became millhands in South Carolina during the late nineteenth and early twentieth centuries were shaped by the paternalism that ruled mill village life. Company directed work and play, discipline and fraternalism, created a workforce that met the needs of cotton mill owners and ameliorated differences among workers. However, the reciprocity that had been carefully nurtured between the mill owners and the workers was eventually eroded as the century gave way to economic depression and labor unrest.

Pinto Beans and Cornbread
Dinnertime in the Village
By Baker Maultsby

Textile mill villages, with their indoor workplaces, rows of houses, and company stores, often gave workers just off the farm their first taste of city life. Still, "city life" didn't always mean a change in diet for the new mill village residents.

Connie Trent, who was brought up in Clifton, remembers getting in trouble at home when she "forgot to slop the hog" on the way to school in the morning. The Clifton village, like numerous others, offered workers space to raise pigs and cows and to grow vegetables. Whether agricultural space was provided as a quality-of-life enhancement or was simply the most efficient way for mill owners to see that their employees had enough to eat, fresh produce was essential to the diets of many mill families.

"Back when we were little, Daddy always had a garden," recalled Betty Gibson, who grew up in Startex. "There was a pasture, and everybody had their own place. We grew 'maters, corn, okry, potatoes. And Daddy would butcher a hog."

Some mill towns, though, had no space for gardening, so workers had to rely on whatever could be bought at the company store. Even in Clifton, Trent remembers eating beans, "taters," and cornbread day after day. And when the supply of pork and beef was running low, some Clifton residents would take off into the woods to hunt squirrels and rabbits. Squirrel dumplings were a favorite in Trent's childhood household. And for dessert, there were no Krispy Kremes or delicatessen pastries to be had. Only "soakie"— biscuits dunked in coffee. Add sugar to the mix, and you had "stickies."

Jeff Crompton's mother Lillian used to make "hoe-cakes," at their home off Saxon Avenue in the 1940s. These were large flat biscuits made in a frying pan and eaten with butter or molasses. For breakfast, she sliced fatback and fried it, then made gravy with the grease and served it with eggs and biscuits. The biggest meal of the week was served at midday on Sunday when the Crompton family typically cooked a chicken from the yard.

Ethalia Roberts, who lived on the outskirts of Clifton, worked as a cook for a succession of mill supervisors. Clifton's top brass, she recalled, "ate good. They had steak, chicken, vegetables. Not the pinto beans like the people on the mill hill."

Journalist and food historian John Egerton says academic research on the diets of mill workers is scant, but he believes mill families who had access to agricultural land ate about like rural Southerners. "They didn't have the best diets in the world," Egerton said, "but they weren't the worst off in the world, either. These were people who by our standards were dirt poor. But poor people had more than we would think."

Still, many Southern laborers didn't get proper nutrition. Prior to 1940, lack of fresh vegetables and meat sometimes resulted in cases of pellagra, a disease that caused skin rashes, mouth ulcers, and diarrhea. At its worst, pellagra could cause death or be mentally debilitating. Dr. Joseph Goldberger, working for the federal government, tracked the disease and discovered its link to nutrition and poverty. He found the disease most common among the

Amanda Brown Kirby of Pacolet Mills and her cows. Many mill families had farm animals and vegetable gardens. (Courtesy, Don Camby)

poorest sharecroppers and laborers who ate steady diets of cornbread, molasses, and the cheapest cut of hog—fatback.

If healthy food was scarce in the village homes, workers' diets weren't likely to get a boost at the mill, either. Although modern industrial plants now feature full-scale cafeterias, for adults and children who worked in Southern textile mills, lunch breaks were a rare luxury before World War II. Often their lunches amounted to leftover biscuits in a bag brought to the door of the mill by a family member. The only food provided was that purchased from the "dope wagon," a mobile cart that sold crackers and soda. (Dope wagon alumni include Trent, who now can be found stirring up laughter at Dolline's Restaurant in Clifton, and the late Frank Hill, the Lockhart man whose jewelry made of bird droppings landed him on the "Tonight Show with Johnny Carson.") According to historian Ed Beardsley, it was not until the 1940s, when textile mills had to compete with the military and its industrial suppliers for workers, that cafeterias became commonplace in Southern factories.

So, as in other aspects of life in the mill villages, dinnertime has undergone a transformation. With mills closed and company stores long gone, village-dwellers, like city folk, eat fast food burgers and buy their produce at Bi-Lo and Publix. As Connie Trent put it, "Now-days, people don't have to chase a rabbit down. They got all kind of meat. I eat good stuff like everybody else."

Lillian Crompton's Poke Sallet

As told by her son Jeff Crompton

Usually you found it growing wild. We lived off Saxon Avenue. My mother would find it along the railroad track. She used to boil that stuff several times and drain it. She'd have to do that to get the poison out. You had to cook it and drain it just right or you'd get sick.

To me it looked like a weed growin' up. You get the tender leaves growin' off of it. It was usually in the spring. That was one of the things we looked forward to. She used to season it with fatback. It was like salad, and it was dark like turnip greens. Then you'd put salt and pepper on it.

Pinto Beans

Recipe from Mary Ellen Lane

First you get a bag or so of dried dark pinto beans and you look 'em real good 'cause when they plow 'em up in the field they got lots of rocks in 'em. Then you get a big pot, scrubbed out with steel wool, and wash the pintos with three clear, clean waters. Then you soak 'em in that same pot you plan to cook 'em in. I've always heard them say that if you soak 'em they won't be so bad to give you gas.

You start out with four or five cups of water. My Aunt Lib used to say add enough water so that you can start off with what you'll need, then you won't have to add new water. 'Cause you know, what's boiling in the beans is what makes that good rich juice.

Then you put in a little salt at the start and you can put in some margarine. For health I stopped using fatback, but with it they sure would taste better. I'd say now a little piece of fatback wouldn't hurt much. With the margarine I usually put in a little Crisco oil too. After that you just let 'em cook. Two, maybe three hours. Pintos is something we never got tired of—even for breakfast.

Ways to Cook Squirrel

Recipe from Connie Trent

Anywhere there's woods, there's squirrels. You can skin 'em, cut 'em up, and stew 'em. If you want to fry it, first you have to skin that thing. Put salt and pepper on it, roll it in flour and fry it up. You leave the bone on there. Fry it slow. Make it good and tender. Cook it slow.

And then you can make gravy with it. You take some of the grease and mix some flour with the grease to thicken it a little bit. Tastes good. Ain't nothing wrong with it. All gravy's good.

Or you can make squirrel up with dough. Squirrel dumplings. You pull the squirrel off the bone. You use flour, butter, and milk to make your dumplings. Boil a pot of water. Put in the meat, and leave it in there 'til it boils. Then put in your dough with some butter. Cook it 'til the dumplings start to get firm. Your dumplings wind up separate from the meat. It's just like chicken dumplings.

Cotton Mill Colic
by Dave McCarn

When you buy clothes on easy terms,
Collectors treat you like measly worms.
One dollar down, then Lord knows,
If you can't make a payment, they'll take your clothes.
When you go to bed you can't sleep,
You owe so much at the end of the week.
No use to colic, they're all that way,
Pecking at your door till they get your pay.
I'm a-gonna starve, everybody will,
'Cause you can't make a living at a cotton mill.

When you go to work you work like the devil,
At the end of the week you're not on the level.
Payday comes, you pay your rent,
When you get through you've not got a cent
To buy fat-back meat, pinto beans,
Now and then you get turnip greens.
No use to colic, we're all that way,
Can't get the money to move away.
I'm a-gonna starve, and everybody will,
'Cause you can't make a living at a cotton mill.

Twelve dollars a week is all we get,
How in the heck can we live on that?
I've got a wife and fourteen kids,
We all have to sleep on two bedsteads.
Patches on my britches, holes in my hat,
Ain't had a shave, my wife got fat.
No use to colic, everyday at noon,
The kids get to crying in a different tune.
I'm a-gonna starve, and everybody will,
'Cause you can't make a living at a cotton mill.

They run a few days and then they stand,
Just to keep down the working man.
We can't make it, we never will,
As long as we stay at a lousy mill.
The poor are getting poorer, the rich are getting richer,
If you don't starve, I'm a son of a gun.
No use to colic, no use to rave,
We'll never rest till we're in our grave.
I'm a-gonna starve, and everybody will,
'Cause you can't make a living at a cotton mill.

Folk singer Dave McCarn wrote this song in 1926 and it was sung widely by Piedmont mill workers. McCarn, originally from Gaston County, North Carolina, had his own string band in the 1920s. Versions of this traditional song have been recorded by numerous artists including the Blue Sky Boys and Pete Seeger.

Young children gather for lunch at the day nursery in Pacolet Mills. (Courtesy Pacolet Elementary School)

3 Improving Textile Town 1910 to 1929

by Katherine Cann

As the nineteenth century drew to a close, six-year-old Ola Greer and her family moved from their farm near Travelers Rest in northern Greenville County to a brand-new house, still unfinished, in Spartanburg County at the Pelham mill village on the Enoree River, about 25 miles away. The Greers, persuaded by "people [who] came through the country trying to get us to work in the mill," were among thousands of families who abandoned exhausted, unproductive farmland and struck out for the mill villages that peppered the South Carolina upcountry. In Spartanburg County, rapidly becoming one of the most important textile centers in the United States, farmers and others seeking employment could look west to textile mills at Victor and Apalache. Or they could head east toward Cowpens and Clifton. To the north and south were cotton mills at Valley Falls and Woodruff. And on the very edge of the city of Spartanburg, the factories at Saxon and Arcadia beckoned.

Driven by desperation and drawn by optimism, the displaced farmers provided the workforce that made Spartanburg the state's leading industrial center in the early twentieth century. By 1909, 25 of South Carolina's 162 textile mills were located in Spartanburg County, and over 8,000 county residents worked in them. In exchange for weekly pay, farmers, who had been suffering the ravages of economic depression for decades, relinquished the independence and poverty of farming for what many sensed might be a better life.

These people constituted the resource most coveted by textile manufacturers and industry promoters. The white population from which the Southern textile mills drew their workers was remarkably homogeneous. Less than two percent of the entire population of South Carolina had been born outside the South Atlantic region of the United States, and 87.8 percent of the white population was born in the state. The 1927 *Handbook of South Carolina*, published largely as a promotional tool, boasted: "Our very own people! Not in another American industry is such a group of workers. Employers and employes [sic] have grown up together, know each other, and have, in great measure, common interests."

From the beginning of the industry in the state, the vast majority of textile workers lived in villages provided by the mill management; by the end of the 1920s, about one-fourth of all Spartanburg County residents lived in mill villages. Before World War I, mill village architecture mirrored the vernacular architecture of the time and was reminiscent both of the New England "saltbox" and the common Southern farmstead. Generally the small frame, uninsulated houses perched on the side of a hill. Narrow dirt lanes crisscrossed the villages in an irregular fashion dictated by the hillside topography, and hedges often afforded some modicum of privacy. A public well in the street served several families; sanitation facilities were nonexistent except for crude surface privies. Lots were sufficiently large to support a garden, and the mill typically furnished fuel free or at low cost. More than 21,000 people, many of them children, lived in Spartanburg's mill villages in 1909, a time when mill village life offered little more than a place to work, a pillow upon

> ***"A new operative from the country naturally goes to a country mill. These people look on Spartanburg as I would look on New York City, as a great big corrupt assemblage of humanity where folks can't raise their children right."***
>
> **—W.J. Britton, Spartan Mill superintendent, in a 1916 interview with author Broadus Mitchell**

which to lay a weary head, and a church where the spirit might find solace.

Textile work was hard and sporadic, the hours long, and the pay low. In 1910, a slight majority of the 9,000 textile workers in Spartanburg mills were women and children whose pay, meager though it might be, contributed to the "family wage" essential for paying rent and buying food. Textile executives insisted that they favored an end to child labor because children were not as efficient or productive as adults. However, they argued, it was necessary to employ children to keep the parents.

The Progressive Movement that began to filter through the United States in the late nineteenth century eventually reached South Carolina and, as a result, in 1903 the General Assembly had passed a law intended to curb child labor in the textile industry. The law prohibited children under 10 from working in the mills. With a permit, children of widows and disabled fathers, and orphans aged 10 to 12, could work in the mill. Six years later, at least 1,710 workers under 16 years of age labored in Spartanburg County's textile mills. No doubt there were many more who worked illegally and whose work was not documented, reported, or observed by state inspectors.

Under pressure from the American Cotton Manufacturers' Association and others, the legislature increased the age limit several times after 1910. Nevertheless, underage children continued working in the mills, often because the parents, desperate for the pay—however small—that a child could earn, simply lied about their child's age. In 1914, state inspectors found a number of violations of the child labor law in Spartanburg County. After a worker at Pacolet Manufacturing Company presented a sworn statement that she could legally work, a state inspector proceeded to the family home "to see the family Bible record, which I found they had. Upon looking at this record I found that there had been made an attempt to change the date of the child's birth. The mother then claimed the child to be over 12 years of age. Upon looking at this record through a magnifying glass it was seen by the old writing that the child was under 12 years of age." The parents were fined $10 in accordance of state law.

The picture was not, however, always clear, and state inspectors could be humane when the occasion warranted. In 1914, the inspector at Clifton discovered a boy he suspected of being underage. He found the child's parents were illiterate and incapable of keeping accurate records, so they escaped the fine, and the mill superintendent sent the child out of the mill. When underage workers did not have the sworn statement required by law, the mill superintendent was deemed responsible and fined accordingly. By 1915, the reported number of child laborers in Spartanburg's textile mills had dropped to 1,379, and the follow-

Glendale mill workers, late 1920s (Courtesy, Charles Hammett)

Susie Smith, age 67, interviewed as part of the Federal Writers Project in 1938:

"If we hadn't found out 'bout the cotton mills, I guess we'd still be in Virginia, less'n we'd been killed 'er something. A man come through there one time and had a letter from the super' at Spartan Mills. He showed us the letter and said that we could get a job in Spartanburg if we wanted to. The old man and Charlie come first and worked at Spartan Mills about two months before they sent for us. We'd saved up a shoe box full of money, so we had plenty to make the trip on. Charlie and his daddy made 'bout two dollars a piece fer a day's work. The old man didn't live long atter we moved here; don't know how many years he's been dead, but it could be counted up.

"Fer a long time I had three boys, and sometimes four, working in the mill. They made about two dollars a day; that must have in 1926. Well, no matter how many worked, we allus jest barely had enough to get along on. Up to 1925 they allus draw'd a little money every pay day, but atter that it was all took up in the store. It went on that way fer a long time, and we didn't see no money a-tall. By the time Doll and Walt got big enough to work, the others had married off, and they never was able to help me no more.

"Doll and Walt has been working at Whitney Mills 'bout six 'er seven years. They make twelve dollars a week a piece, and we manage to keep things going on that. Course they do take up lots in the store, but it ain't like it used to be when a-body couldn't see no money a-tall. I need some shoes now, and ain't got enough money to get 'em. I was getting five dollars a month from the old age pension, but some tattlers went up there and got that stopped."

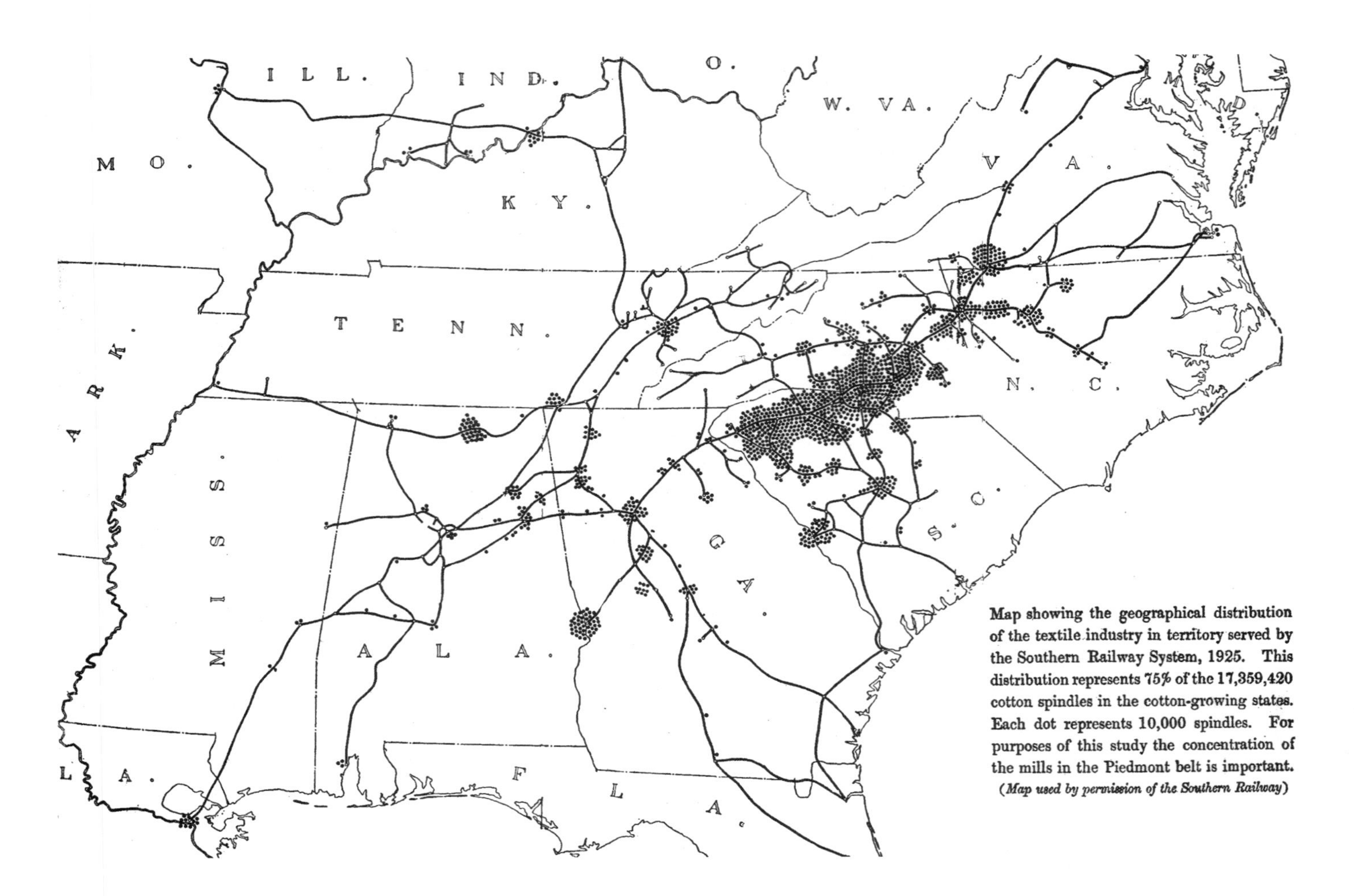

A 1925 Southern Railways map showing locations of Southern textile mills. Notice the heavy concentration in the Spartanburg area. (Courtesy, Bill Lynch)

ing year, a new state law excluded anyone under 14 from textile labor. In 1918, South Carolina officials boasted that the state had made more progress in eliminating child labor than any other state, and inspectors found that most mills complied with the law. In the 1920s young people between the ages of 14 and 16 needed both parental and mill management permission to work in the mill. Still, there is evidence that the practice continued. Children sometimes worked alongside their parents in the mill without compensation. Parents continued to falsify records, and the mill superintendents continued turning a blind eye to workers obviously younger than the law permitted. In all likelihood, the parents' willingness to circumvent the law was a combination of simple economics and the workers' traditional disdain for government interference in family matters.

The outbreak of war in Europe in 1914 heralded great changes both in Spartanburg County and its principal industry. In 1917, the year that the United States entered the war, the city of Spartanburg agreed to permit construction of Camp Wadsworth on 2,000 acres west of the city. The thousands of soldiers who trained there swelled the coffers of city businesses during their leave time. In addition to bringing revenue into the city, Camp Wadsworth meant construction of a highway, Highway 29, connecting the camp and the city. This highway became a major thoroughfare linking Spartanburg with other major textile centers including Greenville and Anderson, and Gastonia and Charlotte. The county's population—83,485 in 1910—had grown to 116,323 in 1929, much of the increase due to the influx of new textile workers, who comprised about 10 percent of the county's population. Construction of the two cotton mills at Woodruff in the first decade of the twentieth century brought so many people into town that the population nearly quadrupled to 2,396 in 1920. During the same period, the population of the city of Spartanburg and the town

Cotton manufacturing required still, humid air, to keep the fibers pliable for processing and to control the static electricity generated by textile machinery. To maintain humidity, mills closed their windows and pumped in steam, filling the factory with dank, dust-laden air. Workplace accidents were not uncommon, as tired and often malnourished workers tended the whirling machines and fought to maintain their concentration. Men, women, and children navigated floors made slippery by the humidity of the mill, guarding the spinning, unguarded cogs and pulleys.

—Author Toby Moore

Young boys in the mill villages enjoyed playing a game called mumblety-peg, or mumble peg. The boys would pound a stick, several feet long, into the ground and place another of approximately equal length on top. One person then hit the tip of the cross-T stick and flipped it into the air. The player knocked the spinning stick as far as he could with a broom handle or another stick. The winner was the one who hit it the farthest. The person hitting the stick the shortest distance had to "root the peg": the other players hammered another peg into the ground and he had to pull it out with his teeth.

—Author John M. Coggeshall

of Cowpens doubled. Such population increases were common in Upcountry towns where cotton mill smokestacks towered above the landscape.

Expansion of the Southern textile industry that had begun in the 1880s continued, although not steadily, throughout World War I and into the 1920s. World War I stimulated unprecedented increases in the value of mill stock. For example, Drayton Mill stock, worth $30 a share in 1916, had risen to $185 per share in 1919. That summer, the South Carolina upstate was in a "frenzy of gambling" in mill shares, and mill stocks were "running wild" as increasing stock values yielded handsome profits for stockholders, such as the 400 percent dividend issued by Inman Mills in 1920, reportedly the largest dividend ever announced by a Southern cotton mill.

The sharp rise in the value of mill stock in 1920 created the impression of "prosperity among stockholders and employees…whereas a few years ago the mills were hard pressed financially." Investors were reluctant to sell, hoping that stock values would climb even higher in 1921. By September 1920, however, the demand for cotton goods had decreased as much as 50 percent. As the year ended, Spartan Mills had 67,000 bales of cloth in storage waiting for better prices. Believing workers preferred a wage decrease to a temporary closing, the company cut wages 30 percent within 90 days. The textile industry thus entered a period of fluctuation, a hint of what came with the Great Depression of the 1930s. Invariably through the 1920s, mill owners and stockholders remained optimistic that good times would soon return.

In the post-war era, the Southern challenge to New England textile manufacturing began to mount, and soon the center of the industry gravitated south. Economic difficulties in New England ultimately led to bankruptcies and the loss of 40 percent of the mills and half the jobs. By 1924, the revitalized cotton textile market, sparked by the closing of several New England mills and a significant increase in orders, meant full employment in the South Carolina industry with many mills operating a 24-hour work schedule. In 1927, South Carolina cotton mills produced 44 percent of all textiles manufactured in the nation; Spartanburg's 34 mills contributed

Three Montgomery brothers, sons of Capt. John Montgomery, formed a textile dynasty in the 1920s. Left, Victory Montgomery was president of Pacolet Mills and, for a time, Whitney Mills. (Courtesy, Walter and Betty Montgomery)

Center, Walter S. Montgomery was president of Spartan Mills, 1902-1929. (Courtesy, Walter and Betty Montgomery)

Right, Ben Montgomery was president of Drayton Mills and Crescent Mills until his death in 1933. (Courtesy, William Gee)

How Textiles Are Made: A 12-Step Process

Step 1: In the opening process, cotton bales are opened upon arrival at the mill and loosened. Modern mills have an opening machine that travels along a line of opened bales, pulling the fibers out.

Step 2: In the carding process, carding machines further loosen and separate the fibers by passing them between two rotating metal drums covered with wire needles. This aligns cotton in a thin web of parallel fibers that is eventually condensed into an untwisted rope called a "sliver."

Step 3: In the combing process, the slivers go to a combing machine that removes smaller fibers and impurities, such as stems. Combing also makes the fibers more parallel.

Step 4: In the drawing process, slivers are made into a thinner strand.

Step 5: In the roving process, drawn-out slivers are fed into a roving frame where they are drawn out further and a slight twist is added. The sliver becomes known as "roving."

Step 6: In the spinning process, the roving is drawn and twisted into a fine yarn and placed on bobbins.

Step 7: In the winding process, yarn bobbins are transferred onto larger bobbins called "cheese cones." This is the end of what takes place in a spinning mill.

Step 8: Some kinds of yarn are dyed or textured in a separate building in preparation for weaving.

Step 9: The yarn goes to a weaving mill, where it becomes a "warp" yarn, which runs longitudinally, or a "weft" yarn, which runs crosswise.

Step 10: To prepare warp yarns for weaving, hundreds of strands are wound from the cheese cones onto a large beam and coated with a starch mixture called "sizing compound" for strengthening. Yarn is then wound onto a loom beam and placed on the loom.

Step 11: The weft, or filling yarn, is fed through the warp. This process was once performed by shuttles, but in modern mills air jets or water jets move the weft across at high speeds, producing a woven cloth (known as greige or grey) on large rolls. This is the end of what takes place in a weaving mill.

Step 12: The cloth goes to a finishing plant where it is bleached, dyed, or printed to produce the fabrics that are then sewn into apparel, sheets, towels and other products.

—John Michener

MEET MR. COTTON BOLL

When we have visitors at Spartan, a guide shows them around. That's why we have included Mr. Cotton Boll, your guide. He will make your trip through the pages more interesting.

Just follow Mr. Cotton Boll through the book!

If you have worked at Spartan for some time, you will know many of the facts in this book, but Mr. Cotton Boll may still point out some things of special interest.

If you are a new employee, Mr. Cotton Boll will help you find your way around. You will feel that Spartan is "Home" more quickly if you keep your book handy. Let Mr. Cotton Boll guide you.

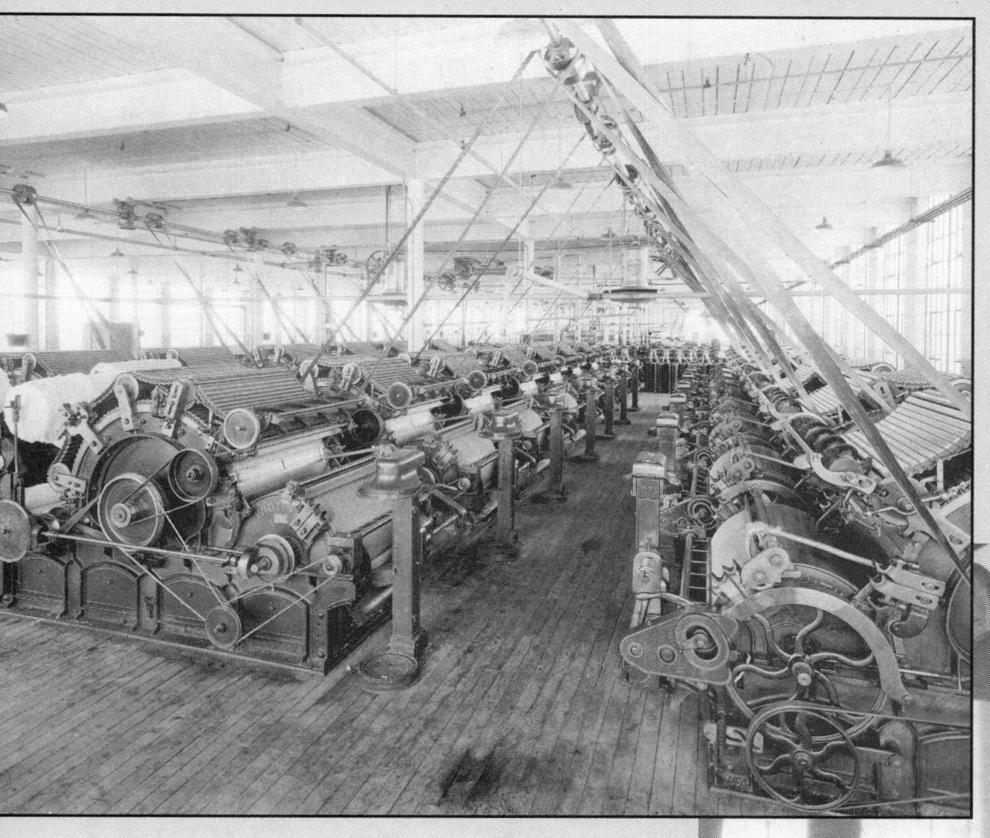

Carding machines, like these employed in the early years of Mayfair Mills, once operated by means of belts attached to the ceiling. (Courtesy, Mayfair Mills)

A twisting machine at Mayfair Mills. Improvements in textile manufacturing ultimately made this process obsolete. (Courtesy, Mayfair Mills)

their significant share.

In the heady post-war economic boom, the character of Spartanburg's textile industry began to change. When the textile campaign had begun in the 1880s, local investors provided most of the capital to build textile mills. For example, Spartanburg's Cleveland and Manning families provided the financial support to organize Arcadia Mills in 1903, and J. B. Kilgore and W.W. Simpson organized the Woodruff Cotton Mill in their hometown. By the 1920s, however, investors, brokers, and marketing specialists from New England became involved in Spartanburg's textile industry. In the process, many textile dollars flowed out of Spartanburg County and into the coffers of companies like Martel Mills of New York, Powell Knitting Company of Rhode Island, and Pacific Mills of Massachusetts.

One of the most profitable years of the local textile industry was 1919, the last year of World War I. That year, 21 local cotton mills, having a total capital stock of $8.9 million, paid out stock dividends of $2.1 million.

—Author Robert Dunn

The infusion of Northern capital into Spartanburg's textile industry led to new construction, remodeling, and renovation. Without question, the most significant event in Spartanburg textiles in the post-World War I era was the construction of a huge bleachery and finishing plant west of Spartanburg. Owned by Massachusetts-based Pacific Mills, the multi-million dollar investment signaled the new era in South Carolina textiles—the large-scale manufacture of high-quality finished textiles. Encouraged by Pacific Mills' success, New York-based Reeves Brothers Inc. constructed a finishing plant at Fairforest late in the 1920s, and Powell Knitting Mills of Rhode Island purchased the bankrupt Model Mill, constructed just after the war by supporters of Textile Industrial Institute, in 1927.

The wave of expansion during the 1920s created a deceptively rosy picture of economic conditions in the textile industry, which directly supported 20 percent of the state's

Operatives at Clifton Mill No. 1, probably around 1915. (Courtesy, Mike Hembree)

Arcadia Mills began construction on a second plant, the Baily Plant, in 1922. (Courtesy, Mayfair Mills)

population. During the "Roaring Twenties" when the national economy flourished, the textile industry weathered myriad changes due to price fluctuations of finished goods, unstable cloth markets, and style changes. While none of these conditions lay within the control of the industry's leaders, all of them exerted enormous impact on the industry and consequently on the thousands who labored in the mills.

Encouraged by the market's turnaround in 1924, South Carolina textile manufacturers embarked on an ambitious expansion program. Unfortunately, the market did not hold. Stocks and profits dropped precipitously, and stockholders began to sell their shares. Nevertheless, textile mills continued to produce more than the market could absorb in spite of dire predictions of the consequences. Few textiles were selling at a profit by 1926. Available work was sporadic, depending on the market. Despite lowered profits in 1926, many mills continued to run on schedule. However, there was sufficient concern that even the South Carolina General Assembly tried to help the industry by organizing a "Wear Cotton Campaign" designed to boost demand by making cotton clothing fashionable. But the effort had little effect. Yet, as late as September 1929, mere weeks before Black Tuesday and the stock market crash, trade in South Carolina textile stocks was more vital than in several years. Most mills were running at capacity, some on a 24-

hour schedule again. Optimism reigned, as many believed that the textile industry was on the "eve of a great revival."

> "Spindles have increased in India from 4,945,000 in 1900 to 8,704,000 in 1929, or 75%; and in Japan from 1,274,000 to 6,698,000, or 426%; and in China from 550,000 to 3,638,000—an astounding growth of 561% ... in the light of such alarming facts, isn't it reasonable to believe that South Carolina is not only losing its advantageous position in cotton manufacturing, but is facing the most serious problem of its history!"
>
> —*Author William P. Jacobs, 1932*

The South Carolinians most directly affected by the volatile nature of the textile industry were the thousands of mill workers. After the war, according to the manufacturers, textile workers enjoyed higher wages, better housing, schools, churches, recreational facilities, and health insurance—all of which came at a cost of millions of dollars to stockholders. Some mills allegedly operated at a loss in order to provide employment.

During the post-war era, the number of women and children who worked in Spartanburg's mills began to decline from about half of all workers in 1915 to about one-third in 1927. Jobs in southern textile mills were segregated by gender. Typical mill jobs held by women included doffing (typically performed by young boys until the advent of child labor legislation); slubber, speeder, and beamer tending; trimming; and inspecting. The most prized factory jobs open to women were frame spinning and weaving, which paid slightly more and demanded a little less exertion. In addition,

Villages Within Villages

The larger mill villages in Spartanburg County often had several different neighborhoods within them. These places were usually named by residents and had their own distinct characters. Here are but a few of them:

Bungalow Town: An area of the Enoree Mill village.

Can Holler: An area of Pacolet Mills down the hill from Walker Street (known as Back Line). It was the location of Bonner's Store, "Pug" Guyton's Barber Shop and "Hunk" Gossett's Store.

Coopertown: An area surrounding Clifton Mills No. 2, named after a family that ran a late 19th-century furniture store.

The 500: An area of the Tucapau village. Houses were numbered sequentially, and this was one of the later additions to the village.

Gobbler's Knob: A nickname used both in Drayton and Glendale to describe a hill above the mill.

Happy Hollow: A name given to E Street on the old side of the Inman mill village.

Keg Town: An area in Pacolet Mills past Brown's Chapel Baptist Church. This was the location of Whitlock's Store.

Red Egypt: An area across the river from the Tucapau Mill.

Shake Rag Hill: A name given to the row of houses on a ridge above Arkwright Mill. When women hung their clothes out to dry, the wind blew them like rags.

Stumptown: An area along Burnett Street where Spartan Mills employees lived. Built in a former forest, this was one of the last parts of the village developed.

Tight Wad: The area in Pacolet Mills where Granite and Cleveland Streets meet.

Vinegar Hill: An area located across the road from Clifton No. 2. This high area was the burial ground for some of the flood victims of 1903.

Reeves and Arcadia built the Fairforest Finishing Plant in 1929 (Courtesy, Herald-Journal Willis Collection, Spartanburg County Public Libraries)

nearly all mills hired women to perform clerical duties. By assigning women to less strenuous tasks, management saved money on wages paid for more complicated and demanding work.

Mill management generally viewed women as "temporary" workers, even though many of them worked as often and as long as they were able. Few, if any, female textile workers worked in supervisory or management positions such as overseer or foreman. However, Molly Spears served as president of Landrum's small Shamrock Damask Mill in 1927. As textile wages in general rose, the number of women textile workers declined. When industry-wide difficulties led to production cutbacks, plant curtailments, and fewer jobs, employers more often than not let women go first, while retaining the men. Consequently, the role of women in many mills was relegated to that of lower paid part-time labor. Women sometimes described their jobs as "hard work, but we loved it because we knew no other."

Textile jobs were also segregated by race as required by state law. Supervisors faced a fine if caught in violation of the segregation law that prohibited workers of different races from working in the same room, but nearly all mills hired "colored help" for heavy work, janitorial service, and dangerous jobs such as stoking and firing the boiler. Black women sometimes worked as sweepers and window washers. African-American workers, however, were not permitted to run the machines and inevitably earned the lowest wages. In 1927,

Flora and Fay Foster at their Saxon Mill village home about 1920 (Courtesy, Caroliniana Library)

A group portrait at Spartan Mills in 1926 (Courtesy, Ed and Mary Ann Hall)

about 4 percent of Spartanburg County textile workers were African Americans.

In most times, textile workers could be sure of two things—wages were low, and they were subject to upheavals in the industry. Factors determining wages included the mill's location and economic circumstances, the kind of job, and the length of service on the job. In the industry's flush days following World War I, average textile wages doubled. But hard times loomed. A 1928 budget prepared by the University of South Carolina sociology department estimated that a family with three children in Columbia required $1,090.92 per year to enjoy a bare minimum standard of living. Average wages in South Carolina's textile industry at that time were $654 for a full year's labor. Figures published in the *Congressional Record* in 1929 showed the $9 weekly wage to be the lowest in the nation. *The Columbia Record* pointed out benefits that, in the judgment of the newspaper, more than compensated for the low wages: "rent—four rooms, with lights and water—$5.00 per month; coal, wood, and water furnished by the mill; free electricity provided by the mill at least one day each week,

This 1920s photograph of Spartan Mills employees includes Homer Burrell, the mill's blacksmith. He is second from the right on the bottom row. (Courtesy, Ed and Mary Ann Hall)

sometimes more." And often, manufacturers justified low wages in the belief that the mills "offer a means of making a decent and comfortable living" and provide employment to thousands who would otherwise be unemployed. It should be noted that determining an accurate average wage is very difficult, due to inadequate bookkeeping and records.

South Carolina manufacturers also pointed to the lower cost of living in the South, which, they insisted, made up for the low wages. In the mid-1920s, steak sold for 25-30 cents per pound at the local Piggly Wiggly, and a pound of fatback, an ingredient essential in southern cooking, cost 11 cents. Flour, a 24-pound bag, cost 85 cents, and a can of tomatoes was 17 cents. A dozen eggs was a quarter, and a pound of coffee, 45 cents. A person could buy a new Ford Roadster for $360, pay for it "on time" and take his family to Sunday dinner where 60 cents bought an ample meal. In the mill villages, however, especially those with a "company store" where workers were often forced to shop, prices occasionally were significantly higher.

The leading South Carolina newspaper, *The State*, bragged on the "remarkable" fact that "working people of comparable economic station nowhere else on earth today outside the United States enjoy living at the level which cotton mill workers regard in the southeast as normal." Whatever the wages, textile workers were among the lowest paid workers in American manufacturing, earning about the half the average wage of all other industrial workers. In 1919, at the height of the post-war textile boom, the average annual wage of South Carolina textile workers was $757; by 1929, it had declined to $652. Some textile companies gave bonuses to workers who demonstrated efficiency and had perfect attendance records, especially in the exhilarating days immediately following World War I. For example, in 1919, W. S. Gray Cotton Mills in Woodruff paid weekly bonuses and an additional 6 percent bonus at Christmas. Other companies merely raised wages.

In addition to low wages, South Carolina textile workers worked long hours. The first law governing hourly employment in the textile industry set the maximum at 66 hours in 1892. A subsequent statute in 1907 lowered it to 60 hours. To alleviate post-war inflation, attempts were made in 1920 to establish a 56-hour work-

World War I caused labor shortages in South Carolina cotton mills, prompting owners to boost black employment—though only in the positions where they could not keep white workers. Black employment climbed from 2,653 in 1909 to 5,237 (10.5 percent of the workforce) by 1919.

The following law was passed by the South Carolina General Assembly in 1915, legally establishing segregation in cotton mills.

SEC. 1272. Separation of employees of different races in cotton textile factories.—It shall be unlawful for any person, firm or corporation engaged in the business of cotton textile manufacturing in the State to allow or permit operatives, help and labor of different races to labor and work together within the same room, or to use the same doors of entrance and exit at the same time, or to use and occupy the same pay ticket windows or doors for paying off its operatives and laborers at the same time, or to use the same stairways and windows at the same time. Or to use at any time the same lavatories, toilets, drinking water, buckets, pails, cups, dippers or glasses: Provided, equal accommodations shall be supplied and furnished to all persons employed by said person, firm or corporation engaged in the business of cotton textile manufacturing as aforesaid, without distinction to race, color or previous conditions. Any firm, person or corporation engaged in cotton textile manufacturing violating the provisions of this section shall be liable for a penalty of not over $100 for each and every offense, to be recovered in suit by any citizen of the county in which the offense is committed and to be paid to the school fund of the district in which offending textile manufacturing establishment is located. Any firm, person or corporation engaged in cotton textile manufacturing violating the provisions of this section shall be punished by a fine not to exceed $100 for each offense or imprisonment at hard labor for a period not to exceed 30 days or both at the discretion of the judge. This section shall not apply to employment of firemen as subordinates in boilerrooms, or to floor scrubbers and those persons employed in keeping in proper condition lavatories and toilets, and carpenters, mechanics, and others engaged in the repair or erection of buildings.

week combined with a wage increase. Two years later, when the General Assembly approved legislation setting the workweek for specified jobs in the textile industry at 55 hours, many mills had already implemented a shorter week. Typically, after 1922, mills in full production ran two 10-hour shifts, five days a week, and a five-hour shift on the sixth day, usually Saturday. South Carolina's 55-hour law established the shortest workweek in the Southern textile industry.

But enforcement of the maximum hour law was lax. Workers as well as management circumvented the law's intent by staying at work longer. Piece workers, mostly women, whose wages were determined by production, not hours, frequently worked longer than 55 hours, and mills sometimes reclassified positions in order to escape the law's requirements.

Night work was common in all of South Carolina's textile mills by the end of the 1920s. Although regulated by state law, mill managers frequently violated the night work regulations just as they did the maximum hour law. As was the case with piece workers, many of the night workers were women. Attempts in the late 1920s to eliminate night work elicited a strong defense from South Carolina textile manufacturers who argued that stopping the night shifts would cause massive unemployment and pointed out that this

Isaac Miller, my father, was born in western North Carolina in 1892. He moved to Spartanburg County, married my mother, Lillie Lee Bivins, and went to work as a fireman for Arkwright Mills during World War I. Later, he was a fireman at Beaumont Mill, where he worked for 40 years until his death April 13, 1956.

My father was firing the boilers at Beaumont Mills during the Depression and World War II. Those were dismal years but we lived through them. At this time, black employees were few inside the mill. Some were firemen, platform cotton truckers, and restroom attendants.

He fired the boilers that would create steam to run the looms in the cotton mill. Every one-hour period, he would have to throw coal in the 10 furnaces to keep an even flow of steam. He had one assistant to roll coal for him. Back then, every cotton mill you saw had a towering smokestack. Billows of smoke would roll out when the boilers were being fired.

When I was in the fourth grade, on my way to Cumming Street School, I would have to drop my father's breakfast off to him about 7:30 to 8 a.m. Each morning, I would be late for school. In the wintertime, the principal, Eugene Rivers, would whip me in my hand with a switch for being late. In the spring or fall, he would make me stay in one hour after school.

Isaac and Lillie Miller had six children: James Benjamin (who died at two months), Deborah, Susan, Jonathan, Grace Helen, and Harold.

—Harold Miller

Shop crew of Enoree Manufacturing Co. in 1921 (Courtesy, Anthony Tucker)

work pattern enabled Southern mills to double capacity without capital expenditures. Supporters of the system drew a pretty picture of the leisurely three-day weekends, Saturday through Monday, that the night workers had and contended that the only real opponents were the New England mills and the manufacturers of textile machinery. Only a few regarded the night work system as needlessly harsh.

Working conditions in the mills were frequently unhealthy and dangerous. Textile workers could not escape the cotton fluff that gave rise to the derisive term "linthead," nor the constant roar of the machinery that often seriously damaged their hearing. A 42-year-old worker at Pelham Mill died after his arm was caught in a shafting wheel that spun him around over 100 times before the machine could be stopped. Accidents such as this one prompted legislative discussion of workplace safety regulations, but many years passed before the General Assembly took such action.

When, as was often the case in the 1920s, textile plants shut down or curtailed production, workers' wages declined, and

TUCAPAU MILLS

No. 311

Name	Shelton Clark
Wages	1760
Rent	250
Mdse.	200
Cash	1310

Received Payment

JUN 1-1929 192

B&W—117622

MILL NO. 2 APRIL 6 1935

MULLWEE CLAUDE

IN ACCOUNT WITH

CLIFTON MANUFACTURING COMPANY

CLIFTON, S. C.

By Labor		
" "		
" "		
Total Wages	18	15
To Rent	1	10
To Electric Cur.		
To Insurance		
To Mdse.		
To Welfare		
Balance Cash	17	05

Mill workers received their pay in these envelopes, called "tickets." If they owed money at the company store, the amount was subtracted and indicated on the front. (Courtesy, Bill Lynch)

> **Almost in the same breath with a person's name is mentioned the shift he works on. One's shift indicates when he has spare time, when he eats and to some extent his prestige since the first shift carries the highest prestige.**
>
> —*Author John Morland*

meeting obligations became even more difficult. In the spring of 1926, Spartanburg mills cut back the workforce as much as 33 percent, citing market paralysis resulting from declining cotton prices. For two months, cloth had been accumulating in mill warehouses. Curtailing the workforce, the textile companies believed, would prevent the surplus from growing even larger and depleting the mills' cash resources.

The entire community, not merely the mill villages, felt the ramifications of economic distress in the mills. The closing of a mill, said a grocer in nearby Rock Hill, even temporarily, "meant a real hardship to our mill operatives." Deprived of their weekly pay, South Carolina textile workers struggled to survive. "...[T] hese mill folks will have to live on credit if they can get it. The small grocers who cater to the mill operatives' business are hardly in a position to carry the load of feeding all these people...Most of us have to live hand to mouth...and we cannot carry...the mill folks over a period of two or three weeks and then wait until they can pay us. It is hard lines for the mill folks, very few of whom have money laid aside."

Sometimes mill families left families behind in the mountains of North Carolina. Elizabeth Ensley, who lived in North Carolina, was visiting her Pacolet family on Milliken Street.(Courtesy, Don Camby)

MILL VILLAGES REFLECTED the personality of the owners, and the paternalistic philosophy of mill management was everywhere evident in the mill community—from houses in which the operatives lived to the local recreation center to the ubiquitous baskets of fruit and nuts at Christmas. Realizing that healthy and contented workers meant steadier profits, many textile companies in the 1920s began remodeling projects that ameliorated some of the saddest features of mill life. The mills' largesse during this era included schools, churches, dwelling improvements, YMCAs, recreation halls, playgrounds, various health programs, and beautification projects. In 1918-1919, Enoree mills "overhauled" all village houses, "putting them in first class condition." That same year, Chesnee Mills began construction of a combined school/community building with an auditorium, classrooms, library, domestic science room, and first-aid room. On the second floor of the "handsome brick building" was a lodge room, roof garden, and dressing rooms, and the basement housed a barbershop and club rooms. At Saxon Mill Village, in the last days of World War I, the mill introduced the concept of "community organization," which called for monthly "town meetings" where the residents

These men built the trolley line to Glendale. (Courtesy, Charles Hammett)

1/31/23

Mr. Stanley Converse
70 West North Avenue
Atlanta, Ga.

Dear Stanley:

Replying to yours of the 25th inst. I agree with you that if you were to undertake the practical side of mill work the best thing you can do after you finish your course in Atlanta is to start in the Card room and get a working knowledge of the details of manufacturing.

It is a pleasure to know that you would prefer if possible to be in this locality and you may rest assured that we will do everything that we can to locate you here when the time comes.

I note that you say you would probably not be ready to go to work until next fall. This would be alright from our standpoint, and if you will keep me advised we will try to make our plans fit in with yours.

With best wishes in which all here join.

Yours truly,

President

Letter from Clifton President Choice Evins to Stanley Converse, who succeeded him as the company's fourth and last president in 1945.

discussed "subjects pertaining to the good of the community" and appointed committees of residents to try to act upon the ideas. That year, the company installed running water in all houses, built a well-equipped and well-lighted playground, and gave new books to the library. Furthermore, a first-aid station in the new community building provided a place where children and workers could receive medical examinations. Other improvements included new desks for the school and repairs to the Baptist Church. Workers at Converse Mill literally "saw the light" in 1919 as the company installed electric lights in all the houses and on the streets. W. S. Gray Cotton Mills and Clifton Manufacturing installed sewerage and water in most workers' houses. And at Clifton, there were two savings banks where about one-third of the village residents kept savings accounts. By 1926, South Carolina officials reported that a majority of South Carolina mill houses had waterworks, electric lights, and sewerage. But not all of them.

Spartanburg County's new mill villages, those built in the 1920s at Lyman, Powell Mill, and Fairforest, were distinctly different from those built earlier. Based on the concepts of mill village designer Earle S. Draper, the new-style villages incorporated gently curving roads and low-density housing. Larger lots somewhat gave the illusion of the space characteristic of workers' rural roots, or that's what the designers intended. The houses, located on varying elevations and at irregular intervals, eliminated the boring uniformity of an earlier day. In addition, many of the new-style villages included green spaces. Not all roads led to the mill, as they once had, and not all houses were in sight of the looming mill itself. Reliance on a "city planner" like Draper suggested that textile manufacturers were concerned about their public

Employees of the mills often addressed the mill owners by their first names. For instance: "Mr. Walter" (Walter Montgomery, president of Spartan Mills) and "Mr. Jim" (James Chapman, owner of Inman Mills).

image, which the mill village invariably suggested. These changes, as well as changes in transportation, marked the beginning of the workers' turn away from the mill and its village as the center of their lives.

To mill owners, the mill village was a positive influence in the workers' personal development. Motivated by the tradition of the mill executive as community benefactor, the necessity to prevent labor organization and simultaneously maintain control over the workers, and the desire to mitigate some of the most demoralizing influences of industrialism, textile companies engaged in various types of "welfare work" in the mill community.

Because education was regarded as an important function of mill management, South Carolina mills furnished a graded school with an elementary curriculum in virtually every mill village. Where there was no school, the mill made arrangements for the residents to attend a local public school, sometimes only after paying a fee. Before the twentieth century, mill schools were crowded, located in inadequate facilities, and insufficiently funded. Early education in mill schools was characterized by a short school year, which varied from village to village. By the 1920s, however, the system had become more standardized, and in 1920 only three of 143 mill schools were in session less than nine months. Mill education benefited from the 1920s improvement trends. At Enoree, the mill hired two new teachers and raised teacher salaries to $100 a month. Late in 1921, mill officials completed the new three-story brick school building as well as a nearby teacherage and principal's home. By the middle of the decade, the State Superintendent of Education reported that the quality of education in the mill schools surpassed that of public education in general.

While their parents worked in the mills, Pacolet children enjoyed organized playtime in one of Victor Montgomery's many welfare initiatives. (Courtesy, Pacolet Elementary School)

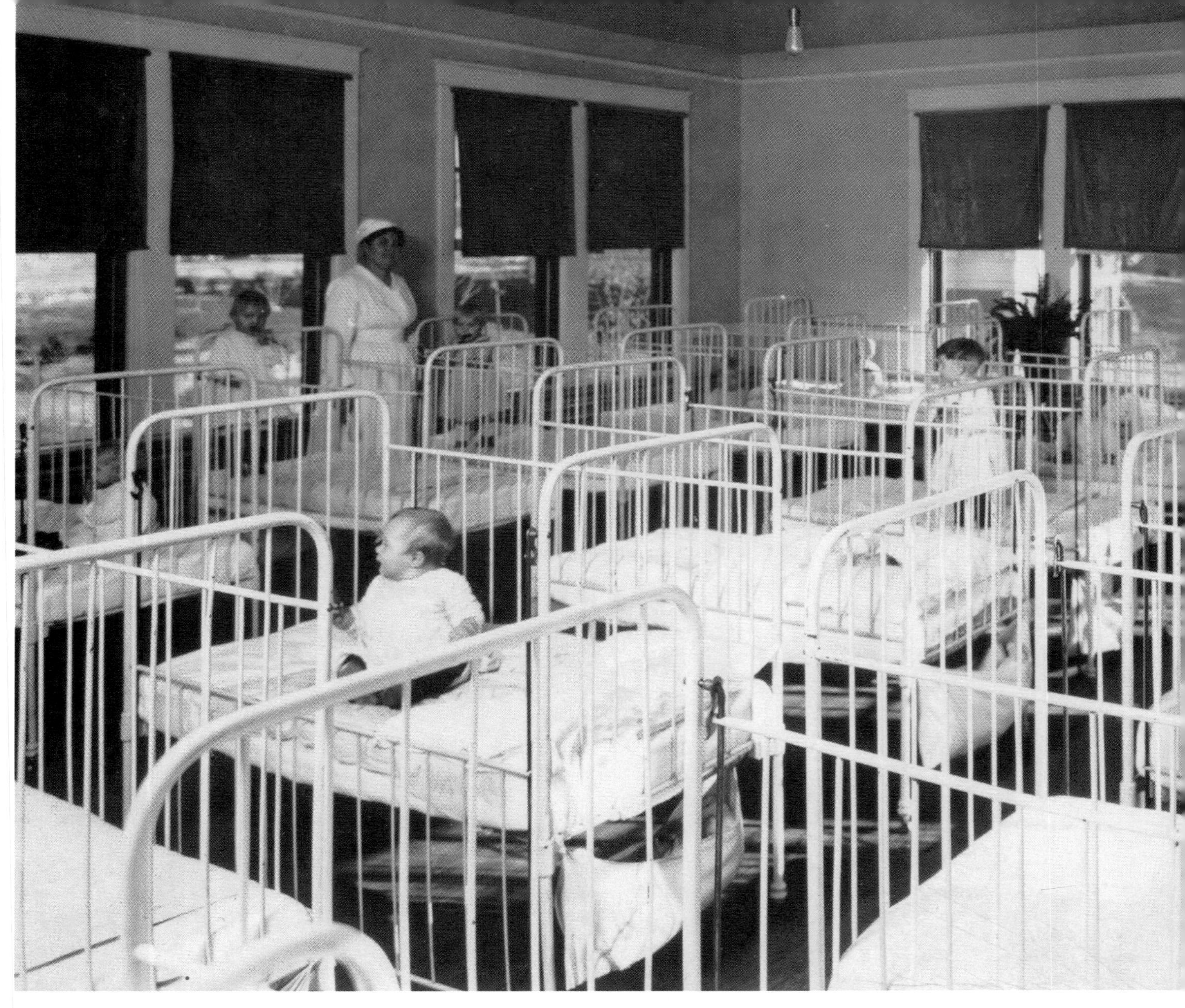

Children in Pacolet in the 1920s grew up in a day nursery with dozens of cribs. (Courtesy, Pacolet Elementary School)

Many textile workers, however, had grown up in a time and place in which education had little value, and many of them could neither read nor write. In 1914, in Spartanburg County—known for the quality of its school system—one-fourth of adult men could not write their names when they registered to vote. Mill village schools would solve the literacy problem for children, but adult literacy was an entirely different matter. Some of the mills organized night school for the workers. Others supported state-sponsored adult education programs and the Opportunity School. In Spartanburg, particularly, a partnership between some of the most prominent and influential textile owners and Methodist minister David English Camak resulted in the establishment of Textile Industrial Institute in 1911. Aimed at providing an elementary education for adult textile workers, TII sought to capitalize on the workers' desire to improve their condition and the companies' desire to maintain a contented workforce. The school attracted students from the mill villages in Spartanburg County. For example, in 1919, individuals from 16 Saxon families were enrolled there, taking advantage of the school's unique work-study program. Due partly to its affiliation with the Methodist Church, TII also attracted students from other parts of South Carolina and even from other states. Shortly after World War I ended, the school built a textile mill to provide a

Fraternal, or secret, orders were popular among the men in mill villages. Meeting weekly, they usually followed a secret ritual. One of the main attractions was the life insurance benefit they offered. Most based their membership on people who embodied good character in the mill villages. In Saxon in 1927, for instance, 150 men belonged to the Woodmen of the World; another 40 belonged to the Masons, 35 to the Junior Order of the American Mechanics, 30 to the Red Men, and 20 to the Knights of Pythias.

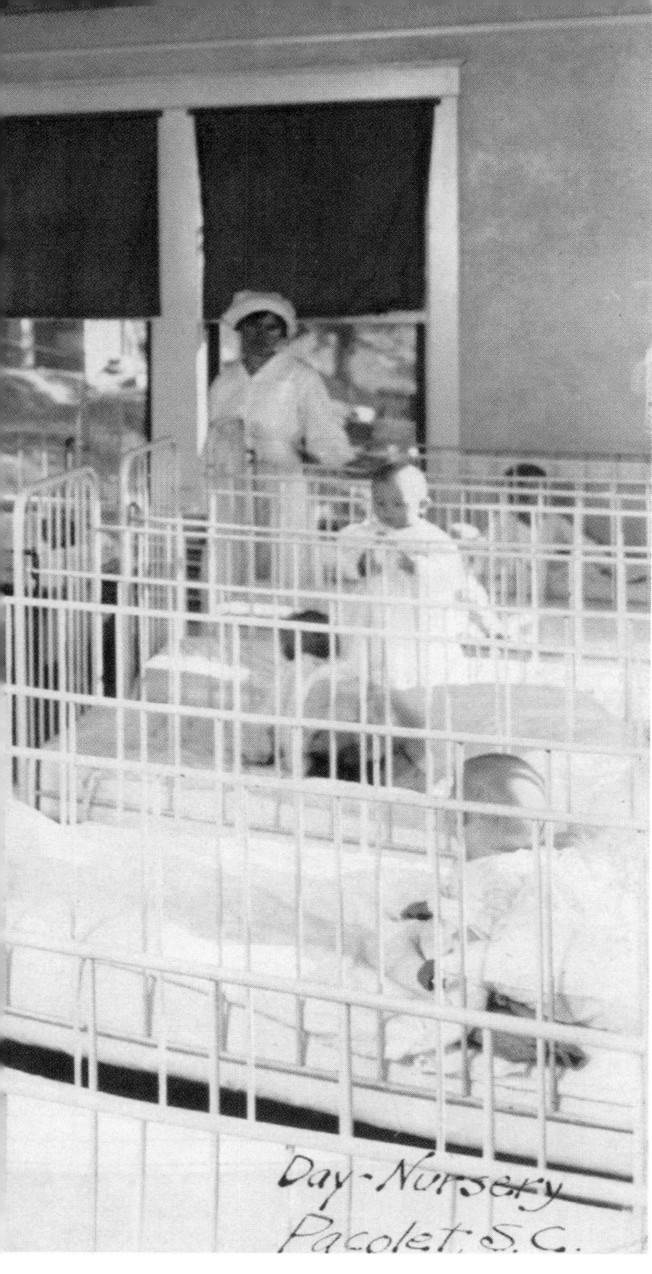

place of steady employment for the students, as well as a source of endowment funds. The South Carolina Commissioner of Agriculture, Commerce, and Industries described this Model Mill as the "best built, best equipped cotton mill in the world, almost with the perfection of a Swiss watch." The Model Mill became a victim of the times and poor management, though, leading to its sale in the late 1920s to Powell Knitting Company of Rhode Island.

FEW SOUTHERNERS IN THE EARLY twentieth century enjoyed good health. Plagued by disease and accompanying lethargy, Southerners were perhaps the least healthy people in the nation. Until the post-World War I era, textile companies paid little attention to the health of their workers, many of whom suffered from the debilitating effects of hookworm, pellagra, malaria, and tuberculosis. The transience of the mill population and the close, stifling, and often unsanitary quarters in which they lived and worked provided ideal conditions for the emergence and transmissions of communicable diseases. In addition, the heat, dampness, and swirling dust that inevitably, and of necessity, accompanied the process of textile manufacturing, contributed to a myriad of medical conditions that plagued many textile workers.

The most widespread disease in the South prior to World War I, hookworm, was transmitted by contact with bare feet and unwashed hands and bodies, conditions all too common both in mill villages and on farms. Stagnant water and open sewers provided ideal breeding grounds for malaria-carrying mosquitoes. The typical Southern diet, bereft of fresh meat and green vegetables, contributed to the incidence of pellagra. While much of the population of the South suffered from these diseases, the concentration of men, women, and children in textile mill villages contributed to a high rate of infection.

The South Carolina State Board of Health, established in 1878, had the responsibility to "investigate the causes, character and means of preventing such epidemic and endemic diseases as the State is liable to suffer from; the influence of climate, location and occupation, habits, drainage, scavengering, water supply, heating and ventilation." As time passed, the state Legislature expanded the board's authority, so that by 1927, it consisted of 10 departments that performed a wide range of services, including the dispensing of medicines and anti-toxins, tracking cases of contagious diseases, controlling mosquitoes, and educating the public about health issues. When invited to do so, the State Board of Health had the authority to inspect textile mills.

During the 1920s, South Carolina joined other southern states in vigorous crusades against the most virulent communicable diseases—malaria, tuberculosis, venereal diseases, pellagra, and hookworm. As a result of efforts at mosquito

When it opened in the mid-1920s, Pacific Mills was the BMW Manufacturing Co. of its time. (Courtesy, Louise Foster)

"I was so small that they had to make me a thing, you know, something I could stand on with wheels on it to reach the roping and do my cleaning up. I'd push that little thing along and whenever the real boss men would come along, they would put me in the waste can. They didn't want the overseers to see me, cause they knew I was too small to be in the mill. They forgot they put me in that waste can, and then the bossmans was done gone. I was hollerin', 'Help! Help!' I was hollerin' because I wanted somebody to get me out of that waste can. [My brother said] 'Sis, I'm sorry. I forgot I put you in there.'"

—Eloise Garner, who went to work at age nine in a local mill

Schoolhouse built for Tucapau children, 1923 (Courtesy, Junior West)

Mr. Steadman, center, and his two sons, were early students of the Textile Industrial Institute. (Courtesy, Spartanburg Methodist College)

control, in 1928, the General Assembly required sewerage facilities in all the state's mill villages. Passing over the industry's objections to such costly changes, the law compelled textile mills to equip employees' homes with toilets. However, the law applied only to textile mills located on creeks with sufficient water and drainage. Many textile mills, including the factories at Saxon and Enoree and W. S. Gray Mills in Woodruff, had already installed sewerage facilities in the villages, but in 1928, as a result of the law, the number of South Carolina textile mill villages with sewer systems doubled.

The nature of textile manufacturing in the early twentieth century made for a hazardous workplace, once described by a U. S. senator in 1907 as "inhuman." While mill owners steadfastly defended or ignored hazardous conditions in the mills themselves, they sometimes made improvements when such would increase productivity—and profits. As the post-World War I prosperity stimulated welfare work in the mill community, it also stimulated improvements in the workplace—new machines and air cleaners and filters, for example. However, even after mills had installed such improvements, a state survey revealed that working in a mill endangered the health of three quarters of South Carolina's textile workers.

The mills encouraged recreational "welfare work" to stimulate community spirit and cohesiveness as well as healthy bodies (i.e. healthy workers). Early in the twentieth century, athletic competitions among the various textile companies, not only

November 25, 1926

Mr. J. C. Evins, President
S.C. Cotton Manufacturers Association
Spartanburg, S.C.

Dear Choice:

I have never felt satisfied about the failure of our mills to observe Thanksgiving Day. When the Saxon Mills was first started, I made them close down for Thanksgiving Day, but when the next season came around, I was told that the mill people preferred to have holiday for the circus, and as labor conditions were acute at that time I complied with the request. If the attitude of the mill people was correctly interpreted at that time, I am sure that it is not their attitude now, for their religious customs and observances are too near akin to those of other county and city churches for them not to wish to make a similar observance of Thanksgiving Day.

But entirely aside from the religious phase of the situation, I feel that it is a very bad object lesson in citizenship for our mills to thus disregard the Proclamations both of the President of the United States and the Governors of the States.

This is no new thought with me, nor possibly with other mill men, and I do not wish to attract any publicity, but it has been on my conscience each Thanksgiving season, and I have notified the superintendents of both our mills [Saxon and Chesnee] that neither one of them are again to be operated on Thanksgiving Day so long as I am in control of the management.

I am advising you of this in a partially personal and partially official way in the hope that in some manner a general observance of the day by the mills may be brought about next season.

With all good wishes,

Cordially yours,

Jno. A. Law

Choice Evins replied to John Law that Clifton Manufacturing also intended to observe Thanksgiving the next year and said he would try to persuade all mills to do the same.

In 1915, about two-thirds of mill families in South Carolina relied on primitive privies for human waste disposal. By 1925, with the advent of village sewer systems, more that two-thirds had indoor flush toilets. Roughly 30 percent still relied on outhouses.

—Author Edward Beardsley

in Spartanburg County but also throughout South Carolina, had become a popular diversion. Competition was reportedly so intense that fans of one mill team feared for their safety when in rivals' territory. In most mill villages, the athletic program was the most organized and developed of mill "welfare" programs. Organized, mill-sponsored leisure activity became imperative with the elimination of child labor, the decline in the number of women textile workers, and additional leisure time resulting from reduced work schedules.

Although baseball was probably the number one recreational activity for Spartanburg County

School children in Glendale (Courtesy, Charles Hammett)

Spartan Mills' medical clinic was housed in the community building on Forest Street. (Courtesy, the Spartanburg Herald-Journal)

The community of Una was founded by immigrants from the mountains who worked at local mills but didn't like the rules associated with living in the villages. During the strikes of the 1930s, many blacklisted workers found refuge there.

As the first wave of strikes began to break over the Southern mills in 1929, the Cotton Manufacturers Association of South Carolina bought large ads in the state's newspapers to present what they called, "The Truth about the Cotton Mills of South Carolina." These ads were an attempt to counter what a growing number of union representatives in the state were telling textile mill operatives. The following are excerpts from those 1929 ads:

About wages:

"The wage of the cotton mill operative cannot be measured by his pay envelope alone. Other services, known as wage equivalents, supplied at the expense of the mill, offer him comparatively free of charge, many advantages that others of similar income can't afford. Comfortable homes are furnished to operatives at a nominal rental of less than $5 per month average. Mills frequently supply laundry service, nurseries, clinics, welfare service and group insurance to safeguard health, life and income. Fuel is generally supplied at wholesale cost, and electric lights, water and sewer services are usually furnished free or at a nominal fee, never at a rate to pay the actual cost. These things, when added to the weekly paycheck, offer a wage greater per individual than is offered by many other types of industry."

About the stretchout:

"In his new capacity as an expert weaver, he is enabled to supervise more looms and thus earn greater wages. In one South Carolina cotton mill, the weaver who formerly alone attended twenty-four looms can now with more ease, less exertion and greater skill, care for a hundred looms with the help of his assistant. With one hundred looms, his work is heavier and his health is in no way jeopardized, thus there is nothing in extended labor to cause suspicion. It is a scientific, a sound and an honest attempt of the manufacturer to meet exacting competition with improved production and a better production."

About profits:

"Net returns on actual invested capital in South Carolina mills during the past two years has averaged little more than five percent. In some recent years, there were no profits, and if the abnormal war period were discarded, the average for the history of the industry in this state would probably not exceed four percent. Earnings in almost every other leading industry have far exceeded those of textiles. Even railroads, long regarded as the invalids of American business, have fared better than the textile industry."

Mill village children, like the Walker brothers of Enoree (shown here in the 1940s), often created their own toy cars called "iron wheel wagons," using discarded gears from the factory as wheels. Steel rods from the same trash heap were used as axels. Other parts were made from scrap lumber, their fathers' old belts, broomsticks, rope, and springs from screen doors. Occasionally, races were held on the steepest hill of the village, and trophies were awarded to the winners. Often, an unofficial contest designated the best looking wagon and the ugliest. These creations were extremely difficult to steer, and brakes were usually non-existent. The noise made by these contraptions was great, and third-shift workers trying to sleep in their pre-air-conditioned houses were sometimes irate! (Courtesy, Anthony Tucker and Shirley Holmes)

mill workers in the early twentieth century, it was certainly not the only one. Some recreational facilities were quite extensive, but, with the exception of the baseball field, intended for more informal activities. Many mills hired community workers, often female, to coordinate recreational activities. In many mill villages, one could find YMCAs (YWCAs were fewer in number), annual community fairs, boys' and girls' clubs, and community contests to encourage home gardening, community beautification and beautiful, healthy babies. In 1924 Victor Montgomery built an amphitheatre in Pacolet to host annual May Festivals and other musical performances.

Textile executives sought to instill patriotism in "their people." During World War I, nearly all South Carolina textile mills, including those in Spartanburg County, reported military service and death among the workers. Saxon Mill proudly boasted that 82 young men from the village had served in World War I. On the home front, mill employees participated in Red Cross fund-raising

In 1920 the Bureau of Labor Statistics reported a study showing that mill operatives were 50 percent more likely than other workers to die before the age of 44, largely due to "wear and tear" on the job.

activities and subscribed to wartime bond campaigns. In 1927 when Charles A. Lindbergh, the United States' greatest aviator, came to Spartanburg shortly after making his landmark flight from New Jersey to Paris, Spartanburg's textile mills gave their workers a half-day off so that they might see Lindbergh on the square and be inspired by him.

While improving mill villages and mills themselves may have expanded productivity, textile companies sought even more ways to boost efficiency. Following the lead of New England manufacturers, many South Carolina textile mills introduced the "stretchout" in the late 1920s. Based loosely on the theories of so-called "efficiency experts" who analyzed work time and motion to determine the pace at which all tenders of the same machines should work, the stretchout set machines at the highest possible speed and increased the number of machines tended by each worker.

Implementing the stretchout was ill-conceived and poorly planned, but management contended that it represented a "scientific, a sound, and an honest effort of the manufacturers to meet exacting competition." Furthermore, they argued, the system enabled "piece workers" to earn more by producing more. Almost from the beginning, workers complained about the stretchout. State Representative H. C. Godfrey of Spartanburg called for an investigation of the new system and proposed a law limiting the maximum number of looms tended to 36. After lengthy, often vitriolic debate, the House defeated the measure, but the debate contributed to the outbreak of strikes in nearly every textile section of the state. Workers at Woodruff Cotton Mill, then part of the Brandon Corporation of Greenville, organized a strike in 1929 that lasted three weeks. Allegedly, the strikers met at the local baseball park and pledged to stay out of work until the stretchout was abolished. Mill owners feared that the strike might spread. Charles G. Wood, Commissioner of the U. S. Department of Labor, came to Upstate South Carolina and urged potential strikers in other mills to stay on the job until efforts at conciliation had been made. Protests against the stretchout, Wood said, challenged the right of mill owners to operate their

Pacolet Mills women, from left: Shirley Longe, Ada White, Nora Snyder, Jennie Bradley, and Myrtle Voultine (Courtesy, Don Camby)

Employees of the Saxon spooling room, 1915 (Courtesy, Sue White)

plants as they saw fit. Strikes, he thought, tended to unite the owners to the disadvantage of the workers. Although some strikes had limited success, at Woodruff the employees went back to work, and the stretchout system remained in place.

TENSION AND UNREST GENERATED by the stretchout surprised textile industrialists who had long basked in the assurance that the "splendid spirit of cooperation" between management and workers resulted in a high standard of living and made workers more resistant to "radical leadership or the whisperings of would be reformers." Southern antipathy towards organized labor was longstanding, and before the 1930s, unionism made little headway in the South. Objections to labor unions included the possibility of violence and subsequent destruction of private property. South Carolinians, particularly those in management positions, believed that unions bred disloyalty, radicalism, and other undesirable traits. Union organizers may have realized that the seemingly abundant supply of ex-farmers desperate for a job made it doubtful that workers could successfully dictate their compensation. In addition, many workers were disillusioned with the unions, which failed to deliver on their promises.

In the midst of the 1929 labor strife, the South Carolina General Assembly appointed a special legislative committee to investigate conditions in the industry, the only committee of its kind in the nation. Vindicating the workers, the committee attributed the 1929 strikes to deplorable living conditions in the villages and to overwork resulting from so-called "efficiency measures" (the stretchout) that exhausted two-thirds of the workers and put the other one-third out of work. Furthermore, the committee failed to blame any of the disturbances on organized labor, preferring to ignore the relatively silent specter of unionism

Boom times after World War I brought the need for more mill houses in Tucapau. This is Chestnut Street in an area known as New Town. (Courtesy, Junior West)

in South Carolina.

In 1929, in the midst of the most serious labor disorder of the century, the South Carolina State Federation of Labor, formed in 1900, held its annual convention in Spartanburg. Delegates to the convention attacked the state Department of Agriculture, Commerce and Industry for failing to enforce the labor laws adequately. Industry representatives called for elimination of night work (especially for women and children), a shortened workday, and higher wages. In addition, they called for doing away with "paternal features." The textile workers present asserted their intent to pursue these goals without resorting to violence. The convention made headlines but accomplished little of substance. Perhaps its most significant result was to alert both management and labor that change was in the air.

Mill executives tended to ignore the industry's perennial problems of low wages, long hours, and unsafe working conditions and to see what they wanted to see—contentment, well being, and

In 1928, the year before strikes began to break out in Southern textile mills, South Carolina had the lowest textile wages in the country, according to the U.S. Bureau of Labor Statistics. South Carolina laborers earned $9.56 a week, compared with $11.73 in Georgia, and $12.23 in North Carolina. In Massachusetts, which was quickly losing its industry to the South, textile workers made $16.47; in Connecticut they made $18.40.

satisfaction among the workers. But others noted the deleterious effect of millwork and mill village life. An Upstate farmer described the mill workers he saw in town on Saturdays: "Their faces were pale...for they worked in the winter from daylight until after dark...[They] made me feel that they had been captured, that they were imprisoned, that they had given up being free. For a long time they felt that way themselves, for they were cotton farmers, born to the wild wind." Contemporary critics of Southern paternalism pointed to the isolation of the workers in the mill villages, harsh working conditions, long hours, low wages, and the "mill complex" that evolved. Mill executives, meanwhile, applauded the system as humanitarian and profitable, for the companies, for the workers, for the state as a whole. Outsiders observed that the system discouraged labor organizing. But still others, some of whom were intimately connected with the industry and its thousands of workers, pointed out that "Southern industrialism left the mill employees unexposed to other occupations and other lifestyles, and, while technically free, in reality the Southern textile worker was a prisoner of the system which controlled so much of his life."

Late in the 1920s, Spartanburg County textile mills produced a huge variety of textiles—cloth for sheeting, fancy prints, bedspreads, damask, hosiery, twills, osnaburg, for example. The largest company was Clifton Manufacturing Company, which operated three mills containing 86,800 spindles and 2,650 looms to produce a large volume of sheetings, prints, and drills. The smallest traditional textile factory, Mary Louise Mill at Mayo, operated only 6,144 spindles for manufacturing yarns only. Even smaller was the Shamrock Damask Mill at Landrum whose 60 looms wove damask bedspreads.

In 1927, *The Handbook of South Carolina* touted the "five outstanding developments" in South Carolina in recent years. All were related to the textile industry. They included improvements in the mills and mill villages, the manufacture of finer grades of cloth, the diversification of the industry by the addition of dyeing and finishing plants, the development of inexpensive hydroelectric power, and the growing independence of Southern manufacturers from Northern selling houses, a situation which had robbed South Carolina of revenue for years. The Spartanburg Chamber of Commerce noted in the mid-1920s the "progressive spirit" of the Spartanburg County mills. The mills, according to the Chamber, paid good wages, provided extensive welfare work, provided superior schools, libraries, community buildings, beautiful parks, and "homes equipped with modern conveniences." In South Carolina, according to *The New York Times*, "the future opens wide the eyes of the imagination," and the gospel according to industrialism reigned supreme.

These postcards illustrated spindles in Spartanburg County. (Courtesy, Bill Lynch)

Mary Irene Gault

Mary Irene Lavender Gault, daughter of Drayton's first and only barber, was born in Drayton in 1913. Though she only worked in the mill for a short time, she has lived within 100 feet of her birthplace for more than 87 years.

I remember when we didn't have any lights in the house, had lamplights, and we'd study our lessons by the lamps. I remember when they put the lights in—I don't remember the date—but I was just little, and people said when it comes up a storm and if you have lights in your house, the lightning would strike. Every time it would come up a storm, I was just scared to death. But then that wasn't true. And I remember when they put the water in the house. I think I was around 12 or 13 years old when they put the water in the house. They built onto the back porch, they just built a little room and the door, and they put us in a commode. No lavatory or anything, just a commode. The seat on the commode, when you got up, that seat would come up and that would flush the water…I remember that just as good as everything. That was our bathroom! The water, we had was a pump right out here in front of this house. You just go out there and pump to get your buckets of water. Had buckets a-settin' on the back porch to get your water. I remember all that…

Mary Irene Gault at her home in Drayton (Courtesy, Mark Olencki)

[We heated the house with an] open grate. The grates are still there, just a grate. And oh, it was cold in the wintertime. And the house was real high ceiling and just wooden walls, and it was so cold. We lived through it and all. That's the only fire we had. Then Daddy had a wood stove in the kitchen to cook on, then later on in life got an oil stove…

The school was back over there on Spring Street. A big old school, and it didn't have any water in it. Had an outside toilet and we'd have to go across the road over there in the cow pasture and to a spring on down in the cow pasture, and I know when I'd go I'd be scared to death of them old cows running around. I remember all of that…

Mr. [Ben] Montgomery, [for] the trash, he'd set some kind of barrels out. That's when I was a little girl and he hired these two black men, and they would come and pick up the trash. We didn't have to pay anything—we just started paying since Milliken closed the mill. He paid for that too. I remember when I was a little girl—they said you was lazy if you didn't have a garden; I know Mama had one—it was a mule and the plow, and they'd plow your garden for you. Mr. Montgomery paid for everything, and you could have your garden and I remember Mama always had a pretty garden right down there…

Saturday night in the community building, they had a picture show, and we went every Saturday night. You'd pay a dime, a Mr. Snyder would run the picture show, and he lived right over there in that house. But we'd get a dime and oh, we just couldn't wait. That's where we'd do a lot of our courtin', you know. We could set there by our boyfriend. And one time this girl, we had a boyfriend apiece, but the boys didn't know it. They was our sweethearts but they didn't know it because we hadn't told them. That's how you did back then…

Mr. Ben Montgomery was just wonderful, and I know [he and] his wife lived up there on Main Street, and every time she'd ride over to Drayton, I can't think of the black man who drove her car—you know they had servants and all—she'd be in the back seat and he'd be in the front seat. And she'd have him stop down there, and Mama would go out to the car and talk to her. Mama thought a lot of Mrs. Montgomery. They was real wealthy people, but he was just so good to everybody.

Mr. Montgomery, when he built the [community] building, he built bathrooms for the women, and my daddy—you'd go in the barbershop and he'd let you have a towel and a cake of soap for a dime. And he'd put the towels in the laundry for one dime. You see, we didn't have no bathtubs. And that's the way we'd go. And the women's was on this side. And the men would go there to the bathtub. We had to go down there to take our bath. We went all the time. The men had to go through the barbershop to go to their bathroom. They sometimes take their own towels, but usually they didn't, and they'd just give my daddy a dime. It was a towel and a little cake of soap, about like that [holds fingers apart]. But you know everybody seemed happy back then. It was a hard time but they didn't know they was having a hard time.

I used to go down there [to the company store] all the time. They had a big company store. Like you'd go down there, order your groceries, well, they'd lay out all your stuff out in the floor and whoever'd run the wagon, they'd have to load that up in the wagon and deliver it to your house. Back then everybody just seemed to be happy.

—Interview by Mickey Pierce

Drayton Mills, 2002 (Courtesy, Mark Olencki)

Voices from the Village

Lee Loftis

Lee Loftis, the son of a New Prospect farmer, went to work at Inman Mills in 1921. Loftis, who was 17 when his family moved to Inman from the countryside, describes his early days in the mill and on the village. Here he describes children playing in the mill and how new hires sometimes got playfully taunted. He was age 97 at the time of this interview.

Lee Loftis outside Inman Mills (Courtesy, Mark Olencki)

We were farming, and it just got to where you made a living and that was it. You raised practically most of your food. You didn't have any money, and there was only one money crop in this area, and that was cotton. You'd borrow money for food and for necessities for the summer until you sold your cotton in the fall. Then you'd pay your fertilizer bill and your food bill. What you had left, you'd buy some clothes for the wintertime. But you didn't have anything left. There was several of us children, and my daddy decided to move to the mill. I went to work the day we moved there at lunchtime. They started me off learning to spin. I didn't do that for a day or two until they put me to sweeping, 14 cents an hour, 55 hours a week, $7.70. I was excited because 14 cents an hour to me was a lot of money. I had been working on the farm at 50 cents a day. I liked it.

That racket, of course, we were not used to that. But the racket in the spinning department where I went to work didn't bother me too bad. In fact you got used to it. You could talk. I never did get used to the weaving. Those looms made so much racket. At that time the spinners ran what you called sides. A frame had two sides. They had two spinners run 10 sides. The warp spinners ran eight sides…

We moved in a four-room house on what we called G Street. After a short while we moved over on F Street in a six-room house because we were a large family. There was about eight of us, I believe, and there was five of us that worked in the mill. That same house was the house I went to keeping house in when I got married in 1925.

That whistle blowed in the mornings. You hit the floor and get up, you know. It blew about 10 or 15 minutes to seven in the morning. Of course, 98 percent of the people lived on the village then …I was always up. They was just running one shift at the plant then. I was down there at six o'clock. We'd go in with boys and girls, and they had what was called wipin' the rovin'. This was cleaning up. You had skewer sticks and creels. This stick, it would run in a little plastic cup, just a small thing, but it would gather lint. It was like music almost there in the morning. You'd go in there—there was no labor laws, you know. You could just hear them sticks paddling up and down. I never could do it as fast as they could. If I was sweeping, I would go ahead and have it good and clean at 7 o'clock so I could play around through the mill until about 8 o'clock. I used that time because I was interested in getting more pay and getting another job, maybe two cents an hour more.

I think mill people is the best people in the world, but there wasn't many of them when I went to work that had a chance to get an education. We had some good men that, say, was fixers and things like that, they didn't have an opportunity for education, and they didn't have much opportunity for much advancement over there. By the time I was 20 years old they were training me to fix. That was the highest pay next to the supervisor…

We ate good. My daddy had a good garden at the mill. You didn't get paid until on Saturday after you'd worked five hours. You'd get off at 12. As you come out, the paymaster would be down there in a little cage. You could smell fish at almost everybody's house on Saturday nights. I guess it was a cheap food and a good food, you know. My overseer used to say, "I look for Saturday nights better than anything because we're going to have fish for supper." The windows was up. You didn't have screen doors or screen windows. Then you had people would raise hogs and cows down there. The company would build barns, a place for people to keep their cows…

In the plant, you'd have a new one come in. They's a lot of people come in from the country and the mountains. And you know, the machines was fastened to the floor. You'd tell these new boys, "Now we gotta move this spinning frame." He's strain hisself to death. And then some of them would get too rough, and boys would drag them down the steps, you know. I remember one who came from the mountains. And there was two of the boys came in there, and I think one girl. And this boy, he was not a heavy fellow, small, but he'd come from the mountains, he was tough. And there was some of them gonna drag him down the steps. He just got over there in the corner and pulled out his knife. He says, "Now, come on! Some of 'ems gonna be a-bleedin' when you go down there!" And they backed off. They didn't drag him down the steps. Strickland boy, I remember that name. Now, you didn't let the supervisors know this. I guess it's something like what goes on in the colleges and high schools today.

There wasn't no jobs but the textiles, and the textiles had gotten going, building, and they couldn't get enough people to operate. They'd go up in the mountains and hire people, large families, and they'd get flat cars on the railroads, you know, and they'd put all them people and furniture on there and even move their cows. I remember the Dixon family. They told me a lot of times about them a-coming down, you know. The company had wagons and would meet them up here at the depot, and they'd carry them down and have a house for them. You could have chickens, and you could have chicken lots. And of course you had to allow people their horses or mules. Their families would go to work, and the man wouldn't do anything but plow gardens.

We was boys, and us bein' boys, 16 and 17 and 18 years old, and they'd be some of us take a notion and we'd quit, want to play ball, and so we'd loaf around a while, maybe a week, a week and a half, play ball. You'd see Mr. Wofford, the overseer. I did that two or three times. Mr. Wofford was a good man. He'd say, "Leroy (they always called me Leroy), you got your play out now, come on back to work!" I'd have a job sometimes that I could hurry up and I could get through by 12 or 1 o'clock. The rest of the day I'd dodge the overseer, and that's the way I learned other jobs. I learned to doff, fix, overhaul. I enjoyed it. I had a good life.

—Interview by Betsy Wakefield Teter

Inside Inman Mills (Courtesy, Mark Olencki)

Ira Parker Pace

Ira Parker Pace's father died when she was very young, leaving her mother with ten children. Ira, born in 1911, went to work for Pacolet Mills. Here she describes her youth and early married life.

I had to stop school at 14. See, my mother was a widow-woman, and there was three of us [children] that when we got 14, we had to quit. What I learned was just that much, nothing much, because we had to quit when we were 14 and go to work. My mama never had worked in the mill before. She had 10 children, so that kept her busy. And my daddy died and left all of us. I think we've been lucky though. She raised, I'll have to say, a good bunch of children.

I did spinning. That's what they called it. The spinning room. I wore a dress and an apron. We had to have an apron to put our cotton in. I wore an apron, most of the time, with a bib on it. But I didn't get dirty all that bad. Then too, I'd change every day if I did. See, my mama done our washing because we didn't have a washing machine at that time. Isn't that awful?

Ira Pace (Courtesy, Mark Olencki)

We were poor people, and Mama was a widow-woman, and after we had a full week in the mill, she'd give us a little money. All of us tried to work everyday so we'd get a little money. They'd be coming round in the afternoon, just about time we got ready to go home, and they'd give us our money for the week. I made maybe $18 or $19 a week. We carried it home to our mother, see, 'cause that was the rule. We had to carry it home and give it to her, then she'd give us some back. And the movie wasn't but 10 cents. You could go to the movie right down here for 10 cents. And oh, that was something!

Well, we went to the movies, and, you know, bought what we wanted. You know they sold things down there to eat and to drink. And we had an ice cream parlor in the company store, they called it. And we'd gather up down there, you know, us and our boyfriends.

There were big rock steps. Huge. To go up in the hall. See, the hall was a place—in that hall we had 300 or more people in one room, at the movies. When you'd first go in, the barber shop was there. You'd go back in there, and that was the men's area—the pool room was there and I don't know what all.

The first house on this line here, it's torn down, it was the girls' club. There was a porch on yonder side, and we'd get out there and flirt with the boys. They'd sit down on the steps, and we'd flirt with 'em. Anybody could go to the girls' club. Especially people that lived here. Everybody loved the club. We had a good time. We could just go down there and get on the couch and read if we wanted to. Oh, we had a time. But they tore down that house.

Mama was the first one on Pacolet to get us a self-playing piano. I've still got it. I've got it right now sitting in my living room. But it won't play, and it's out of tune. We had a self-playing piano, and I had all kind of boys just coming here just to sit and play the piano. They'd buy rolls for it.

[Leroy and I] were both raised here, and he was just one of them old guys that [would come over]. He worked in a machine shop, with the mill. He worked down here, learned down here. After he learned up and everything, finally they sent him to the new mill. No. 2. And they sent him up there, and he worked there until he retired.

We was living right close to the mill, on that line, and me and him went to courting over there. Finally, my mama died. We had planned a wedding about two months before she died, and I said, "Hey, my mama needs me and my money. And I'm not marrying you or nobody cause she can't live alone and I'm staying here." And I told him. So I stayed.

And he said, "We will get married, won't we?"

I said, "Yeah, we'll get married." But I put him off three months after she died. Well, I was so hurt, and didn't have no mama.

I was almost 18 when I got married. And 18 months later I got pregnant. Then I got sick on my stomach. Well, that just goes with it. And it was so bad to have to be sick on your stomach and work, so he [Leroy] made me quit. He told me, said, "You ain't going to work now, and throwing up like you've been doing." But I wanted to keep working.

We lived right on top of the mill almost, right on the very back line. Not the back line but almost. We lived pretty close to the mill, and we liked to sit and watch people going in and out. We could see people going in and out on the gangways. When my last baby was born, I was sick about all night, then I had him just about when people was going to work. Leroy raised the window and hollered, "Guess what folks—I got a big baby boy up here!" He was so proud of that baby.

—Interview by Jennifer Griffith Langham

At home in Pacolet (Courtesy, Mark Olencki)

Ola Smith

Ola Smith was born on a cotton farm in southern Cherokee County in 1918 and came to live in a black mill village called Marysville (known in more recent years as Maysville) on the outskirts of Pacolet Mills in 1925. Built by Victor Montgomery, this group of about 20 houses was named after an early African-American resident, Mary Brown Knuckles, and provided housing for the black families who did manual labor and domestic work in the village and around the mill. Here, she describes attending the two-room Marysville School (built in 1915) and the prejudice she encountered in the village because she was black.

My mama married a man from this side of the river, and so that brought her over to Pacolet Mills. [My stepfather] worked at Pacolet Mills at that time. They were outside workers. You know, there wasn't any blacks in the mills. At that time he was doing waterhouses [bathrooms], cleaning in the mills. But other than that, they cleaned streets or whatever there was to do on the outside. They first lived in a company house across the river until they could get a house in the village they called Maysville. I was about seven.

There was no trees, no nothing. It was just four rows of houses there, and you could see from one end to the 'nother. I remember one big tree about middle ways of Brown Street out there. You could see everywhere...They said in the beginning, it was just the house on the corner. Maysville used to be called Gilliamtown, and this one man, he was named Gilliam, lived in this house on the corner out there. They said that was the beginning of Maysville. I guess the school building probably came along about that time too.

Ola Smith
(Courtesy, Mark Olencki)

We played out in the sand. There was a lot of sand around at that time in the village...We used to build frog houses over our feet, put sand on our feet. In the evenings we would play out like that. And ring games. I guess we played some ball games too. I remember that. I don't remember being hungry or nothing. I remember we always come home at dinnertime from school and eat lunch. I was happy with it. I can't remember no bad times...

My mama carried me to school and I would go in every morning, hollerin' just as loud as I could, but I soon got used to it. I guess I wasn't used to the children when I first come over. 'Cause where I lived [before], it was a farmhouse. At the time, when I started, it was just two rooms. Two more rooms were built after my children started school. I really liked it. We had our teacher, Mrs. Daisy Davis. She would teach us by piano, by music. She would sing the alphabet. She would play and we would sing A, B, C, D and all those...They had more than one grade in the room. I think it was up to fourth or fifth grade in the first room, and the other side was bigger children. We had those pot-bellied heaters. I don't know if it was coal or wood. It was a real natural fire, I mean! They had folding doors in between, and when it got warm, they would close the doors so that one class wouldn't disturb the other one...

We had to go down to the mill sometimes. All the things that was happening in the city was down there, where they used to call "down at the flats." They didn't say "down at the mill." It was a store, a drug store, a company store. The company store was grocery and dry goods. They had the little ice cream parlor thing. Downstairs, they finally built a girls' club, but that was only for the white part of the community. The blacks didn't have nothin' like that to go to. We could get groceries. We didn't have a place to sit and eat ice cream either. We had to come back outside and eat it. But we was glad to go in there and get it...

As a child, after I got large enough, when we'd get out for lunch, our daddies worked down at [the mill]. They would have lunch hour from 12 to 1. And we would have to carry dinner down to the mill. There's a place there called the boiler room where they got all the heat and steam from the building and things. We would carry lunch down there. And so coming back, if we didn't want to have to fight, we had to get out in the street with the cars and come up, and walk up, because white school children would catch hands 'cross the sidewalk and if we didn't push them out of the way or something like that...we'd just get off the sidewalk. They were rough on us! I call it rough anyway. It was like that all the way, but we didn't have a lot of business down there, just to take lunch. When I'd go to the store, I'd always be with my mother. Never had to go to the store by myself.

I would stay over here with my mama all the school days and on Fridays I would want to go to my grandmama's, back across the river. The first house over there, they had dogs. And we had to walk down in the road because those little boys would sic their dogs on us. They would do all kind of mean things at that time. I don't know if their parents was working or what, but no one ever did make 'em do right! When I was over at my grandmama's we had to come over here to the flats and get ice. They had icehouses at that time. This man that [worked at] the ice plant...he was what you called "very prejudiced." In fact, he didn't like black people at all. At least that's what I thought about it. They would have ice, the big blocks laying out, and some of them would be cut in different sizes. As they lay out there it would melt down. So when me and my cousin would go to get ice, he would always kick the smallest piece that was on the floor out to us, and we would take a string and tie it around the ice and put a stick through and both of us would carry it like that. Now that part was bad. I remember that. I noticed it after I got old enough to know the difference.

Marysville was one of several black mill villages in Spartanburg County, and one of the few that retained its identity into the 21st century. A black village at Lyman survived, while others at Converse, Valley Falls, Spartan, and the Cliftons were demolished. Black families were evicted from their villages at Drayton and Whitney when the villages were sold off in the 1940s and '50s.

—Interview by Betsy Wakefield Teter

Outside Marysville School (Courtesy, Mark Olencki)

Mill Village Religion
Weaving the Community Together
by James Dunlap

In 1850 Spartanburg was a sleepy crossroads village with only four mainline churches—Baptist, Methodist, Presbyterian, and Episcopal. After the Civil War, the Carolina Piedmont experienced an unprecedented industrial boom. In his 1907 book, *The Cotton Mills of South Carolina*, Columbia banker and civic leader August Kohn listed no less than 26 mill village churches in the 23 industrial communities around the burgeoning Hub City.

In this report Kohn noted, "The cotton mill operatives, as a class, are of a decidedly religious temperament." In Spartanburg, this was an understatement. The most prominent feature of any mill village, second only to the mill itself, was the church. Church services and revivals provided social and entertainment outlets as well as religious activity. In Spartanburg mill villages, churchgoing was a way of life.

The mill village church was part of the textile industry's original infrastructure. Owners such as Edgar Converse and John H. Montgomery built churches alongside cotton mills, houses, company stores, recreation halls, and health clinics. Their churches often were products of their own religious faith. In 1900 historian J. B. O. Landrum described Converse as "benevolent and liberal in his contributions to all worthy subjects that came before him." Before his death in 1889, Converse frequently attended "the religious occasions which he loved to encourage among his people." John Montgomery was "for nearly half a century a consistent member of the Baptist Church" and was "foremost in the support of...all the claims of [church] missions and charity."

The Men's Bible Class at Saxon Methodist Episcopal Church in the 1930s. Their teacher was "Grampa Satterfield." (Courtesy, Kenneth Burnett)

The textile workers responded in kind. They took up collections and usually shared the financial burden for pastors' salaries and church construction. According to Kohn, "the mill folks, as a rule, are very generous in their support of church affairs" and "in proportion to their means...very liberal towards their churches." In the mill village church, the owners and operatives together created a lasting and powerful institution.

In same year that Montgomery established Pacolet Mills at a place called Trough Shoals, he also helped pay for the construction of Pacolet Mills Methodist Church. The company also paid for separate Baptist and Methodist parsonages and a duplex office building shared by both churches, according to current Pacolet Mayor Elaine Harris. In 1927 Montgomery's son, Victor, contributed to the renovation and enlargement of the church, renamed Montgomery Memorial Methodist Church. Just inside the door remains a bronze plaque in honor of "John H. Montgomery, statesman—instrument of God."

Other mill owners repeated the pattern across the county. Inman Mills founder James Chapman, for example, constructed a village of New England-style saltbox houses arranged on a neat square grid and immediately provided a chapel for use by all denominations. Attendance at this Sunday School reached 200 by 1906; in 1909 Chapman provided funding for the construction of Inman Mills Baptist Church. Fifteen years later he paid $8,000 and the congregation another $8,000 for the construction of Aldersgate Methodist Church. He not only funded the churches, his mill heated them, too: he ran steam lines from the main mill to provide warmth to both the Baptist and Methodist churchgoers.

Author and social worker Marjorie Potwin observed a similar situation at Saxon Mills Methodist Church. The owners and congregation shared the cost of the building, but church members paid the entire cost of the pastor's $1,800 salary. At the same time the 308-member congregation also raised $1,195 for various outreach and mission programs.

Mill village churches tended to be evangelical and conservative, emphasizing temporary suffering on earth and promise of heavenly reward. Potwin noted, "There is no vague perplexity or academic questioning." The doctrine, she wrote, was "puritanic" in its condemnation of card-playing and dancing but, in her estimation, "the prayer of spinner or weaver seem to bring Heaven very close to those to whom its glories mean so much."

Until the Great Depression most observers praised the religious nature of the textile operative. During the New Deal, however, union organizers entered the South and sometimes found evangelical Christianity a key impediment to successful organization of the Southern cotton textile industry. In the years surrounding the 1934 General Strike, union leaders charged the owners with using churches as instruments of social control. Cotton mill workers, they maintained, were intimidated from the pulpits of company churches.

Sociologists studying the issue also denounced the relationship between owners and the mill village church. Operatives, they said, fell easily into line because of a fatalistic religious attitude reinforced by company paternalism. The leading critic of mill owner paternalism was Yale sociologist and theologian Liston Pope, whose 1942 *Millhands and Preachers* is still considered the definitive study on textile religion. In Pope's view, the mill owner saw himself as the fatherly provider for the needs of the workers. In return for mill-funded churches, the owner demanded absolute loyalty in the struggle against outside forces such as labor unions.

Not all Spartanburg County mill churches fell into the pattern described by Pope, however. Members of Saxon Baptist Church published a pro-union newspaper and openly defied Saxon Mills owner John A. Law during the period surrounding the 1934 strike. One of the most outspoken and active local ministers was the Rev. David English Camak, founder of Textile Industrial Institute. In the early 1920s, he allied himself with former student Olin D. Johnston, then a young Democratic legislator working for improved mill conditions. By 1923, Camak and TII had parted ways. The official reason was "poor health," though Camak continued to serve the Methodist Church outside of Spartanburg for more than two decades. In his memoir, *Human Gold from Southern Hills*, Camak says Methodist bishops told him there was "no demand" for him in city pastorates.

Churches generally survived the strife of the mid-1930s, though the turmoil often pitted churchgoers against each other. During the years of World War II, church was integral to mill life. Clifton, for instance, claimed three Methodist churches, three Baptist churches, one Presbyterian, one Wesleyan Methodist, and one Church of God.

Added to the volatile mix of evangelical religion and labor unionism was the influx of Pentecostal churches into the mill villages after World War II. These new churches also focused on the hereafter and served to blunt even further the trade union message. They also drew thousands of members away from traditional Baptist and Methodist churches. The Church of God, unknown in 1900, moved across the landscape "like a mighty army." This brought another dimension to the controversy surrounding mill village religion that concerned "sober churchmen" like Gordon Blackwell, who studied the issue as a young man and later served as president of Baptist-affiliated Furman University.

In Spartanburg mill owner and operative alike were only recently removed from the hard times of the dirt farm or mountain hollow. They shared a world view that credited the cotton mill system with freeing the South from a backward agricultural past. The mill village provided an ordered, productive, and mutually beneficial arrangement rooted in Christian beliefs.

The cotton mill, "in spite of its shortcomings…gave my grandfather and many others the opportunity to ascend out of extreme poverty into a better life," said Ruth Trowell Watson, a third-generation resident of Pacolet. Her grandfather, Maynard Lancaster, left a Spartanburg County farm and, at age 13, took a job at Pacolet Mills. "The desire to work hard and live by the Ten Commandments was all that was required," she said.

The mill village system passed from existence in the last half of the 20th century, and by

2002 most of Spartanburg's numerous cotton mills were shut down. Many mill churches, however, remain viable institutions. Pacolet Mayor Harris laments the closing of the mills and the passing of a unique way of life. Still, she maintains, all is not lost. The churches, she said, hold together what remains of the Spartanburg County cotton mill community.

The Gift of Literacy
Camak's Textile Industrial Institute

by Katherine Cann

A Methodist minister who cared deeply about improving the lives of mill village residents founded a new kind of school in 1911 that aimed, through a liberal arts-based curriculum, to transform textile workers into stable, productive citizens. David English Camak's vision—the Textile Industrial Institute—provided an education for hundreds of adults who labored long hours in local cotton mills.

Born in rural Winnsboro, South Carolina, Camak attended Wofford College. Later, as minister to churches in several Upstate mill villages, Camak determined God's plan for his life. TII's purpose, "to find, train, Christianize and place men and women to help do the thinking for the five hundred thousand cotton mill operatives of the South," was the embodiment of that plan.

Hoping to make the vision a reality, Camak initially approached officials of the Methodist Church who showed little interest. Finally, somewhat to his surprise, Camak found backers among the textile magnates of Spartanburg. One of them, Walter S. Montgomery, gave Camak $100 and provided a house in the Spartan Mill village where classes could be conducted. The first session in 1911 opened with one male student and two female students, but the enrollment grew, and during the first year, 45 students attended the new Institute. By 1913, the Institute had moved to a building near Saxon where a campus eventually grew.

Initially, TII offered only primary grades, but enrollment demands necessitated the addition of a high school, a total of 11 grades, by the 1916 school year. Students in the primary grades could enroll at any time during the year and advance at their own pace. The students were adults; the average age in 1914 was 22. Understanding that the need to work was the major obstacle to education for the textile mill population, Camak devised an innovative schedule, rotating weeks at work in the mill and in class. Students from rural areas who had no textile experience worked on the same "week about" schedule in the school's kitchens, in its fields, and with its livestock.

With local support, Textile Industrial Institute constructed the Model Mill that opened near the campus in 1920. As conceived by Camak, the Model Mill would produce cloth to be sold to increase the school's revenue and give students "hands on" experience with a variety of textile processes. A type of gingham known as "character cloth" became the school's signature product. Samples of "character cloth" can be seen today at the Spartanburg Methodist College library and at the Spartanburg Regional Museum of History. The Model Mill, however, failed because of the recession in the textile industry that began in the 1920s and the inexperience of the managers. In 1923, Camak left TII and in 1927, Powell Knitting Company of Rhode Island purchased the Model Mill.

The first home for TII. (Reprinted from Human Gold from Southern Hills)

In the 1920s, Textile Industrial Institute became affiliated with the Board of Missions of the Methodist Episcopal Church, South, and in the 1960s, began a continuing relationship with the South Carolina Methodist Conference. In 1927, the school added a Junior College Department to complement the high school and grade school departments. Gradually, as public education in South Carolina improved, the lower and high school divisions were dropped, and the school became a full-fledged junior college. The name was changed to Spartanburg Junior College in 1942 and to Spartanburg Methodist College in 1974. Today the college serves a diverse population drawn from a number of states and several nations.

At a time when education in South Carolina was limited both by lack of facilities and lack of will on the part of the people, Textile Industrial Institute provided a singular opportunity, and to many offered a means of escape from a harsh life spent among the bobbins, spindles, looms, and cotton dust of a textile mill.

Power from the Hills
John Law's Lake Campaign

by Norman Powers

Power was the problem.

Other difficulties had fallen to the march of progress. New fertilizers introduced after the Civil War had made Spartanburg County's depleted cotton fields productive again. (It was no coincidence that Captain John Montgomery, founder of Pacolet and Spartan mills, had been a fertilizer salesman.) Transportation of raw cotton to mills and of finished fabric to manufacturing centers had been solved by the late 19th century with the coming of the railroads and Spartanburg's transformation into the "Hub City."

The Green River Valley before it was flooded for Lake Summit (Courtesy, Walter and Betty Montgomery)

Workers lured away from rural hardship by steady wages in the mills had arrived in such numbers by the turn of the century that entire towns to house them dotted what had once been pasture and field. But with nine Spartanburg mills in operation by 1900, it was the raw power to turn more and more looms, much more power than the Spartanburg Rail Gas and Electric Company could provide, that was the challenge.

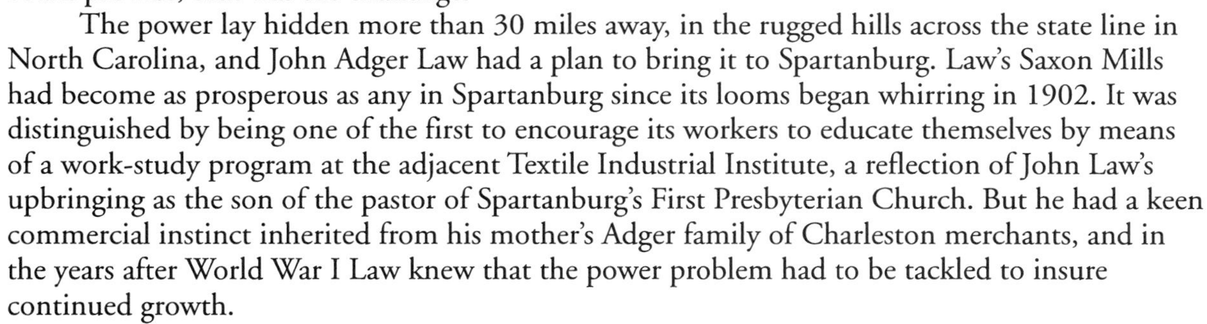

The power lay hidden more than 30 miles away, in the rugged hills across the state line in North Carolina, and John Adger Law had a plan to bring it to Spartanburg. Law's Saxon Mills had become as prosperous as any in Spartanburg since its looms began whirring in 1902. It was distinguished by being one of the first to encourage its workers to educate themselves by means of a work-study program at the adjacent Textile Industrial Institute, a reflection of John Law's upbringing as the son of the pastor of Spartanburg's First Presbyterian Church. But he had a keen commercial instinct inherited from his mother's Adger family of Charleston merchants, and in the years after World War I Law knew that the power problem had to be tackled to insure continued growth.

Having spent summers in the breezy hills around Hendersonville, North Carolina, Law was familiar with that city's success in damming the Big Hungry River as early as 1904 to supply hydroelectric power. It didn't take long for Law to come up with the idea of a "Manufacturers Power Company" after discussions with five other mill owners, among them Walter S. Montgomery. Montgomery was the son of Captain John, so the business of running a mill was in his blood. Like Law, Montgomery realized the difficulties ahead without a reliable and industry-controlled power source and saw the potential in the Piedmont's rivers and tributaries.

In a series of rapid deals, the Manufacturers Power Company was reorganized as the Blue Ridge Power Company by absorbing Hendersonville's Power and Light Company. The new company bought hundreds of acres of pristine woodland along the Green River in Henderson and Polk counties in North Carolina and, just after World War I, set about solving the power problem once and for all by building two dams to harness the power of the river for Spartanburg mills. One dam, constructed in Mill Spring, North Carolina, west of Columbus, reflected Law's influence in the name given to the lake it formed, Lake Adger. The 324-acre body of water formed by the second dam, 15 miles upstream and at a higher elevation of some 2,000 feet above sea level, was named Lake Summit.

The site was chosen carefully. It lay between the tiny hamlets of Zirconia and Tuxedo, where the river had long ago carved a modest valley through the hills. Even better, the Spartanburg to Asheville railroad passed along one edge of the valley after climbing the formidable Saluda Grade, providing an efficient means to get construction equipment and materials to the site and to carry away the thousands of tons of excavated rubble. In 1920, water from the new lake began spilling over a 254-foot long, single-arch dam into a flume, eight feet in circumference. (The present flume, which can be seen from the old High Bridge on Route 176 between Saluda and Hendersonville, is a replacement, but it follows the same route as the original). The flume narrowed to less than five feet before entering the power plant in Tuxedo, sending the Green River against the turbine blades with tremendous force, and electricity flowing south to the mills.

The new power source, however, soon proved just as attractive as a family retreat, even after cheaper steam turbines replaced hydroelectric power and Lake Summit's lines and powers plants were sold to a subsidiary of Duke Power in 1929. Removed from the tourists in nearby Saluda and accessible only by a dirt road, the tranquillity of Lake Summit became a family tradition for the Laws, the Montgomerys, and railroad cars full of nieces, nephews, in-laws, and cousins. In 1919, as the lake and dam neared completion, both men began construction on comfortable summer homes on adjacent parcels overlooking the southwest shore of the lake and joined by a common driveway. The timber was cut to order by Spartanburg's Clement Lumber Company and carried by rail to the lake's opposite shore, where it was loaded onto a barge and floated to the construction sites. Margaret Law, John's younger sister and a well-respected artist, built her own home and studio on a nearby lakefront lot, where supplies were ferried to her by boat and winched up to her porch in a basket. Even Law's personal secretary, Marjorie Potwin, settled into her own home next to her employer's during the summer months.

Others followed, especially after Law formed the Lake Summit Corporation and encircled the lake with its present public roadway. One regular summer resident was Spartanburg pediatrician Dr. Lesesne Smith, founder of the Spartanburg Baby Hospital. Dr. Smith's Lake Summit private sanitorium provided relief for the "summer complaint" of diarrhea and fever that claimed scores of Spartanburg children during the hot, humid Piedmont summers. The sanitorium's location proved so beneficial that Dr. Smith used it for an annual Southern Pediatrics Seminar, held each summer from 1921 to 1958.

Today, some 120 homes surround the lake, many of them converted to year-round residences. Still chief among them, however, are the Law and Montgomery houses, their common driveway a testament to the shared vision of the two men, and the lake they overlook a reminder of the power problem they both solved.

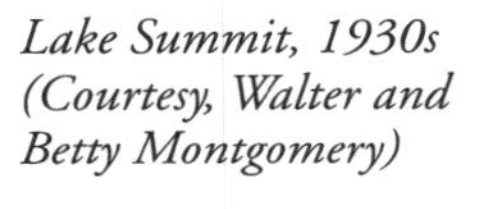

Lake Summit, 1930s (Courtesy, Walter and Betty Montgomery)

Company Stores
One-Stop Shops for Mill Workers
by John Messer

Established primarily for the mill worker, company stores provided very early one-stop shopping. These stores, prevalent from the early 1900s until late 1940s, were considered a "perk" for mill workers, but often served entire communities, including area farmers and small town residents.

These stores were located adjacent to the mill, or in the mill office area, and sold food, clothing, hardware, furniture, livestock feed and supplies, seeds, home garden implements, fertilizers, wood, coal, and ice (usually made on premises). There were also cooking pots and pans, wash pots and tubs, laundry, coal-burning heaters, fruits, nuts, candy, ice cream, sandwiches, griege and finished cotton fabrics for home-sewn clothing and draperies, and even caskets. Of necessity within walking distance of mill village homes (in the early years, private automobiles were a distinct luxury), company stores usually provided free delivery on purchases that were usually made on credit.

The company store at Inman in the early 1900s (Courtesy, Jim Everhardt)

During the entire lifetime of company stores, it was safe for school children to walk unaccompanied from the furthest reaches of the mill village to deliver the shopping list to helpful clerks in the company store, who not only filled the order but arranged for or made free delivery. Deliveries were made by horse-drawn wagon and later, flatbed Model T truck.

It was a treat for village children to go to the store, for the smell was original, unique, and inviting. Old-timers can still close their eyes and imagine the smell of fresh fruit, nuts, candy, hog "shorts" and other livestock feed, the meat market with fresh ground sausage and cured hams, pickle barrels and salt fish, hot dogs and hamburgers with chili and onions, and even new overalls. All these scents mingled together to form a pleasant and unmistakable smell. The treat was magnified by company store candy counter clerks who often saw to it that children got more than their money's worth for tightly clutched nickels and dimes.

Being a company store employee was considered a "plum" job, since there was only one shift per day—during daylight hours, whereas most mills, except in cases of curtailing or reduced hours/days, operated around the clock after the 1920s. Also, company stores had no flying lint, prevalent in the spinning room, no overhead central shaft machinery drive belts anxious to seize and remove intruding fingers or long hair, nor the clackety-clack din of cam looms in the weave room. Pay scale for company store employees was also

The company store in Arcadia, 1910. Among the men pictured are Vance Johnson, second from left; J. C. Epting, third from left; R. L. Doggette, fourth from left; and B. W. Johnson, store manager. (Courtesy, Mayfair Mills and Fred Epting)

higher than that of mill workers. Male company store employees usually wore shirts and pants, unlike male mill workers who almost universally wore overalls.

Some have criticized the company store as an attempt by mill owners to entrap workers in debt, preventing their seeking employment elsewhere. Stores did sell on credit, and there were always those who, for whatever reason, found themselves in debt. The percentage of those "owing their soul to the company store" likely was much lower than the current percentage of blue-collar workers owing thousands in credit card bills or in hock to check-cashing companies.

Workers who had charged more at the store than their weekly earnings were given a yellow statement by the paymaster on payday, usually Thursday or Friday, instead of an envelope containing cash. Company stores also issued money, which could be spent only in the company store. Some mills issued paper money called "jay flips," which were issued in $1, $2, $3, $4, and $5 booklets. Booklets could be purchased on credit, with the amount deducted from net cash in subsequent pay envelopes. Someone needing ready cash could purchase a $5 booklet on credit, and sell it for $4 cash, losing $1 in the process but having access to immediate cash. Company stores also issued coins called "flukums," "loonies," or "bobos" in the same denomination as current coins.

Company stores have long since vanished, along with a majority of the textile plants, but pleasant and precious memories remain with many who shopped there, worked there, or just hung out on the immense front porches.

Top, The company store at Pacolet Mills was one of the grandest in the county. (Courtesy Pacolet Elementary School)

Bottom, Salesman Charlie Holland waits for customers at the Spartan Mills company store. (Courtesy, Walter and Betty Montgomery)

1942 Spartanburg County Food Prices
Mill Stores vs. Other Stores

	Company Stores (*Average in cents*)	Supermarkets (*Average in cents*)	Independent Stores (*Average in cents*)
Eggs, dozen	**17.5**	**15**	**18.25**
Sugar, lb.	**6.75**	**6**	**7.75**
American Cheese, lb.	**32.5**	**39**	**36**
Fatback, lb	**16**	**17**	**17**
Cornmeal, peck	**35**	**39**	**34**
Potatoes, lb.	**4**	**3.5**	**4**

1942 Survey done by William Hays Simpson for his book, *Life in the Mill Communities*

Mill Baseball
Our Boys of Summer
by Thomas K. Perry

From 1885 to 1955, baseball flourished in the mill villages throughout the Southeast, and nowhere was the level of skill and intensity to win greater than in the Upstate of South Carolina. Big, strapping country boys, who swapped open fields for a steady wage in the deafening roar of the cotton mill, regained a measure of freedom with their exploits on the diamond.

Our heroes emerged from such beginnings, and generation to generation they came, to entertain and inspire. Spartanburg boasted excellent teams from Drayton, Saxon, Glendale, Whitney, Converse, Arkwright, Spartan, and Arcadia mills, but the early years were dominated by the teams from Tucapau and Beaumont.

The Tucapau boys made a name for themselves by venturing outside the mill village for competition, losing to Wofford College 4-3 in April, 1906. The next year they pulled off a near miracle, defeating the Spartanburg team of the minor league Carolina Association 2-0 (May 8, 1907), a hard loss to swallow for major league hopefuls. They were even known to play the likes of the Chicago Bloomer Girls, starring the famous pitcher Maude Nelson. Exciting as it must have been that day (May 12, 1910), the scorekeeper neglected to turn in the score.

Even when battling among themselves, the mill kids posted memorable efforts. Beaumont's Mayberry struck out 28 men in a 21-inning, 1-0 win over Inman Mill; his mound foe, Nix, fanned 24 in a losing cause. Both men pitched complete games.

When the contests were at Wofford Park, the band of W. A. McSwain entertained the crowds with popular songs of the day and marched ahead of the teams from the outfield to home plate in pre-game ceremonies. The game's popularity encouraged the mills to field strong teams, which meant recruiting, so Spartan Mill plucked Montague Nichols from the U. S. Naval Academy for the 1914 season. Drayton chose to stay local, getting Wofford College's Arthur Hamilton that same year.

Arcadia's 1935 team was the champion of the Eastern Carolinas League. (Courtesy, Louise Foster)

Fans came to the ballfields by trolley, horse and buggy, Tin Lizzies, and flatbed trucks. Sun glinted off mica-flecked infields, their rough surfaces contributing to untold errors, and sometimes the outfields had to be seen to be believed. In a 1914 game against Saxon, Arkwright's right fielder O'Dell made the play of the day. One witness said, "He ran back on a ball that looked impossible to catch, ran down into a cornfield, and caught the fly just off the silk of a big roasting ear." Saxon center fielder Johnson was lauded in like fashion the next year, after his two brilliant catches "while running in a plowed portion of the field" preserved a 2-1 win over Spartan Mill (August 8, 1915).

Mills overcame the derision of "cow pasture ball" as owners upgraded their facilities in the late 1910s and 1920s. Young men played on, honing their skills, and mill village reputations were built on the team's success. Bolt of Clifton had a strong pitching streak in 1920: a 9-0, no-hit win versus Glendale (May 1); a second no-hit gem, 4-0 over Saxon (May 8); and a 6-0, one-hit win

Drayton's championship baseball team, 1932 (Courtesy, Louise Foster)

against Saxon (May 15). Winning programs offered boosts to employee morale and community pride, and owners deftly used both as productivity tools for their workers.

The 1930s became the true glory days of textile baseball in the quality of play, spectator interest, and fierce loyalty to the village heroes. Game day was a festive atmosphere when Harold Sullivan and his Spartan Mills Band rode through the village on the back of the company's flatbed truck, loudly announcing the upcoming game at the park just off Preston Street.

Folks whooped it up for the kids of the American Legion team, a familiar entry into the Spartanburg County League, who used the experience against stiff "linthead" competition to win the 1936 National American Legion Championship. Glendale's Buster Hair delivered overpowering mound performances in a trio of 1937 wins. He struck out 20 in a 7-0 win at Clifton (April 17), 17 in a 2-0 duel with Fingerville (April 27), and 23 in a 4-0 victory over Spartan (April 30). Pacolet's Ernie White, soon-to-be St. Louis Cardinal, fanned 155 in only 126 innings during that 1937 season.

The introduction of night baseball added more excitement. Lyman defeated Hickory, North Carolina, 4-1 (May 7, 1937) before 2,000 fans, then edged Pacolet 3-2 in Eastern Carolina League play the next evening in front of 2,500 paying customers.

The boom years came to a close amid the noisy rumblings of World War II. Folks still came to the parks, much as they had during the Depression, to share their fears, to forget the horrors of war, to laugh for awhile and share the latest village gossip. Though we did not quite understand, textile baseball had changed forever. As soldiers returned to the mills and to their beloved game, something was missing. Owners sold the houses on the mill villages, and that fierce community pride was lost. With post-war prosperity, it was easier to buy automobiles and televisions, and the world was truly at our doorsteps. There just wasn't time to cheer the local boys to victory, and the game we loved slowly died away.

Old-timers' gatherings were started to keep the memories alive, and Ernest "Powerhouse" Hawkins led Spartanburg's effort. An exceptional player for Arkwright Mill throughout his long career, he was the traditionalist of the textile leagues, bold when it came to declaring how great the old guys—including himself—really were. The reunion drew dedicated followers for more than 25 years and sadly ended with the death of the great "Powerhouse."

Our boys of summer gave us a bit of innocence and reminded us how simple and joyful was the game we called our own. Their gift was enough to last a lifetime.

Olin Johnston
Champion of the Textile People

by R. Phillip Stone II

Olin D. Johnston rose from humble beginnings to become governor of South Carolina and a United States Senator, but he never forgot the textile workers who placed him in high office. Whenever he met textile workers at shift changes, his hand outstretched, he would announce, "It's Olin D. in person."

Born near Honea Path, on November 18, 1896, Olin Dewitt Talmadge Johnston and his family moved from working on their rented farm to the Chiquola Mill in Anderson County while Johnston was still young. He entered the Textile Industrial Institute in Spartanburg in 1913, graduated in 13 months, and entered Wofford College in 1915. While a student, Johnston worked as a proofreader for both the *Herald* and *Journal* and held a number of other jobs in the city. After his sophomore year at Wofford, he entered the Army and fought in World War I. He served as a sergeant in the 42nd Division, the Rainbow Division, in France.

Upon his return from the war, Johnston resumed his studies at Wofford. Graduating in 1921, he entered graduate school at the University of South Carolina, where he took a master's degree in political science in 1923 and a law degree in 1924. While in law school, he made his first attempt for public office, winning a seat in the state House of Representatives from Anderson County in 1922.

When he graduated from law school, he married Gladys Atkinson and established a law practice in Spartanburg. Gladys Johnston became one of his closest political advisers. In 1926, Johnston ran again for the Legislature, this time representing Spartanburg County in the house for two terms. Johnston's major accomplishment in the labor-unfriendly Legislature was a relatively weak act that required mill owners to install sewers in mill villages.

Johnston entered the race for governor in 1930 as the champion of the state's textile workers. He won more than 58,000 votes in the primary and led his nearest opponent by 15,000 votes. In the runoff, however, he lost by fewer than 1,000 votes to fellow Spartan Ibra Blackwood, who won an overwhelming (and suspicious) majority in Charleston County. Undeterred by this loss, Johnston ran for governor in 1934 as an ardent New Dealer. He campaigned for an end to the textile stretchout, for maximum hours and minimum wage legislation, and on a pledge to use the government for the betterment of the masses. Running in the middle of the 1934 textile strike, he had to walk a fine line between supporting the strikers and making himself attractive to town folks, farmers, and other groups of voters. At large stump meetings he would speak in generalities, and only when he was alone with mill voters would he talk about empowering the working class. When the votes were counted, he had defeated Cole L. Blease by a resounding 35,000 votes.

Olin Johnston about 1960 (Courtesy, the Spartanburg Herald-Journal)

For South Carolina, Johnston represented a new type of governor, one who was an unabashed economic liberal and who supported laborers in their quest for a greater share in the economy. When he took office in January 1935, he declared the era of "ring rule" in South Carolina was over. For most of the state's history, men from rural Lowcountry counties held the reins of government, and they opposed many of Johnston's initiatives. One of his primary goals was passage of a package of laws friendly to textile workers, though out of a series of over two dozen bills, only one passed both houses of the General Assembly. The rest died in the Lowcountry-dominated state Senate. Even the one that passed, the Worker's Compensation Act, was weakened by the senators.

Late in 1935, the rejection of much of his legislative program and his desire to wrest control of the highway department's enormous budget and patronage from his political opponents prompted Johnston to make a drastic move. When he failed to gain the resignation of members of the highway commission, he declared the highway department to be in a state of insurrection and ordered the National Guard to take control of the department. In what was perhaps the most famous fight between a governor and Legislature in South Carolina history, Johnston got the $3 automobile license tag that he had demanded in his campaign, but he lost the power to appoint highway commissioners and permanently damaged his relationship with legislative leaders. Ultimately, Johnston's supporters learned that while they could amass enough votes to elect a sympathetic governor, they could not overcome legislative obstacles to their goals.

Still, Johnston's administration saw some successes, including the creation of the state labor department and an industrial commission to administer the worker's compensation law. Under his leadership, South Carolina embarked on an ambitious rural electrification project. With the personal support of President Franklin Roosevelt, South Carolina's program was a pilot for the later federal Rural Electrification Administration.

In 1938, Johnston sought to capitalize on his term as governor and his close relationship with President Roosevelt by running for the U. S. Senate seat held by Ellison D. Smith. Johnston lost a hotly contested race to the long-time senator. The defeat was as much a referendum on the Roosevelt administration and the president's meddling in state politics as it was on Johnston's administration. In a move unprecedented in South Carolina politics, Johnston ran for governor again in 1942, winning a narrow victory. However, the end of the New Deal and the outbreak of World War II meant that labor issues took a back seat to defense matters in the second Johnston administration.

Johnston again challenged for Smith's Senate seat in 1944, and this time, he defeated the elderly Smith. As a senator, he retained his labor sympathies, voting against the Taft-Hartley Act in 1947. He remained a New Dealer throughout his term, though his liberalism in economic matters did not translate into racial liberalism. Like many politicians of his time in the South, he used racial demagoguery to shore up support among white voters. However, he declined to bolt the Democratic Party to support the Dixiecrats in 1948, and his political following was so strong that he helped deliver Spartanburg and Anderson Counties to Harry S Truman in the election, the only two counties in South Carolina not to support Strom Thurmond's presidential bid.

He won three more elections to the Senate, defeating Thurmond in his 1950 bid and a young Ernest Hollings in 1962. So loyal were his supporters that his daughter, Elizabeth Patterson, was able to tap into his political base in her successful congressional races during the 1980s.

Johnston died in Columbia on April 18, 1965, and Vice President Hubert H. Humphrey and at least 18 of his Senate colleagues attended his funeral at Spartanburg's Southside Baptist Church. The funeral procession from Spartanburg to Honea Path attracted thousands of his supporters and friends. They remembered him simply as "Olin D.," the man who knew their needs and aspirations so well because he was one of them.

Marjorie Potwin
Shaping a Brave New World in Saxon

by Alice Hatcher Henderson

In 1916, Marjorie Potwin, a slender, young redhead from a textile town in Connecticut, arrived in Spartanburg to work as recreation director in the Saxon and Chesnee textile mills of John Adger Law as part of the new cooperative extension program established by the federal Smith-Lever Act. Within the next 10 years, she not only completed a master's degree in labor economics at the University of Chicago, a Ph.D. from Columbia University, and wrote a book entitled *Cotton Mill People of the Piedmont*, but she also became the powerful community director at the Saxon Mill and the controversial companion of John Law, a bank and cotton mill president and—especially shocking to the community—a married man with six children. Their relationship remained the subject of gossip for years and eventually ended in Law's divorce from his wife and marriage to the young Yankee.

Marjorie Potwin was admired by some and detested by others. She received praise from those who supported her energetic efforts to turn Law's mill villages into imitation New England textile communities. Regular school and work attendance, personal and housekeeping cleanliness, and devotion to task completion were the characteristics she wanted to teach the mill village inhabitants. For those workers who were not goal-oriented and displayed little interest in "improving" themselves, she had only contempt—a feeling that they reciprocated. During the textile mill strikes of 1934-36, one of the union's demands was that she be dismissed from her job. At that time, Saxon Mill's personnel policies and work rules were the subject of frequent complaints to the Cotton Textile Labor Relations Board, and there was little evidence that Marjorie Potwin's academic study with liberal labor economist Paul Douglas had brought compassion and fairness to Saxon.

Marjorie Potwin's book, published in 1927, provides interesting insights into her own thinking. In many ways, she seemed more sympathetic and less prejudiced against mill workers than other Spartanburg residents. She called attention to the tremendous changes the families had been forced to make in their move from the more relaxed schedule of subsistence farming in the North Carolina mountains to the long hours in the mills. She emphasized again and again that, among the mill workers, there were individuals with ability and ambition who showed great potential and outstanding human qualities. Her references to African Americans who worked in the mill villages as maintenance men, laundresses, and cooks were generally positive. She cited the strong mill village interest in religion, politics, clubs, music, and sports as evidence that the workers were more than subhuman lintheads.

On the other hand, she was unable to conceal her contempt for families that wanted to keep their farm animals in the yard rather than in the pastures provided for them, that used their bathtubs to store coal or potatoes, or who didn't get their children to school regularly. In her book, she complained, "they seem to lack studied self-direction." For their part, many mill workers deeply resented her moral double standard. In the mill villages, adultery was punishable and could lead to job loss, but she appeared to flaunt the rules. She must have been aware of their attitude and stated in *Cotton Mill People of the Piedmont,* "As they see it, right is right and wrong is wrong."

Her missionary-like zeal to bring about a brave new world in the mill villages is apparent in her writing. John Adger Law is enthusiastically portrayed as a benevolent despot, bringing better health, recreation, and educational opportunities to his workers. She cites the summer vacation camp for mill children that he established at Lake Summit as an example of his attempt to improve the lives of his workers. She refers to the textile workers' rising standard of living and access to better education and more consumer goods, but says little about low wages, the pressure to speed up, and unsafe working conditions.

COTTON MILL PEOPLE OF THE PIEDMONT

A Study in Social Change

BY

MARJORIE A. POTWIN, PH.D.

NEW YORK

COLUMBIA UNIVERSITY PRESS

LONDON: P. S. KING & SON, LTD.

1927

Majorie Potwin's book (Reprinted from Cotton Mill People of the Piedmont)

Upon Law's retirement, the newly married couple moved to Lake Summit, which had been built by Law and John Montgomery as a hydroelectric project to provide cheap power to the textile mills. Ironically, he had already built a family home on the same lake for his first wife and children. When Law died in 1949 at the age of 80 while on a visit to Marjorie's family in Connecticut, the *Spartanburg Herald* ran a front page story on the death of this community leader, a man who had served as president of two banks, two mills, the national Cotton Manufacturers Association, and on the boards of Converse, Wofford, the Kennedy Library, the Piedmont and Northern Railroad, and the United States Chamber of Commerce. The names of his first wife, six children, parents, and brothers were all listed. However, the obvious omission was a reference to his wife at that time,

Marjorie Potwin Law. Apparently, she was still too controversial to be mentioned.

Marjorie Potwin Law lived at Lake Summit for 14 years after her husband's death, serving as vice president and general manager of the Lake Summit Company. She was known as an active businesswoman and supporter of the arts. She died at her Lake Summit home in 1963 at the age of 71. Her funeral notice was carried in the Hendersonville paper but not in Spartanburg. She was probably the first woman in Spartanburg to earn the Ph.D. degree, her book on the Saxon mill village was on the shelves of most of the leading libraries in the country, but in deference to the morals of the community, her life and death passed without comment by the Spartanburg papers. She had married the "king of the mill hill," but she had not become the "queen."

Pacolet Mills
A Southern Utopia

The following article was published in Nation's Health magazine in 1927 by Dr. R. G. Beachley, deputy state health officer in Hagerstown, Maryland, who spent time in Pacolet Mills in the mid-1920s as Spartanburg County Health Officer. Here he lauds Pacolet management for improving health conditions so much that the village death rate was lower than that of the county or state as a whole. Indeed, Pacolet was considered by many as one of the best-kept mill villages in the South.

Located in the Piedmont section of South Carolina is one of the most ideal cotton mill villages, from the standpoint of health and beautiful surroundings, to be found in the United States, or even in any foreign country. The village is built around the mill and houses the employees of the Pacolet Manufacturing Company.

Nothing has been left undone for the comfort of the mill workers and their families. The homes furnished the employees are not the ordinary form of shack or small frame house so often seen in the mill villages of the South, but each family occupies a detached bungalow built along most modern lines, with attractive architectural designs. In passing through the village one might believe he was in a suburb of a modern American city to judge from the appearance of the streets, homes, and surrounding yards. Each house has a front yard which is kept in excellent condition, shrubbery being supplied by an expert gardener who maintains a spacious hot-house and nursery for this purpose. Needless to say, the interior of each home is modern in every respect, as to heating, water, electric lights, and sewerage.

Young children gather for lunch at the day nursery in Pacolet Mills. (Courtesy Pacolet Elementary School)

Some years ago the mill authorities employed a trained nurse to organize and establish a system of welfare work among the employees and their families. The work has contributed wonderfully to the health of the people, and especially to the low death rate from communicable diseases, as well as the low infant and maternal death rate. In fact, the general death rate is as low as could possibly be expected. In any case of sick-

ness among the employees or their families, when it is desired, the nurse gives whatever service might be required and arranges for special medical attention, such as hospitalization or operative treatment.

One of the most unique developments has been the establishment of a baby day nursery, where mothers who are employed in the mill may leave their children during the day. This unusual feature of mill social service work was developed by an additional building being added to the nurses' home, connecting by a large passage. The babies are brought in when the mothers go to work. Every child gets a full bath daily and is cared for scientifically, having plenty of recreations, rest hours, and proper amount of nourishment. This day nursery is under the supervision of the nurse and doctor. A baby clinic is held each week and any child who is brought to this clinic is examined and put on treatment and diet, free of charge. Both the doctor and nurse are present at these clinics.

During the past six months a physical examination clinic has been organized, the object being to give a thorough and complete physical examination to each adult in the village in an effort to further reduce the death rate and protect the health of the employees. The people are very responsive to this new idea of an annual physical examination and, needless to say, this clinic is a step forward in industrial medicine and is obtaining excellent results.

In addition to the careful medical and hygienic conditions given the employees, a large community house is provided for recreation, with reading rooms, club rooms, baths, kitchens, and dining room. The president of the mill calls this community house the "Matrimonial Bureau," for in this building the young ladies of the village give their parties and dinners and act as hostesses, under the direction of the club leader, in entertaining their

Top, New homes built by Pacolet Mills in the 1920s were considered the best mill housing in Spartanburg County. (Courtesy Pacolet Elementary School)

Bottom, District Nurse Belle Fuller served the community of Pacolet Mills for years, traveling from one place to another on horseback. (Courtesy, Pacolet Town Hall)

young friends, and the fact that a number of weddings have been solemnized in this attractive building leads one to think that the president may not be entirely wrong in his statement after all.

A special building is provided for the men, with pool rooms, bowling alleys, gymnasium, picture shows, and banquet room with kitchen, for their recreation.

Two fine churches are located in the village, and the community stands on record as one that has complete harmony between the churches, and between the churches and the community.

A community council is another interesting organization in this village. The council is composed of a representative from each department in the village and holds a supper meeting once a month, when the activities for the month are planned and discussed. This council is responsible for seeing that the plans are carried out. They have one mass meeting each month, one or more general parties for the young people, and discuss all the meetings and all developments from the standpoint of the community.

A community aid society is composed of a committee made up of two ladies from each street in the village, whose duty it is to find and report any cause requiring temporary help. If a case is reported by a committee woman, it is then investigated by an investigating committee, and, if advisable, help is given, but it is given only through the woman bringing the first report.

A loan closet is maintained, filled with hospital supplies, electric fans, gowns, sheets, and any second hand clothing that may be brought in. These supplies are loaned, or, in the case of clothing, donated to the needy family. This has been one of the most valuable organizations in the village, as it helps to maintain a spirit of good fellowship and an interest in each other among the entire community.

Classes in first aid and home hygiene are given by the nurse and her assistants. At the completion of these courses regular graduation exercises are held, making it a real social event, which greatly helps to insure perfect attendance.

A physical director is employed, who conducts classes on physical education in the schools.

The organization of a Sunset Ball League contributed much to the pleasure of the mill during the summer. Teams from every room in the mill played a short game every afternoon between supper and dusk.

A recent health survey of the mortality and morbidity of the Pacolet Mill village shows that there were no deaths during the past year from typhoid fever, diphtheria, scarlet fever, infantile paralysis, malaria, measles, tetanus, or cancer. Likewise, there were no deaths of mothers or babies during childbirth, or its complications. The general death rate of the village is much lower than that of the State or county.

From observing the conditions in this cotton mill and its community one cannot help but be moved by the possibilities that can be brought about for these mill workers, and it is the earnest hope of the health officials throughout the South that Pacolet may set a standard to which all cotton mills might be raised.

This house was built when Pacolet Mills was still known as Trough. (Courtesy, Don Camby)

AN OPERATIVE'S RESIDENCE, Trough, S. C.

Curing the Disease of the Poor
Spartanburg's Pellagra Study
by Gary Henderson

To pellagra patients healed by Dr. Joseph Goldberger's work in Spartanburg, he was a hero. To politicians, physicians, and mill owners embarrassed by his research, he was a radical.

By 1910, thousands of Southern mill hill residents were exhibiting signs of pellagra, sometimes known as the "sickness of the four D's"—diarrhea, dermatitis, dementia, and death. People who contracted the disease sometimes continued to function with butterfly-shaped skin sores as their only symptom. But often victims progressed to a point when effects of the disease began to destroy them; about 40 percent of them eventually died. Children and non-working women were particularly hard hit.

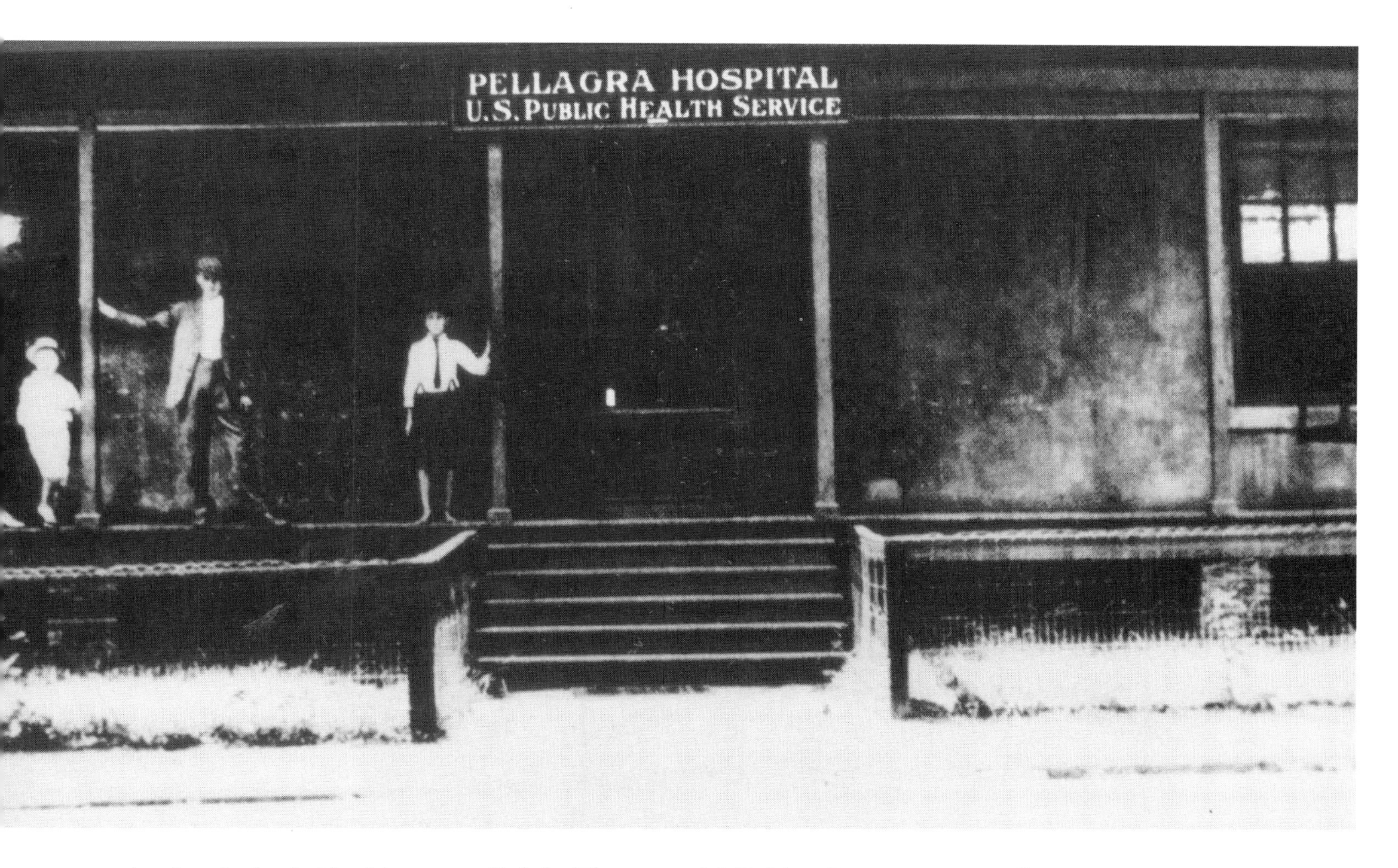

The U.S. Public Health Service operated the Pellagra Hospital in a building at Spartan Mills, 1916-1921. (Courtesy, Louise Foster)

A privately funded health team, called the Thompson-McFadden Commission, was first invited to Spartanburg in 1912 to study the problem. During four years of research, they identified at least 300 local cases of the disease but came to the faulty conclusion that pellagra was mostly likely spread by germs and was caused by improper sewage facilities. The Spartanburg County Medical Society was closely involved with the study; Wofford College offered its chemistry lab, and Spartan Mills made a building available at the corner of College and Forest streets for a pellagra charity hospital. (After three years of studying the Spartan Mills village, the commission concluded it was "one of the worst pellagra districts in the South.")

But it wasn't until Goldberger arrived in Spartanburg that the true cause of the disease was identified. Goldberger, an epidemiologist who had run health studies in other areas of the country, joined Dr. George Wheeler and his staff at the pellagra hospital to conduct tests that focused on textile workers and their families. His initial research showed that pellagra had nothing to do with the lack of sewage systems—and everything to do with the income levels of the victims. "Dr. Goldberger realized immediately there was a direct relation to how much money a person earned and pellagra," said Charles Gershon, an Asheville urologist who's writing a book about Goldberger and the work he did in Spartanburg. "This was a disease of the poor."

What made Goldberger controversial was his claim that poor meals on the dining tables of poverty-stricken textile workers and sharecroppers were causing a scourge of pellagra to sweep across Spartanburg County and communities throughout the South. Goldberger, the son of Slovakian immigrants, was appalled by the living conditions he saw on Spartanburg County's sharecropping farms and in the area's mill villages.

"He saw very quickly the textile workers' and sharecroppers' diets consisted of grits, salt-cured meat (fatback), molasses and corn bread," Gershon said. "There was little money to buy anything else, like fruits and vegetables. The average textile worker was making about $12 a month at the time. They bought their food at the company store because they didn't have the money to go anywhere else. (Some of) the company stores didn't carry fresh fruits, vegetables, and other things that are part of a balanced diet."

Cotton, grown to supply the textile mills, was "king" in the South when Goldberger came to Spartanburg. What disturbed him was seeing cotton rows planted up to the porches of sharecropper shacks, with no room left to grow the foods needed to stop the pellagra epidemic. In Inman, a major cotton-growing area, the disease was rampant.

Goldberger was so sure that the germ theory was untrue that he tried to infect disease-free volunteers by having them consume a tablet made in his lab from the blood, body secretions, and fecal matter of pellagra patients. Goldberger even injected himself, his wife, Mary Farrar Goldberger, and Dr. Wheeler with blood he'd drawn from victims of the disease. None of the volunteers contracted pellagra, leading Goldberger to pursue his notions about lack of vitamins and minerals in the diet even harder.

His research clearly showed that the incidence of pellagra declined with an increase in wages. When South Carolina textile wages rose during World War I from an average of $311 a year in 1914 to $757 in 1919, there was a corresponding drop in pellagra. But hard times returned in the 1920s. Textile paychecks shrunk, and by the end of the decade, pellagra cases reached an all-time high.

Goldberger's theories challenged the economic and social structures that were the cultural backbone of Spartanburg and the South. Pointing out the problem made him seem even less credible to local leaders. "The whole thing might have been more acceptable if Goldberger had been a Southerner, not a Jew," Gershon said. When President Warren G. Harding called for a federal report, Spartanburg Congressman Jimmy Byrnes denounced the effort. A representative of the local Chamber of Commerce wrote a letter to the editor of *The New York Times* complaining that Spartanburg's reputation had been besmirched. E. B. Walker's letter, printed July 30, 1921, said: "There is no poverty or famine in our cotton villages...The operatives are paid good wages, and there is no reason, except ignorance, for their not buying food that would give a well-balanced diet."

Tests Goldberger conducted on a group of prisoners in a Mississippi jail were further evidence that pellagra was related to poor diet. Typically, the inmates at the jail were fed foods raised on a farm they worked. Over a period of months, fruits, vegetables and dairy products were replaced with items similar to the ones Goldberger had observed on the kitchen tables of the textile workers and sharecroppers. All of the volunteers in the Mississippi study contracted pellagra.

Goldberger's son, Joseph Jr., a retired physician, was 19 when his father died of cancer on January 17, 1929. "I never thought much about his work then," Dr. Goldberger, now aged 91, said from his Texas home. "People used to say, 'He was a great man.'" He later came to realize the importance of the medical research his father conducted. He places the pellagra studies in Spartanburg near the top of the list. His father left a private medical practice to work in public health. He died penniless, with no pension, he said.

Today, niacin and vitamins in packaged foods and improved diet have all but wiped out pellagra, except in some developing countries. Charles Gershon said Goldberger's work in Spartanburg and other areas of the South was a vital contribution to medicine and the health of Americans. "It was one of the great stories of public health in this country."

Adapted from an article in the Spartanburg Herald-Journal. *Reprinted by permission.*

Lyman, South Carolina
A New Kind of Company Town
by Toby Moore

By the time Boston-based Pacific Mills arrived in Spartanburg County in 1924 to build a new mill and village, South Carolina had witnessed at least 70 years of experiments in the design of company towns. Experience had not yielded much improvement. While the state's first mill village, antebellum Graniteville, featured solid buildings made from stone, the cotton textile boom of the post-war period was marked by hurriedly designed and cheaply constructed towns. Workers found these often drab and crowded villages an improvement over the squalid rural housing of the turn-of-the-century South, but not by much.

The typical late 19th-century mill village consisted of rows of wood-frame houses, perched on brick or stone pilings, on large, usually grassless lots. Unpaved streets, many no more than paths, turned muddy or dusty according to the season. Behind the houses would run an alley, and then another row of homes, or fields for grazing, or simply woods. Most houses had garden plots, chicken houses or even pigpens, and outdoor toilets. Rural carpenters copied traditional architecture, and designs varied, at least until D. A. Thompkins of Charlotte codified the vernacular designs in his *Cotton Mill, Commercial Features*, published in 1899. From a handful of drawings, engineers could erect hundreds of homes in a matter of months.

Pacific Mills, which would name its village Lyman, after its president, had in mind a different sort of town. Flush with cash following World War I, and smarting from rising criticism of living conditions in the villages, cotton textile companies had begun to pay more attention to the quality of the places in which their workers would live. These later villages, some of the last company towns ever built in the United States, would benefit from new ideas in city planning, a profession still in its infancy.

Drawing from the ideas put forth by Frederick Law Olmsted, his sons, and other pioneering landscape architects, these city planners would conceive not only such leafy suburbs as Myers Park in Charlotte and resorts like North Carolina's Lake Lure, but also new industrial towns such as Kingsport, Tennessee. It was the designer of Kingsport, John Nolen, who brought to Charlotte

The village of Lyman, 1930s (Courtesy, Herald-Journal Willis Collection, Spartanburg County Public Libraries)

a young city planner named Earle Sumner Draper in 1915. Draper would eventually design nearly 150 mill villages and expansions for companies up and down the Piedmont, including Pacolet Mills in Spartanburg.

Draper sought to marry modern ideas of how a town should work with the appreciation of nature espoused by Olmsted, who had designed New York's Central Park. Draper opted for sweeping curves in his streets, instead of the traditional grid, and urged mills to upgrade their houses and facilities. Along the streets, city planners made space for sidewalks and shade trees, and left open space for parks. Variety, not monotony, was their goal.

Winslow Howard, whose family moved to Lyman from Brandon Mill when he was eight, remembered that the houses were painted different colors. "We lived in a shack in Greenville...To go into that thing [in Lyman] and to see the plastered walls and lamps hanging down, turn them on over by the door instead of jumping up and grabbing at a string to pull it, you know," Howard recalled. "That didn't happen around here."

In 1954, when Pacific Mills sold the houses in Lyman to its workers, the Greenville real estate agent handling the sale called Lyman "undoubtedly the best property of all textile villages which our company has handled." The town was laid out "with modern ideas of residential planning as contrasted with the villages built 25 years prior...the houses themselves have central halls and bathrooms were in the original design rather than an addition to the back porch." Pacific Mills backed up the architecture with better schools and better maintenance, and the village gained the reputation as one of the nicest villages in the Piedmont, "at the top, along with Dunean in Greenville," the real estate agent reported.

"They tried to give you just what you needed, which we weren't used to," Howard remembered.

This last era of village building reached its apogee at another Draper design, the mill and village in Chicopee, Georgia, outside Gainesville. Built in 1927 by the Chicopee Manufacturing Company to provide surgical gauze to the Johnson and Johnson pharmaceutical company, Chicopee is often cited as the finest mill and village to have been built in the South. "This work has been done by Southern engineers and contractors, using Southern labor and materials and concentrating the benefits of the entire investment as far as possible in Southern territory," a company publication touted at the project's birth. The company installed underground telephone and power lines, sewers and water lines, none of which were standard in Piedmont villages. A "model dairy farm" provided milk, and "all public playgrounds will be in charge of competent directors and special provisions will be made to regulate the play of younger children."

The last burst of mill village building exemplified by towns such as Lyman and Chicopee did not last long, and efforts to improve living conditions did not reach most villages. The social aspirations of planners such as Draper often foundered on the profit demands of the mill companies. By the mid-1920s, cotton manufacturing had preceded the rest of the country into recession. From 1926 to 1932, the industry turned a profit in only three years; fewer than half of all southern mills paid a regular dividend. The years after Lyman's construction were marked by severe labor violence, and the model villages designed by the landscape architects did not find themselves immune to the strife. Lyman itself was occupied by two companies of National Guard troops during the General Strike, after its closure by a flying squadron. Within a decade of the construction of Chicopee, the industry had begun selling its houses and dismantling the mill village system. The towns survive, however, as testament to the brief but fruitful coupling of old-fashioned corporate paternalism and the most modern of city design.

The New England Textile Collapse
A Southern Migration Begins

by Katherine Cann and Betsy Wakefield Teter

Competition, overproduction, and labor problems plagued the New England textile industry in the 1920s, encouraging investors to seek alternatives. The decline of the New England industry resulted from competition from Southern mills that produced the same goods at lower costs.

Initially, the South's ascendancy was in the manufacture of coarse yarns and medium-grade goods. Following World War I, however, Southern mills produced larger quantities of finished and fine cloth, directly competing with New England goods. The competitive advantage of northern mills was further eroded when New England state legislatures bowed to the pressures of textile workers by establishing maximum hour laws, prohibiting child labor, and adopting laws to halt the exploitation of women workers. Furthermore, wages increased dramatically during World War I, over 200 percent in Massachusetts. Labor costs in the South, where protective legislation of this nature was virtually unknown, were much lower. In addition, a rash of textile strikes immediately after World War I led many New England textile companies to declare bankruptcy.

In the wake of declining profits, New England textile executives and investors found South Carolina and other southern states, where business connections were well established, appealing. In addition, southern legislatures offered various inducements. Hydroelectric power was inexpensive, and many plants were relatively modern, stocked with new machinery. New Englanders bought such local companies as Tucapau Mills, Valley Falls, and Arcadia Mills. Reeves Brothers of New York established a beachhead in Spartanburg County with the 1923 purchase of W. S. Gray Mills in Woodruff and construction of a finishing plant in Fairforest in 1929. Massachusetts-based Pacific Mills built one of the largest plants in the Southeast at Lyman.

The cornerstone of South Carolina's textile empire was the same thing that had once given New England a monopoly—a large pool of unskilled workers. Southern factory workers had the reputation of being "pure-bred American stock from the mountains [who]...appreciate opportunity...[with] a great native intelligence and quick to learn." The tractable Southerners worked longer hours for less pay than New England workers and were not inclined toward organization. In 1924, the difference in wages between New England and the South reached

Arlington Mills, Lawrence, Massachusetts, in the 1910s. (Courtesy, the American Textile History Museum)

its greatest gap when New England workers made 65 percent more than their Southern counterparts.

In October 1929 the *American Wool and Cotton Reporter* published a litany of New England mill failures and slowdowns, an estimated 300 in all. Fall River, Massachusetts, "once the richest cotton manufacturing city in the United States," was operating at less than 25 percent capacity. "Thousands of operatives have been thrown out of work, hundreds of thousands of spindles have been scrapped or abandoned." Among the Fall River manufacturers out of business or in bankruptcy were Chace Mills, Conanicut Mills, Globe Yarn Company, Anacona Company, Hargraves Mills, Mechanics Mills, Pocasset Manufacturing, Tecumseh Mills, Weetamoe Mills, and Sanford Spinning.

Nowhere was the devastation worse than Lowell, Massachusetts. "The Bay State Cotton Corporation of Lowell, a fine reinforced concrete mill, set up at a cost of about $4 million, with nearly a thousand wide looms for the production of sheetings, a mill that was meant to be the lowest cost wide sheeting mill in the country, is absolutely empty, the machinery has been sold, the payroll is non-existent," the magazine's editors reported. "The mill is for sale for about $2 million and there are no buyers." At Appleton Mills in Lowell, 2,000 operatives were out of work and 500 tenement houses stood empty. "Week by week wages of $40,000 are unpaid."

Strangely, however, despite the boom years to come in the textile Piedmont, the stage had been set for Spartanburg—which had once billed itself as the "Lowell of the South"—to repeat the story of its adopted namesake. Sixty years later, the New England textile collapse would happen all over again.

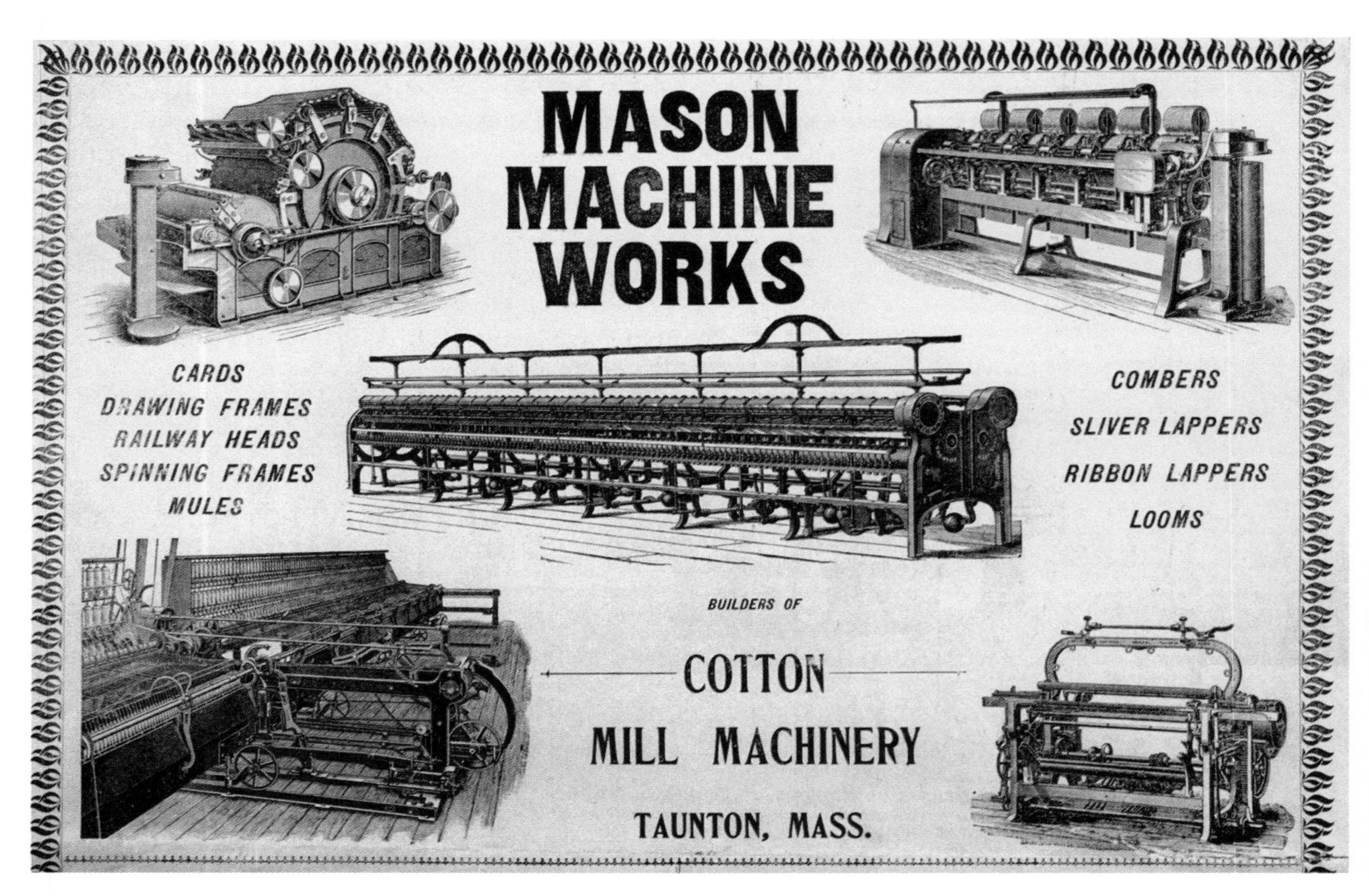

New England companies also provided machinery for local mills. (Courtesy, Spartanburg Regional Museum)

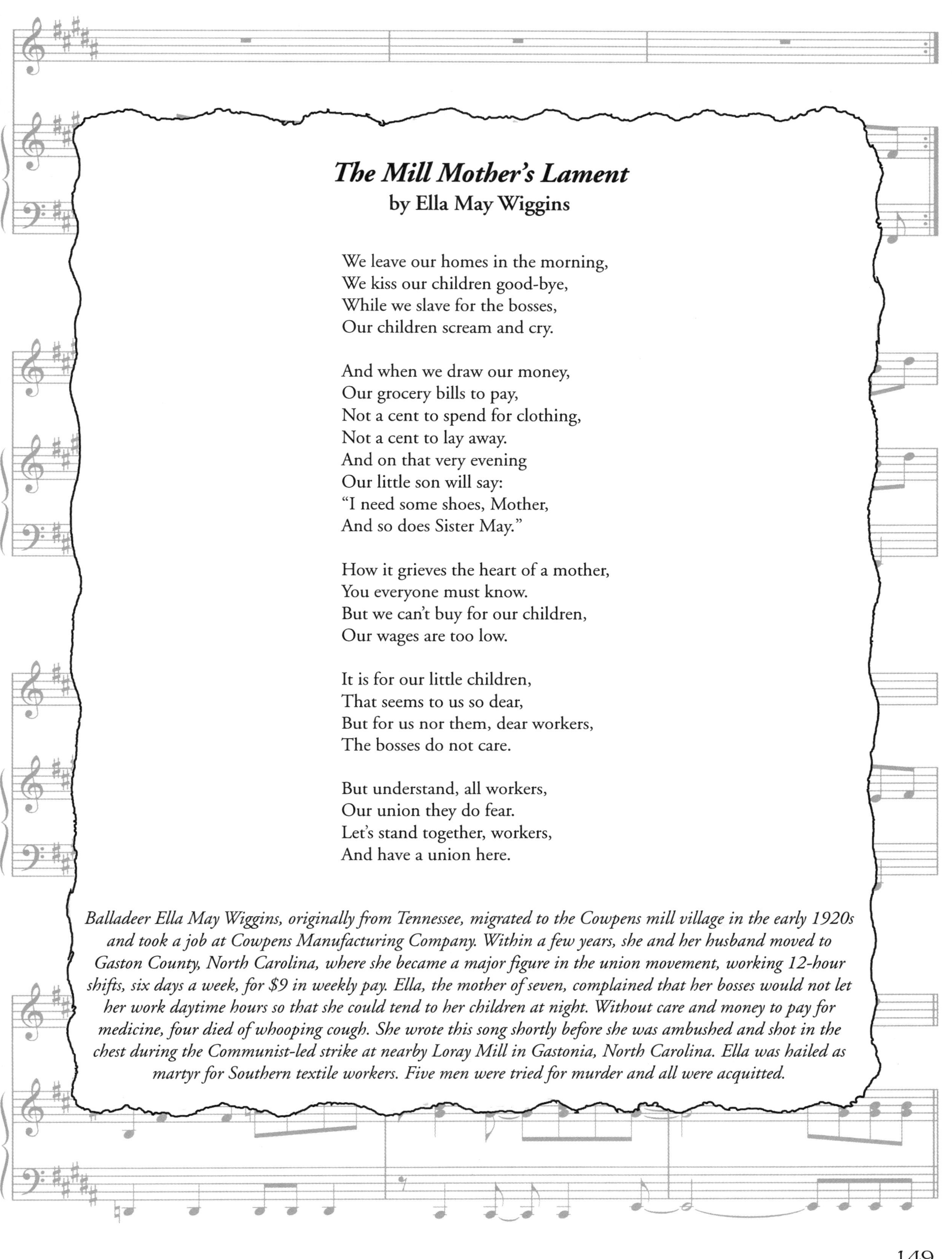

The Mill Mother's Lament
by Ella May Wiggins

We leave our homes in the morning,
We kiss our children good-bye,
While we slave for the bosses,
Our children scream and cry.

And when we draw our money,
Our grocery bills to pay,
Not a cent to spend for clothing,
Not a cent to lay away.
And on that very evening
Our little son will say:
"I need some shoes, Mother,
And so does Sister May."

How it grieves the heart of a mother,
You everyone must know.
But we can't buy for our children,
Our wages are too low.

It is for our little children,
That seems to us so dear,
But for us nor them, dear workers,
The bosses do not care.

But understand, all workers,
Our union they do fear.
Let's stand together, workers,
And have a union here.

Balladeer Ella May Wiggins, originally from Tennessee, migrated to the Cowpens mill village in the early 1920s and took a job at Cowpens Manufacturing Company. Within a few years, she and her husband moved to Gaston County, North Carolina, where she became a major figure in the union movement, working 12-hour shifts, six days a week, for $9 in weekly pay. Ella, the mother of seven, complained that her bosses would not let her work daytime hours so that she could tend to her children at night. Without care and money to pay for medicine, four died of whooping cough. She wrote this song shortly before she was ambushed and shot in the chest during the Communist-led strike at nearby Loray Mill in Gastonia, North Carolina. Ella was hailed as martyr for Southern textile workers. Five men were tried for murder and all were acquitted.

A typical mill house in Clifton.
(Courtesy, Converse College)

4 Textile Town in Depression and War 1930 to 1949

by G. C. Waldrep III

The 1930s really began in Spartanburg, as they did around the country, a few months early: on October 29, 1929, when the New York stock market crashed. To be sure, few in Spartanburg immediately understood how widespread the crash's effect would be: the *Spartanburg Journal,* for instance, ran news of the crash on its front page, but larger headlines were given to other stories: the ongoing search for a downed airliner in New Mexico, "Rain Greets County Fair Opening." Ironically, a third article given more space than the Wall Street crash in that day's paper reported the American Federation of Labor's announcement of a new nationwide organizing campaign. Still damp from the downpour at the fairgrounds, mill owners and textile workers alike had no way of knowing that for the next 10 years their lives would be shaped by two forces: the crash's legacy on the one hand (the Great Depression), and the battle to unionize the mills on the other.

Unlike other large American industries, such as coal or automobiles or steel, Southern textiles had always been a decentralized concern, with dozens of companies and hundreds of mills competing within the same market. Cloth woven at Spartan Mills was not especially different from cloth woven at Clifton or at Valley Falls. Textile products, moreover, were cheap: the cost of a shirt or a towel as opposed to, say, a Ford Model T. The short-term profit margin for even the most efficient mill was always precarious. Mill owners complained regularly that staple supply (raw cotton) always outweighed product demand. Nor were South Carolina mill owners known to be a very cooperative bunch, even among themselves. By the 1930s the Montgomery-Milliken interests controlled more mills in the Upcountry than anyone else, but if they attempted to raise the prices of their products, a swarm of smaller, family-owned mills stood ready to rush into the gap. The fact that most mill owners and managers knew each other well did nothing to lessen this cutthroat competition for market share.

The sudden slump of mill orders following the crash did not surprise Southern mill owners, but the extent of the Depression—as it invaded every nook and cranny of American life and dismantled the economies of Europe—did. The results, by 1931, were massive layoffs. Some mills reduced their workforces, often by as much as half, while others attempted to keep as many men and women working as possible by running the machinery only a few days a week. Saxon's workforce by the mid-1930s was half of what it had been in 1929; workers at the Cliftons remembered boarded-up, abandoned company houses off and on throughout the decade. Some mill owners publicly claimed that

When the banks closed in 1933, Inman Mills was one of the few plants to meet its financial commitments. Barney Bishop, secretary and assistant treasurer of Inman Mills, had prepared for such a financial crisis. In the years preceding the collapse, Bishop had withdrawn from the banks more cash than was needed each week to meet the payroll. The money was stored in the company vault, and by 1933 he had accumulated enough to pay the workers in cash until the banks opened.

—Author Bobby Dean Jackson

Two men were put to death in the electric chair in June 1930 for the dramatic robbery of Drayton Mills paymaster Earle Belue. Paul Johnson and Ray Coleman, who became known as the "blunder bandits," tried to steal $10,00 from the popular young paymaster in a "daring daylight crime" December 15, 1928, as he transported the weekly payroll from the bank to the mill. Belue, 30, was shot in the head in a crime that shook the village for many years.

they were running their mills on a purely humanitarian basis—"to put food on the tables of our operatives," they often said—but they had their own reasons as well: totally laid-off workers were not very loyal workers. When the economy picked back up, these mills hoped to get the proverbial jump on their competitors by having a larger workforce ready to go. Brighter times were always just around the corner.

The economy did not pick back up, and by 1932 Spartanburg was making a serious bid for the title "hardest hit city in the nation." Textile wages had always been low. They had in fact been falling across the South since 1920. Some Spartanburg mills had been running "short time" since 1926. Textile workers were used to privation: as historian Bryant Simon has noted, they simply "did not...have to fall as far to hit rock bottom," compared to workers in other industries. "Rock bottom" meant lacking the necessities of life, such as food and clothing. By 1932, many Spartanburg millhands were there.

The key problem, to many textile workers, was not wages exactly, but the dreaded "stretchout." A Drayton worker, J. Luther Campbell, remembered it this way:

When I went to Drayton, it wasn't too long after I went that they put these minute men on, timing the time it took you to do anything. They checked you, they checked you out: checked all the spinners. They come around with a time clock. They timed you. If you went to the bathroom, you was timed how long you

The Drayton Mills spinning room, 1931 (Courtesy, Barbee Sinners)

A typical mill house in Arkwright, about 1930 (Courtesy, the Spartanburg Herald-Journal)

> *"I remember when the mill stood idle for 10 weeks—no one had any money and not much credit. The company stated it would extend credit in the amount of $2 per man and wife and 25 cents extra for each child. Everyone thought it was a gift and was very disappointed when we learned we had to pay it back."*
>
> **—Leo Kirby, born in Pacolet Mills in 1911, as quoted in the *Spartanburg Herald***

was in there. They timed you putting up an end. They timed you setting in a stick of roving. They just timed you, period.

We knew what was happening. We knew they was going to add more to you—we could see what they was doing...They could see how much you was doing in eight hours. Why, it wasn't long when they come around that if you was running ten sides, you run twelve: stretchout system, that was it.

As the Depression tightened and markets dwindled, mill owners in Spartanburg and elsewhere increasingly turned to the stretchout as a way of squeezing more product—and therefore more cash flow—from their workers.

On their part, managers derided "stretchout" as an excuse workers used to resist all changes in work assignment, job classification, or technology, whether real or imagined. Competition in the industry was fierce, and it was real; the market for textile products—both national and overseas—was shrinking rapidly. Some managers, especially in the smaller mills, saw the choice as between stretchout on the one hand and bankruptcy on the other. Others simply felt that the stretchout was an

Alfred Willis took this photograph of Clifton weavers. (Courtesy, Converse College)

Spinners pose for a photograph at Spartan Mills. (Courtesy, Walter and Betty Montgomery)

unavoidable part of industrial progress. As Robert W. Bruere of Roosevelt's Cotton Textile Labor Relations Board responded to South Carolina Senator James F. Byrnes when Byrnes expressed concern about the stretchout,

> **When you consider that every introduction of improved machinery or technical methods is likely to be interpreted as stretchout you will appreciate the optimism of your hope. The protest of the workers in the textile industry against the stretch-out is as old as the introduction of the power loom, and will, I fear, continue so long as the inventive genius of man is applied to textile production.**

In Bruere's view, textile workers were simply behaving unreasonably. Bruere and his allies in the industry never quite came out and called textile workers *lazy,* but this was implicit in many of their statements, and it infuriated workers. As an outraged spooler from Clifton No. 1 wrote in a letter to President Roosevelt, "You will find very few people at the mill who are not willing to do all they can do. You will find plenty of people working so hard doing their *very very best* to run the work that is put upon them...why a race horse would not be allowed to go as fast as he could go for eight hours."

Spartanburg's textile workers, exhausted, anxious, and hungry, were among Franklin Delano Roosevelt's most ardent supporters. Always politically active, textile workers flocked to the polls in record numbers in November 1932 to elect him to the presidency; as one mill worker from Converse later put it, "Roosevelt, he was our king." Mill owners were not so enthusiastic. They sensed, correctly, that Roosevelt would try to restore economic prosperity by regulating American industries. Among themselves, Spartanburg's mill

> **The workers have to eat their dinner or supper when they can, usually when they have "caught up to" their machines. These meals are brought to the mill by children, wives or husbands. It is a common sight to see a small child carry a dinner pail and a fruit jar of iced tea to the mill gate and leave it there for his dad or mother.**
>
> ***—Author John Morland***

> We have found you are working your help unmerciful at the Enoree Mills, S.C. espicely the Spinners and Cleaners, weavers and Card Room help, it is up to you to reduce this load and you and your superintendent must put an end to this slavery, we do not mean to cause any trouble among the help, only expect you and your superintendent to look after this and this must be done by the first of September 1933 or you will WISH YOU HAD, and to make it better for you, you must change your superintendent at Enoree Mill, S.C.
>
> YOUR first and last warning.
>
> *—Anonymous letter sent to Riverdale Mills from Spartanburg Aug. 15, 1933*

men had discussed various means of solving the "overproduction" problem. In January 1931, Victor Montgomery of the Pacolet Manufacturing Company even went on public record advocating the abolition of night work, a reduction in hours for all mill workers, and a doubling of employee wages (to maintain consumer purchasing power). All three of these ideas, in principle, were among those later adopted by the Roosevelt administration as potential cures for the industry's ills. Montgomery and his colleagues were never able to agree on any self-regulation, however. The only thing that seemed to unite them was their adamant opposition to government regulation of any kind.

Roosevelt's National Industrial Recovery Act (NIRA)—the first key piece of New Deal legislation—went into effect across the country on July 17, 1933. Among other provisions, the act mandated shorter working hours, inaugurated the nation's first guaranteed minimum wage ($12 for a 40-hour week), and set up a government agency—the Cotton Textile Labor Relations Board (CTLRB)—to handle worker grievances. Textile workers, of course, were delighted: higher wages and an eight-hour day finally seemed within reach. Mill owners, on the other hand, were nervous: as S. M. Beattie of Greenville's Woodside Mill said in a speech to the South Carolina Textile Manufacturers Association a few days before the act took hold,

> **It has been said that when Columbus discovered America, he did not know where he was going; when he landed he did not know where he was, and when he got back he did not know where he had been.**
>
> **Now it seems to me that our present situation, under the "New Deal," is somewhat similar, when we are suddenly confronted with the necessity of carrying on under conditions that are new and strange to us all.**
>
> **Our industry is one of the oldest, and by reason of the nature of its growth, one of the most individualistic, with about a thousand plants and many hundred executives; with mills widely scattered; and turning out a great variety of products, operating under conditions which differ as greatly as do the Mexican and Canadian borders; with many if not all of us firmly believing in the rights and privileges of the individual.**
>
> **To suddenly change all this and introduce a new order, where the majority rules through a single representative, with power to enforce compliance, is indeed a revolutionary change…**

Mill owners one and all turned to the stretchout as the only legal way around the NIRA's provisions.

Across Spartanburg County, jubilation changed to horror as workers began to comprehend the industry's united response to NIRA. In his weekly radio broadcasts, Roosevelt urged Americans to write to him, and Spartanburg's textile workers did—by the hundreds. Unable to make a living on their hardscrabble farm, Mrs. Carl Langford and

> ***"When I went to work in 1933, I had a pretty good college education in a cotton mill. The mill was pretty lenient with everybody back then. A woman could work, and she could come out in the yard and nurse the baby, if she had a baby, then go back to work. Maybe it wasn't the good old days that some people think, but we thought it was. We liked our neighbors and the neighbors liked us, and they would help one another."***
>
> **—Mill worker J.C. Fowler**

her two daughters had sought employment at the Enoree mill, where all three worked six weeks filling batteries. Ten days after NIRA went into effect Mrs. Langford explained her situation to her congressman:

> **It was very hard for us. We worked eleven hours at night and only received pay for seven hours work. They paid us $1.05 a night. We went to work at 6 o'clock, stopped at eleven for lunch, started back to work at 11:20 and worked until 5:20 in the morning. We kept up with the work they put on us...**
>
> **We did our best, thinking when the 8-hour law came on we would be fixed but on the 17th of July we went into work as usual and the second boss came around and told me they were going to put so much on us that he wouldn't need us any longer.**

"We feel sorry that the change was made," Mrs. Langford concluded; "as it was we were making $5.25 a week and as it is we are not making a penny." The gulf between the Textile Code's promises and its actual results left workers dazed. "I have been reading all the Papers about the new textile code and was greatly pleased with it for I understood that hands were not to be put on more work," S. E. Knightson of Arkwright complained five days after the Code went into effect, but in fact "it has caused these mills to almost double work."

Nor did the Cotton Textile Labor Relations Board (CTLRB) do much to help. The state board for South Carolina was officially composed of three members, one for "industry" (Greenville's J. E. Sirrine), one for the "public" (a professor from Clemson College), and one for "labor" (Furman B. Rogers of Spartanburg). The state board received hundreds of complaints from beleaguered workers, and the national board sent a handful of agents into Spartanburg County to investigate, but in the end, nothing was done: While Congress had passed legislation creating the boards and the Textile Code, it had passed no laws for enforcing the boards' decisions. In other words, mill owners were more or less free to violate any portion of the Textile Code they wished, even when the CTLRB ruled against them. Furthermore, while NIRA and the Textile Code provided for better wages and

Vernon R. Burnett fixing a loom at Spartan Mills (Courtesy, Kenneth Burnett)

"I think my daddy was a making around $5 a week, something like that ... when he got to work a week, because then mills were so uncertain he didn't know if he'd get two days, three days, or a week, or whatever, and that went on for quite a while, that wasn't just for a few weeks. My Dad came to me and he told me, 'Son you're gonna have to drop out of school and help me.' He said, 'I'm doing all I can but I'm not making ends meet.' I understood. I knew, so I told him okay. So he goes to this overseer of the spinning department, and he talks him into letting me go into the spinning department. Just to make myself available in case a job came open. There wasn't any jobs...I was there in that spinning department for six months before I was ever placed on a payroll of any kind. And I remember when they first put me on the payroll; they gave me a little part of the job and gave another girl the other part. So the first day I worked I made 25 cents and that went on for a few weeks. I mean I was there, ten hours a day, and you might say it was on the job training and I made 25 cents a day."

—Olin Hodge

By-laws of the United Textile Workers of America, 1932 (Courtesy , George W. Moore Jr.)

CONSTITUTION AND BY-LAWS

United Textile Workers of America

REVISED
SEPTEMBER, 1932

UNITED TEXTILE WORKERS OF AMERICA
ORGANIZED AT
WASHINGTON, D.C.
NOV. 19, 1901

AFFILIATED WITH A. F. OF L.
ORGANIZED AT WASHINGTON, D. C.
NOVEMBER 19, 1901

LEDGER NUMBER

shorter hours, neither said anything at all about workload. The "stretchout," however unbearable, was legal.

GALVANIZED BY THE STRETCHOUT and increasingly disenchanted with the government's performance, Spartanburg County's textile workers joined the tide into the United Textile Workers of America (UTWA) in late 1933 and 1934. The UTWA had been in and out of Spartanburg since 1929, when they had lost a strike at Mills Mill's Woodruff plant; they lost a longer, and more publicized, strike at Arcadia in 1932. Under Roosevelt's New Deal, however, the union stood a much better chance of success. For one thing, Spartanburg's textile workers were in more desperate straits than they had been in the past. For another, Roosevelt had himself said on numerous occasions that if he were a worker, he'd be a union man. Finally, one of the other provisions of the National Industrial Recovery Act was the much-heralded Section 7(a), which guaranteed workers the right "to organize and bargain collectively through representatives of their own choosing." For the first time, joining the union not only seemed like a good idea—it seemed like a safe one.

Between the early fall of 1933 and the summer of 1934, the UTWA organized 29 local unions in Spartanburg County, covering all the mills except for those at Arcadia and Woodruff (where workers had just lost strikes), at Chesnee, and in a few smaller, isolated mills (Crescent Knitting, Fingerville, Mary Louise). Five of the local unions were specifically for African-American workers. The organizing drive was public; for instance, the local unions in and around Spartanburg advertised their meeting times and places in the newspapers and in the city directory. The Spartan Mills union even met on company property, in the mill village's community building. Joining the union was a fairly simple act: a worker would purchase a union card from an organizer, and pay regular monthly dues, usually 10 cents. But the risks of doing so were far from simple, as Spartanburg workers knew. Managers at a few mills, like Cowpens Mill, began discharging known unionists immediately. Most mill owners, however, seemed to take a "wait and see" attitude.

This same pattern was repeated all over the textile South. Some textile centers, such as Durham, North Carolina, and Gadsden, Alabama, were quickly and solidly organized by the union. In others, especially those where unions had been badly beaten before, the union's progress was uneven. Workers in Danville, Virginia, and Greensboro, North Carolina, were still reeling from strikes of their own. The same was true for Spartanburg's industrial rival in the Upstate: the UTWA had difficulty convincing Greenville workers, whose own sporadic strikes since the 1890s had always ended in disaster. With Greenville workers wavering, Spartanburg became the center of union activity in South Carolina. Not only were workers in most Spartanburg mills solidly organized by the summer of 1934, they were joined by equally solid local unions in the outlying towns of Newberry, Gaffney, Blacksburg, Union, and just across the state line in Rutherford County, North Carolina. On August 19, 1934, the UTWA even added a Spartanburg worker to its national executive board: Charles W. McAbee, head of the local at Inman Mills.

Electricity, which came to most mills after 1910, brought about the second shift. At that point, two shifts worked 11 hours a day. The National Recovery Act, instituted during the Great Depression, bought about a third shift when it limited the maximum workweek to 43 hours.

Pressure within the union grew steadily over the summer of 1934, especially as it became clear to workers that the government (through the CTLRB) was doing nothing to resolve long-standing disputes over the stretchout and other grievances. In July, the UTWA's state body—the South Carolina Federation of Textile Workers—met for three days in Spartanburg. Frank Walsh of the United Garment Workers opened the convention and set its tone by declaring, "Slavery has not been abolished in the South; the color of the slaves has only been changed." The convention endorsed resolutions calling for a 60-hour workweek, a 33 percent wage hike, elimination of the stretchout, a general reduction of machinery speed, recognition of the UTWA, and re-employment of workers fired because of union membership. In the end, sensing that none of these objectives would come to pass short of an all-out struggle with mill owners, they went on record pledging their support to UTWA officials "whenever it might be necessary to call a general strike." What workers in Spartanburg and elsewhere in the South did not know was how ill-prepared the national union was to support a strike of this magnitude. National UTWA leaders felt they had only two choices: either to call a general strike and risk disaster, or try to hold workers in the mills and lose all credibility. They chose to strike. The zero hour was set for midnight on Saturday, September 1.

Since no mills were running on Sundays in Spartanburg at the time, the real showdown came Monday morning. The *Spartanburg Journal* estimated that approximately two-thirds of the county's

A receipt for union dues at Startex Mills. (Courtesy, George W. Moore Jr.)

United Textile Workers of America

8/24 1935

Receipt No. 98

Received of G. W. Moore Secretary of

Union No. 2070

The sum of $2.30 Two Thirty 00/100 Dollars,

being amount received for dues

J. E. Lavender Treasurer

Printed on Union Made Paper

A Depression-era group of employees of Arcadia Mills, which became Mayfair Mills in 1934. (Courtesy, Mayfair Mills)

14,000 textile workers stayed home, leaving 16 plants idle and 13 running. Those figures would change almost daily over the next three weeks, however, because of the activities of the so-called "flying squadrons." These squadrons, formed of striking textile workers, would travel in caravans to mills that were still running, surround them, and try to convince those inside to come out and join the strike. They took their show on the road, principally to mills in Greenville County and nonstriking mills in the Spartanburg area (like Arcadia), but sometimes as far as Anderson and Ware Shoals. The intent on the union's part was solidarity—to show workers inside the mills that joining the strike was safe, that they no longer had to fear their bosses. Unionists saw their work on the squadrons as a kind of joyful recruiting drive. Needless to say, anti-union workers inside the mills were more frightened than encouraged by the "invitation." And mill owners, with the support of local sheriff Sam Henry and the National Guard, saw the squadrons' activities as a direct and illegal assault on their property.

> ***"A general strike at this time is indicative of a lack of appreciation and gratitude."***
>
> **—Democratic Governor Ibra Blackwood of Spartanburg**

Some mills in Spartanburg were solidly pro-union before the strike was called, some were less solid, and in some the union's presence was very weak. The worst showdowns came at Lyman and Greer, where the union had only a foothold and was unable to shut down production for more than a few days. National Guardsmen with machine guns were deployed by the governor at Lyman, Greer, and Cowpens, a shock that many participants never forgot. Strikers picketed the plants. They received some economic assistance from the national union, whose treasury was quickly depleted. By the strike's second week in Spartanburg, a stalemate seemed to have prevailed. Owners at mills like Spartan, Tucapau, and Inman had no hopes of reopening their plants so long as the strike was on: the unions in those plants were solid. On the other hand, unionists admitted that they had failed to carry the strategically important plants at Lyman and Greer, as well as Chesnee, Arcadia, both mills at Woodruff, and smaller mills across the county.

On the national level, President Roosevelt began negotiating a nationwide settlement with the UTWA almost as soon as the strike had begun. The UTWA's leadership, knowing they could not afford to support the strikers financially for very long, was anxious to settle. The UTWA formally ended its General Strike on Monday, September 24, 1934. The proposed settlement, labeled Roosevelt's "personal appeal" by the *Spartanburg Journal*, had circulated by radio and by word of mouth in Spartanburg County on the previous Thursday and Friday. It recommended that the president ask workers to end the strike and that manufacturers take back strikers without discrimination; that the inefficient CTLRB be abolished and a new "Textile Labor Relations Board" be set up to handle worker grievances; and that no further machine load changes be made by manufacturers until a federal commission could be appointed specifically to investigate the stretchout. Despite the fact that the report provided not a single concrete redress of any of the myriad grievances that had provoked the strike, UTWA leaders accepted it in good faith.

With unswerving conviction that Roosevelt would never betray them, Spartanburg's textile workers also accepted the settlement gladly: the *Journal* reported hearing one striker say, "We'll do anything for the president," while a picket at Lyman promised that "the president is going to play fair with us." On Saturday night, Francis Gorman—the UTWA's national vice president and strike coordinator—telegraphed every local in the country with news of the union's "complete victory." That night, Spartanburg's workers organized their largest demonstration yet. With pickets removed from area plants, strikers from all over the county gathered at 8 p.m. at Morgan Square downtown. They marched behind an American flag to the Central Labor Union Hall, where a mass meeting was held to celebrate the moment; several carloads then departed for celebrations in other communities. Just before midnight, a second impromptu parade with "blaring horns and happy faces" passed along Main Street. At the Central Labor Union's mass meeting, L. E. Brookshire of Greenville captured the optimism:

Since the response to the call has been so marvelous it is my opinion that it is going to instill into these workers a new confidence

William Henry Campbell, a Clifton Mills employee, in 1930. (Courtesy, Ellen Nelson Hodge)

Wartime spinner at Beaumont Mills (Courtesy, Walton Beeson)

and a new sense of their importance as the major element of society in South Carolina and the South. From this time on, organization is going to be the stabilizing factor in handling the affairs of the workers and in all forms of negotiations with their employers.

"The union movement has come into southern mills, and it has come into them to stay," Brookshire concluded. In the words of the UTWA's Special Strike Committee, Southern textile workers had at last won "an end to the stretch-out." "We have taken every trench," the committee's final report concluded.

In fact the aftermath of the strike was a nightmare for textile workers across the South, including Spartanburg. For one thing, mill owners may have been divided over how to deal with the union before the strike, but afterwards they were united in their opposition. At Lyman, "They let in the non-union people but not the union workers," local union president J. H. Stone told the *Spartanburg Journal*. "We were forced back across the street at the point of bayonets, men and women being treated alike in this respect." On top of post-strike discrimination, mill owners continued to expand the stretchout, trusting that the Roosevelt administration would do nothing to enforce the "settlement" Roosevelt himself had brokered. They were correct.

The most bitter part of the strike's aftermath were the evictions. Without any way of protecting

According to an Associated Press reporter who witnessed the assault on Dunean Mills in Greenville, the flying squadron from Spartanburg County consisted of 625 textile workers in a motorcade of 105 cars and three trucks. The lead vehicle flew an immense American flag.

—Author G.C. Waldrep III

> **In such towns as Charlotte, Columbia, Spartanburg, Dalton (Georgia), and Huntsville (Alabama), union support approached 100 percent.**
>
> *—Author Janet Irons*

their jobs, former textile workers were, of course, expected to vacate their company housing. Management at Powell Knitting began pressing eviction notices almost as soon as the strike was over, and soon other plants, like Lyman, Cowpens Mill, and Tucapau, began to follow suit. At Tucapau, for instance, 21 families—including all of Local 2070's officers and several of its shop stewards—were targeted. G. Walter Moore, the local's secretary, later wrote to Secretary of Labor Frances Perkins in Washington in a futile attempt to explain how crucial evictions were in the textile South. "You are possibly aware of the fact that the cotton mill corporations own the houses in which the operatives live and always when there is an employment contract made, there is a rent contract made incident to the employment of the workers," Moore began. "It goes further...the corporation gives the power to employ special officers to police these cotton mill villages; and they usually control the schools and often the churches in the community." Moore declared, "Unless you are here, you cannot conceive of the petty oppression, coercive discrimination worked upon the union employees through the agency of the corporation, under these labor conditions." Under state law, Moore reminded Perkins, "occupancy [of a mill house] is but an incident of employment." Upon discharge from the company, for whatever reason, mill employees and their families were required by law to vacate their houses at the mill's pleasure. Moore reminded Perkins that the federal labor agencies had "taken the position that they have nothing to do with the rental contracts and have steadily refused to lend their power and aid to hold the workers in their homes." Some of the displaced workers found housing nearby; most of the Tucapau evictees, for instance, lived in Wellford, at least for a while. From there, they scattered. As long as the mills owned the homes of their employees, the union's position in Spartanburg would always be weak.

At plants where the union had never been strong, it melted away without leaving much of a trace—except for bitterness on the part of workers who felt that they had once again been betrayed. In many Spartanburg mills, however, local unions came out of the strike as strong as they went into it. Between 1935 and 1937 both the city and the county were rocked by strikes—several of them long and violent—as mill owners provoked conflicts with surviving unions in order to wipe them out. The most memorable of these strikes came at Saxon (1935-36), Spartan Mills (1936), and Startex (1936-37). None resulted in union contracts. A few local unions limped on without any recognition from their employers, but as the 1940s approached, it looked as if Spartanburg's textile industry would remain effectively union-free.

DESPITE THE HARDSHIPS OF THE Depression—including the strikes—life carried on in Drayton and the Cliftons, at Beaumont

Sleeping Spartan Mills' pickets made the cover of the South Carolina NEWSVIEW magazine in June 1936. (Courtesy, Fred Parrish)

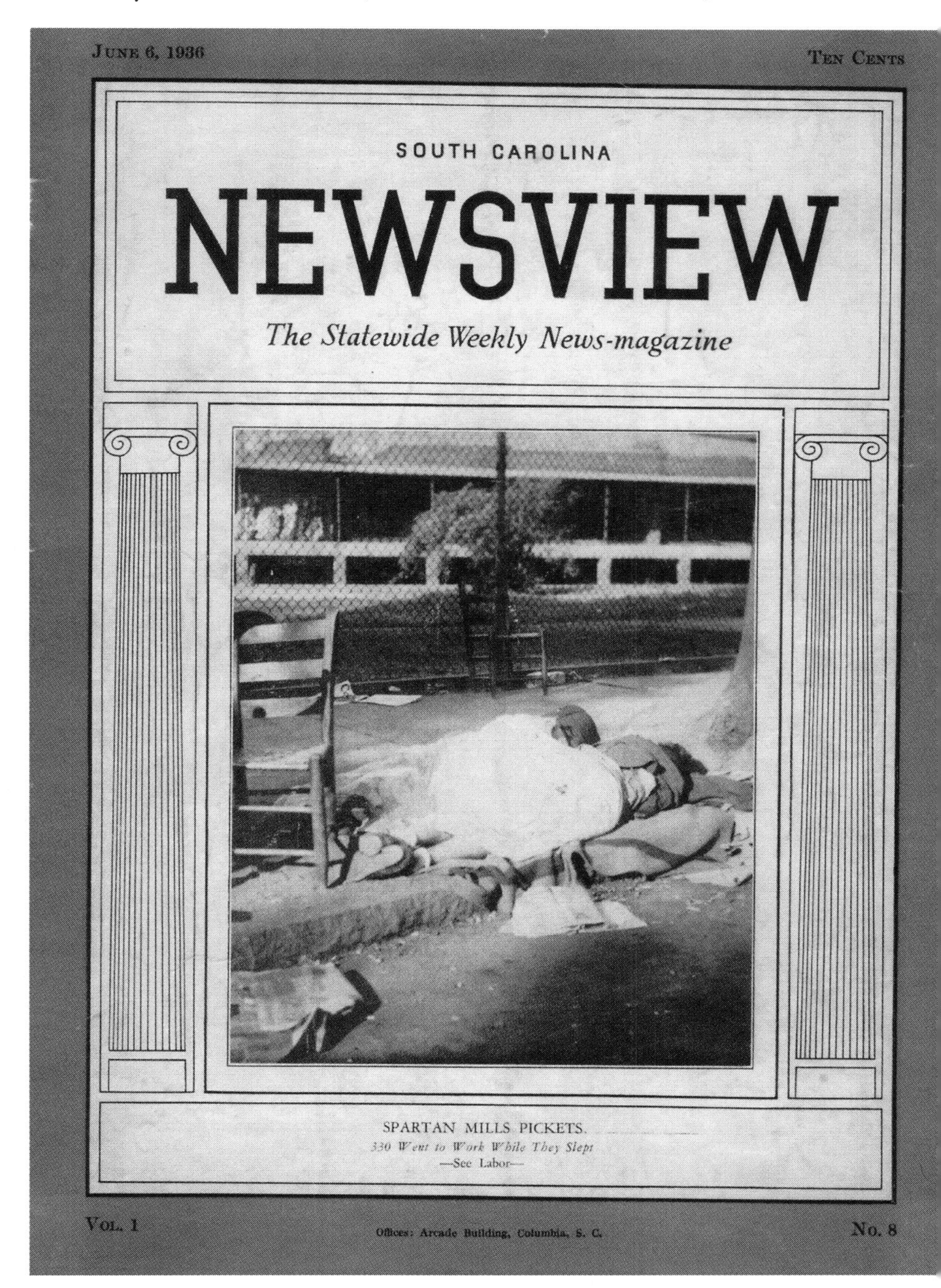
JUNE 6, 1936 — TEN CENTS

SOUTH CAROLINA

NEWSVIEW

The Statewide Weekly News-magazine

SPARTAN MILLS PICKETS.

330 Went to Work While They Slept

—See Labor—

VOL. 1 — Offices: Arcade Building, Columbia, S. C. — NO. 8

Harold Hatcher's Textile Campaign

In the mid-1930s, Harold O. Hatcher served as research director for the Congregational Church's Council for Social Action, headquartered in New York City. As part of that job, Hatcher toured textile mills in the Southeast, including some in Spartanburg County, before publishing his *Textile Primer* in April 1936. The *Primer* was published "to give the average man some insight into single industries." This brochure ended up in the hands of workers in Tucapau.

His 10-cent booklet called for the establishment of a minimum wage law and a new federal program to use excess textile capacity to produce clothing for people out of work during the Great Depression. "It is suggested, therefore, that the government buy or lease textile equipment and put unemployed people to work producing for those who are not able to buy," he wrote.

Among the information Hatcher included was Census data about salaries for textile company officers—at an annual average of $7,250 in 1929, the highest in all of American industry. He also pointed out that officers in 31 of the larger textile companies received salaries in 1934 averaging $36,000 (compared to average mill wages of $850 annually).

Hatcher, who also had a long career working for civil rights in Indiana, ironically ended up back in Spartanburg in 1969 when he went to work for a federal anti-poverty program. In the 1970s, he began creating what would later become Hatcher Garden and Woodland Preserve, a popular public garden on Spartanburg's west side.

Gardener Harold Hatcher's Textile Primer from 1936 (Courtesy George W. Moore Jr.)

Hon. Franklin D. Roosevelt
Washington, D.C.

Dear Mr. President:

Under date of November 6, 1936, I wrote you a letter in regards to the conditions existing at Startex mills Tucapau, S.C., which was extremely bad.

I also informed you as to the steps the Company was taking to get the union people out of the mill village, and that they were evicting the people from their houses and throwing their household goods into the streets of the village.

Now the Mill Company has gone further than that they have gone over court orders and over the courts of the State and are throwing the peoples households goods into the streets.

As I informed you about the relief we were getting at that time, We are getting practical no relief at all now, Some of the people have no coal or wood, Children are actual going hungry, if there is anything you can do toward getting WPA for the people so that they can work, it will be greatly appreciated.

With kind regards,

G. Walter Moore,
Secretary of Local #2070, U.T.W.A.

Moore was one of 83 Startex employees dismissed for union activities between July 1936 and January 1937. Moore never returned to Startex Mills. A representative of Roosevelt wrote back to say that Washington was not in charge of doling out WPA jobs, local officials were.

and Jackson Mill. Mill villages were their own self-contained communities, and although textile workers never had much in the way of earthly goods, they shared what they did have with one another. At the Cliftons, "We was all in the same boat. One didn't have any more than the other." At Tucapau, workers shared because "they didn't have anything else." Everywhere, as at Spartan, textile workers "found out they needed one another." Later, after the passing of the mill village era, it was this sense of the collective that textile workers mourned: in the words of a Spartan Mills worker, "the knittedness, the closeness of the people, and…the compassion people had for one another."

A few companies like Spartan, Saxon, and Pacolet Mills had experimented with "welfare work" in the 1920s, social programs for workers in their off-times. These ranged from Boy Scout troops to health clinics to company-sponsored brass bands. Most, however, left workers to their own devices. Without cars or money, visiting became the most prevalent recreational form: "That's the only entertainment you had," a Clifton resident later recalled. "You didn't have anything to do at home, you'd go to your neighbors', or somebody on the village, just sit and talk with 'em. My mama and daddy did it. We did it, the kids did it," everyone did it. On the weekends, visiting might flower into "socials" or "parties," especially for younger folks. Or it might take on a religious hue, as at Saxon, where workers gathered for prayer meetings and front-porch hymn sings. Sometimes the women would gather separately, sometimes the men; sometimes all would come together, to hear a favorite radio program perhaps or for reassurance in bad weather. Regardless of circumstance, it was the sheer act of visiting that bound mill villagers together. As a retired Saxon worker later put it, "We *visited* each other, we *talked* to each other, we were *concerned* about each other."

The two most important organized social functions on the mill villages of the 1930s were textile league baseball and the churches. Textile league baseball provided mill villagers with a secular communion of sorts. By the 1920s, the Southern textile belt supported hundreds of mill teams in a dozen or more privately-operated leagues. Team rivalries were fierce. Crowds were huge. Baseball was the only weekend entertainment going at Clifton No. 2, a worker remembered, "unless you'd

Joe Allen, left, and two friends play cards in Pacolet Mills. (Courtesy, Don Camby)

go over there when they had that theater over at No. 1, or go to town," neither of which happened very often. The Cliftons were typical: "When you'd hear that music start on Saturday evening, everybody just about would go to the ballgames," another Clifton worker recalled. As historian Thomas K. Perry has noted, "textile baseball was a way of life. It gave a sense of legitimacy in a society content to see the 'lintheads' remain invisible citizens somewhere far away on the wrong side of the tracks."

Mill village churches have been controversial since mill owners first sponsored them in the late 1800s. Some argued that sponsorship of churches proved the mill owners' basic concern for their workers' welfare, while others charged that mill owners simply wanted to use "company religion" as yet another means of worker control. Probably the truth lay somewhere in between. Either way, by the 1930s church activities were vitally important in most textile communities. At Tucapau, "People went to church then...*Everybody* did." At Clifton No. 2, "My mother and daddy would take us to church every Sunday morning and every Sunday night, and every Wednesday night." And at Saxon, "Hardly ever did that church bell ring...without the people was ready to walk in." Little other than tradition and polity separated Baptists from Methodists; mill workers regularly visited in each others' churches and even scheduled their summertime Bible schools so that all children, regardless of denominational affiliation, could attend. While "churches" existed, "church" was the social reality. Religion stood at the institutional center of mill village life.

It was a life, and mill people made the most of it. In their own villages, textile workers could think of themselves with pride, as "the best people in the world," as they so often claimed. Off the villages, however, was another story. Mill workers were still seen by others as a separate social caste. Most better-off whites and even many blacks derided them as "lintheads," which textile workers deeply resented. In 1936, for instance, when the Wofford College literary magazine published a poem describing textile workers as poor, illiterate, ignorant, and degraded, its author was only giving voice to popular and deeply-held views. Outrage on the mill hills ultimately made its way even to the South Carolina legislature. Workers at Beaumont—where the student had supposedly spent time "observing" in order to write the poem—threatened violence against him; nearly 60 years later, surviving Beaumont residents remained bitter

Less than 10 percent of Spartanburg County's textile workers were members of a union by 1936.

at the memory. Textile workers remained a people apart.

ON SEPTEMBER 1, 1939, Adolph Hitler's Germany invaded Poland. In the days that followed, most other European countries either allied themselves with Hitler's regime or (in the case of Britain and France) declared war in turn. Hitler responded by invading France on the one hand and by bombing Britain on the other. World War II had begun.

Although Americans had been following Hitler's rise to power for years in their newspapers, the invasion of Poland was the first real signal many had that war was coming. Spartanburg's two daily papers had always given prominent space on their front pages to news from Europe, but with the invasion of Poland and the

Pacolet Mills doffer Thurl Camby and his two sons, Don and Jennings, on Cameron Street, also known as "Short Line." (Courtesy, Don Camby)

Hon. James F. Byrnes
U.S. Senate
Washington, D.C.

Aug. 5, 1935

Dear sir:

We are writing you in regards to relief, here at Saxon Mill. As you may already know we are on strike for our just rights, and we are justfied in the particular case. And we know you have the influnce with the Hon. President of the United States. And we know it is in his power to give relief in a case of an emergency, and we are urgeing you to get in touch with him and get a special appropriation for this particular case. And we as an organized unit are requesting you to see him at once, as we need relief, at once, and not due consideration.

Hoping you will act at once and advise us at once. With kindest regards we remain.

Yours truly

J.M. Mills chair shop comm.
E.M. Sisk
B.F. Cousin

Saxon Local Union No. 1882

> *"We'd play our own little games. Sounds crazy now but, gee, a lot of people used to chew tobacco back then. On every ply of tobacco, you had a tobacco tag. One of the favorite games of us boys was to gather these tobacco tags up and have a pocket full of them at recess. Go out there, set down on the steps, lay our tobacco tags down and blow on 'em. If I was able to blow mine upside down, I got his!"*
>
> **—Harold Hatchett**

attacks on France and Britain, the tone of their reporting changed. Both the *Herald* and the *Journal* took it for granted that the United States would ultimately be drawn into the conflict. In Washington the Roosevelt administration moved swiftly in the early days of September to set up various defense boards and programs as well as to reactivate the draft. By far the most important local sign of this national transformation was the construction of Camp Croft.

In fact, Spartanburg owed a great deal of its partial economic recovery during the late 1930s not to textiles, but to the construction industry. Federal grants for local projects began to flow into the county in 1936. Unemployed textile workers—or at least adult male textile workers—signed on for work with the Soil Conservation Service (SCS), the Civilian Conservation Corps (CCC), and the Works Progress Administration (WPA). As local historian Vernon Foster noted, in 1937 alone, "WPA projects completed in Spartanburg County included the first Mountainview Home, National Guard armories in Spartanburg and Lyman, the American Legion home in Duncan Park, and a school." Then, in May 1938, the United States Housing Authority announced a mammoth appropriation of $800,000 for "slum clearance" in Spartanburg. Ultimately, this money was divided between redevelopment of the African-American neighborhood on South Liberty Street and construction of an entirely new subdivision for working-class whites on the Wofford Street extension. These two projects—Tobe Hartwell Courts and Hub City Courts—took three years to finish, employed hundreds of men, and pumped a vast amount of cash into the local economy.

Tobe Hartwell and Hub City were nothing, however, compared to Camp Croft. On September 20—less than three weeks after the invasion of Poland—a delegation of Spartanburg officials and businessmen traveled to Washington to investigate the possibility of establishing some kind of military

The Spartan Mills band, 1943 (Courtesy, Kenneth Burnett)

Blanche Kirby Greene rides her bike on Milliken Street in Pacolet Mills. (Courtesy, Don Camby)

base in Spartanburg, only to find that government officials were already considering the idea. On November 8, 1940, the War Department formally announced plans to establish a new training facility on 20,000 acres east of the city.

As Vernon Foster has written, "The construction of Camp Croft represented the most massive single building project in the entire history of Spartanburg County—an entirely new city, almost half the size of Spartanburg itself, to be ready for occupancy within a period of only three months." Over 600 buildings were planned, including 284 barracks, a hospital, "eleven mess halls, sixty-six latrines, seventy-one company offices, twelve classrooms, six guard houses, three fire stations, three theaters, two service clubs, and a guest house. Buildings planned for the Quartermaster Corps included thirteen warehouses with a total of 140,000 square feet of space, a laundry, a bakery, a cold-storage plant, a vehicle-repair garage, and a utility building. Six chapels, nine barber shops, ten athletic fields, four boxing arenas, three swimming pools, two libraries, and the post exchange were built later for the new training center." The cost was estimated at $7.5 million but ultimately passed the $10 million mark. Thousands of men were employed in the construction. What little unemployment remained in Spartanburg evaporated overnight.

Back in the textile mills and villages, the period between late 1939 and America's formal entry into the war in December 1941 was one of fits and starts. The gradual recovery of production in 1938-39 slumped again in early 1940 when wartime conditions effectively cut off American mills from lucrative export markets around the world. By midyear, however, as war preparations

Jesse Dean Fowler, spinner at Spartan Mills (Courtesy, Lisa Caston)

> *"Reeves Brothers has supplied our armed forces with vital cotton fabrics in excess of the billion-yard mark, which, if measured in terms of uniform cloth, would make 160 million uniforms, if stretched into a band 36-feet wide, would run 40 thousand times around the globe, or if stretched in one continuous length would reach 562 thousand miles—a distance equivalent to a round trip to the moon."*
>
> —Reeves brochure, "Out of Research," 1943

Official Presentation

THE

ARMY-NAVY

TO THE MEN AND WOMEN

OF

MILLS MILL

WOODRUFF, S. C.

FOR EXCELLENCE IN THE

PRODUCTION OF WAR MATERIALS

SATURDAY, MARCH TWENTY-SEVENTH

NINETEEN HUNDRED FORTY-THREE

Mills Mill in Woodruff received the coveted "E" Award in 1943. Other local mills receiving such awards were Drayton, Beaumont, and Fairforest. (Courtesy, Bill Lynch)

continued in Washington, mill owners realized that government contracts couldn't be more than a few months away. They were right. By the end of the year, Spartanburg County's textile mills employed more than 15,000 workers: the highest number ever recorded up to that time.

Beaumont was one of the first Spartanburg plants to receive government orders, and its case is representative. In the summer of 1941 the plant began "conversion" to a defense facility. By the fall of 1942, two-thirds of the machinery was new. The plant's capacity had been doubled; the payroll had risen from 750 workers to 1,430. In 1943 and 1944 the company completely overhauled the Beaumont village, adding 135 new houses. Meanwhile, management was actively organizing carpools and bus routes to bring in new workers from the surrounding countryside. As a company publication put it, "Through rain, and there has been rain, through mud, and there has been mud, red, deep and sticky mud, these buses make their routes way out in the county, bringing the employees to work and taking them back home. The Textile Transit Company, in connection with the Beaumont Company, worked out routes and schedules to benefit the greatest number of employees possible not served by other means of transportation."

Another wartime change at Beaumont was the allocation of jobs between men and women in the mill. Traditionally some textile jobs had been reserved for men (carding, fixing), some for women (spinning), and some could go either way (weaving). The shortage of young men caused by the draft forced manufacturers to place women on jobs that had previously been considered too heavy or otherwise inappropriate for them. According to management, "Today the women of Beaumont are running almost half of the jobs in our plant with the certainty that they will run more. We cannot over-estimate the importance of the work done by them. There is no question but that Beaumont could never have accomplished what it has for Uncle Sam" otherwise. Similar changes in employee demographics—the expansion of women's jobs in the plants and the hiring of off-village employees—occurred at nearly every mill in the county.

In terms of labor-management relations, the most obvious wartime intrusion was the National War Labor Board (WLB). Ironically, the government was the source of mill owners' restored profit margins *and* a new source of federal intervention. The War Labor Board was created to oversee the American economy in light of the strikes and layoffs of the 1930s: Roosevelt wanted nothing

> **Production at the two Spartan plants increased from 9 million pounds of cloth in 1938 to nearly 16 million in 1942 as the war effort cranked up.**
>
> —*Spartan Mills annual reports*

As scores of Beaumont workers headed off to war in 1942, the mill began producing a slick, monthly newspaper called "The Beaumont E," named after the award the plant received for war production. This newspaper, edited by Mary H. Phifer, kept the soldiers in touch with what was happening at Beaumont and rallied the mill's employees to keep up their hard work. Each month, the "E" ran letters from Beaumont soldiers in Europe, the southern Pacific islands, and military bases around the United States. Here are but a few of the hundreds that were printed in the "E" during 1942-1945.

Hello, Beaumont!

How is everything going? Hope you are turning out plenty of duck. Sorry I can't help you out, but I have a job to do here. When I get across the pond I hope I can kill a Jap for each of you, then I can come back to dear old Beaumont and get my job back. Hope to see you all soon, and please write to me.

Yours,

Woodrow Wilson
May 26, 1943

Dear friends at Beaumont:

I have just received your letter and was very proud to know that the friends of mine at Beaumont haven't forgotten me and that everyone is doing fine.

As for myself, I am OK and doing fine.

I'll be glad when I begin receiving the paper again. The boys in my battalion also missed it.

I would like to add that I will be happy when the day comes that we boys can take our jobs back at Beaumont.

Keep it coming and we will keep it rolling.

I would like to thank everyone at Beaumont for the paper and keep it coming.

Sincerely,

Pvt. Dorsey
March 8, 1944

Dear Mr. Montgomery:

Your letter of May 28th received and I want to express my appreciation to the employees of Beaumont for the cigarettes.

I don't see why they honor the boys in service so much. Why don't they pitch a party in honor of themselves? They deserve it. I think they have done a job that only the true American could do.

I want to thank you again for the smokes. They are always appreciated by any man overseas.

Sincerely yours,

P.A. Cogdell
June 5, 1943

Target Tokyo was the theme of this monthly issue of the Beaumont "E," a newspaper produced for mill families at Beaumont. (Courtesy, Sarah Koon)

to impede the flow of defense products from the factories to American troops. The regulatory functions of the WLB weren't really very different from those of the CTLRB or its successor, the National Labor Relations Board. What was different was the social and political context—and the unprecedented wartime powers the WLB had to enforce its decisions.

The War Labor Board's chief agenda was to keep production flowing within the framework of existing labor laws. Anyone or anything that obstructed this agenda became a target. If mill owners refused to negotiate with a union that had legally won recognition in a plant, the WLB was there to apply pressure on management. Beaten but not destroyed, local unions in nine Spartanburg County mills attempted to use the WLB's leverage to force mill owners into more cooperative postures. The first official union agreement in the history of the textile industry in Spartanburg County arrived on December 10, 1941—three days after Pearl Harbor—when Stanley Converse of the Clifton Manufacturing Company signed a contract with Textile Workers Union of America Local 325. By 1943 contracts covered other mills whose unions had survived the ravages of the 1930s: Spartan, Startex, Inman Mills, Fairmont. Company officials could still fight the union—many did—but only within the constraints of the Wagner Act of 1936, and beneath the watchful eye of the WLB. Before the war and later, after it, mill officials could still do more or less as they pleased, knowing that by the time the legal battles were all resolved, almost any union movement could be eviscerated. During the war, however, any mill owner who repeatedly violated the Wagner Act could expect no such latitude. In neighboring Gaffney, the Hamrick and Milliken-Montgomery interests had been trying to rid their mills of unions and unionists for years. Suddenly they found themselves under direct orders from the War Labor Board to quit resisting the Wagner Act, recognize their unions, and get on with production. When the Hamricks and Millikens refused, the War Labor Board promptly had mill officials evicted from their Gaffney offices. It handed over day-to-day control of the Gaffney mills to the U. S. Army, which ran them until the close of the war.

Workers found that this sword cut both ways, however. At the outset of the war, both the AFL and the CIO signed no-strike pledges at the national level. But in the mills, old grievances that had been glossed over during the last days of the Depression—chiefly concerning workloads—reignited. Spartanburg's textile workers considered themselves as patriotic as anyone, but they had no intention of quietly submitting to what they considered mistreatment. Mill owners and managers were tempted by the superheated economic climate to implement yet another round of the stretchout. The results were "wildcat" strikes, occurring without union approval and, in many cases, with no union anywhere in sight: 28 recorded in Spartanburg County textile mills between 1940 and 1945. All but one were settled quickly with varying degrees of WLB pressure. The Inman Mills strike of 1944, however, had deep roots, dating back to the 1930s; the leaders of the Inman Mills local union also had considerable negative experience in dealing with government boards and officials. Furthermore, the Inman Mills local union was torn between the AFL and the CIO, then two separate labor organizations. After seven weeks, the WLB unilaterally terminated the jobs of all who had heeded the strike call, some 324 men and women in all. "The mere recital of the facts of the case is sufficient to show that these workers have no one except themselves to blame for their present predicament," the WLB concluded. "They have deliberately taken their jobs and their seniority away from themselves." However justified workers like those at Inman may have felt, wartime strikes deeply eroded local unions' sense of legitimacy and pitted unionists against one another. The Inman Mills union lasted until 1949, but only as a shadow of its former self.

The close of the war did not bring the deep depression that mill owners and mill workers alike feared. Instead, the American economy continued to grow, and the Southern

An advertisement from the Spartanburg Herald-Journal (Courtesy, the Spartanburg Herald Journal)

At a ceremony in Beaumont on October 3, 1942, Walter S. Montgomery Sr. accepted the community's first "E" Award for excellence in war production from the U. S. government. The following is the text of his speech, courtesy of the Spartanburg Regional Museum:

In full recognition of the part *each and every employee* of Beaumont has had in making this day possible, I am proud to accept on behalf of these *war workers* this *great* honor that has been bestowed upon them.

The remarks by our honored guests are challenging and inspire us to *greater* accomplishment. Our country is experiencing the most trying time in its history and exacts *our* utmost from each of us.

We have worked long hours and *we* feel that we have done our jobs *well.* That our Army and Navy know our job is well done is evidenced by the "E" in this flag before you now. The "E" in our flag signifies that the Beaumont workers have fulfilled *grave* responsibilities to our country. The "E" in our flag *also* signifies grave responsibilities which the workers of Beaumont *shall* fulfill in the future as their part in the victory which must come. We have accepted and shall complete our responsibilities to our country through the united efforts of *every man and woman* in *this organization,* all working together as an efficient and harmonious whole.

We are proud to have it said that we have *done well* in our duty to our country and to our fighting forces and, with God's help, we will continue to do a job of *excellence.* The men and women of Beaumont *shall not fail.*

Spinner at Spartan Mills
(Courtesy, George Mullinax)

textile industry entered its second great period of expansion and profit. This time, thanks mostly to the legacies of Roosevelt's New Deal (including the minimum wage), mill owners took workers along with them for at least part of the ride.

FROM THE "COTTON MILL BOOM" of the 1880s and 1890s down through 1940, the company-owned mill village had been the defining feature of life and labor in Southern textiles. The people who lived in the village worked in the mill, and the people who worked in the mill lived on the village. Every village had its exceptions: individuals whose work kept them outside of the mill (village maintenance men, company store clerks) as well as a handful of men and women who commuted to work from farms or suburbs in the immediate neighborhood. Overall, however, the villages and the mills were inextricably linked. One unintended byproduct of the sudden prosperity of the 1940s was the destruction of that relationship. By 1949, with what seemed like incredible swiftness, the company-owned mill village was on its way out.

In fact the process had begun long before. The original purpose of Southern mill villages had been to attract and house workers at water-power sites far removed from established population centers: as at Clifton and Trough Shoals (Pacolet Mills) in the 1880s. Later, mill owners used their villages to retain a reservoir of manpower; later still they realized that mill villages could also be useful means of social control. Publicly, however, mill owners always lamented the cost and trouble the villages gave them. In effect, every textile company operated as two firms: one that produced thread or cloth, and another in the business of real estate management. Some mill owners went so far as to complain that the money they made from the one went back into the other at a loss. This was an exaggeration, of course, as any visitor to a typical bare-bones mill village could see. Prior to the 1930s none of South Carolina's mill owners questioned the necessity of the village system, whatever headaches it caused. As late as the 1940s, Stanley Converse of the Clifton Manufacturing Company not only made preferential hiring from his three mill villages the rule, he also ordered that all company supervisors live where they worked.

Although South Carolina textile manufacturers had been

Mary Phifer was one of Spartanburg's busiest women during World War II. She produced the Beaumont E and had a regular radio program where she interviewed textile workers. (Courtesy, Walton Beeson)

This editorial, exhorting workers to come to the mills on Sunday, appeared in Spartanburg's Textile Tribune *March 20, 1942.*

As To Seven Days A Week

The cotton mills of this section are producing war materials—materials the boys at the front must have if this war is to be won by us. And let it be very seriously noted that there is an IF staring every one of us in the face.

Some of the mills are operating on a seven-day week schedule because the government is calling louder and louder for the materials which these mills are making. It is reported that at some of these seven-day week plants there is a tendency for entirely too many employees to stay out of work on Saturdays and Sundays, thus slowing down production. There has never yet been a pleasant war and this one is no exception. It's tough and going to get tougher.

Now there may be some people working in textile plants who conscientiously object to working on Sunday. Well, just remember this: the next time you are called to work on Sunday and refuse to do it when your work is needed for war goods, that over yonder in Japan the textile workers are working EVERY Sunday and every other day of the week; not just 8 hours a day, but 12 or more, making goods that will go to supply the Jap soldier that the American boy who used to work right there in your mill must face and FIGHT. And remember, too, that Japan has some big cotton mills and has boasted that she will outproduce America. So the textile worker in America is fighting this war just as surely as the man at the front with a rifle...with the exception that it is peaceful and quiet on the home front.

The goods you are making are needed NOW, not the next week or next month or next year. Time is short, the hour is late, the demand is urgent.

And just a word about the religious side of the question. If the Christian religion is worth anything it is worth fighting for and it is worth working on Sunday for. IF we win this war we shall continue to have the Christian religion, but IF Japan licks us then we will have the Japanese religion forced upon us. And how do you like the Japanese religion? You know...how a little Jap woman centuries ago claimed to have a vision; that in that vision, the sun entered her womb and she conceived a child and gave birth to a baby boy whom, she was told in her vision, was to rule the entire earth and the heavens and the waters and all that in them is. And that "religion" has been handed down to each succeeding Jap ruler to this good day, hence the emblem of the Rising Sun and the sun worshippers.

If a Christian can't muster enough courage to work on Sunday to keep that kind of religion from being spread over the earth, much less forced upon him, he's not very much concerned about the religious conditions of the world.

Unidentified Spartanburg County man tends a loom during World War II. (Courtesy, George Mullinax)

The county fair, begun in 1907, was a favorite fall activity of local textile workers. Expanded and renamed the Piedmont Interstate Fair in 1946, the annual carnival drew visitors from mill villages throughout the region. Carolina Scenic Coach Lines ran special buses three times a day from Pacolet Mills, "Fare: Regular Fare add 10 cents."

discussing the "problem" of the mill village since at least 1932, the first major sales of mill village property in the South took place in the immediate aftermath of the General Strike. Harriet Herring, a sociologist who interviewed a number of those involved in these early sales, later argued that at least some sales were motivated by disenchantment on the part of mill owners and managers. Whatever faith they still possessed in the paternalistic ideal (the community as one big happy family, under the benign rule of the mill owner or superintendent) had evaporated during the labor conflicts of 1933-36. These companies, according to Herring, no longer felt any responsibility for the "welfare" of their workers. They shed their villages like old skins.

In Spartanburg County, Drayton and Whitney fit this pattern. In December 1935, officials at Whitney commissioned a formal survey of their village property. Sales of individual homes began on July 3, 1936. Drayton surveyed its village on December 11, 1936, and began selling lots in early 1937. That Drayton and Whitney were the first to sell was not coincidental. Both were suburban mills, drawing an increasing number of workers from nearby rural communities (such as Fairview Heights and the Cannon's Campground neighborhood) as well as from Spartanburg proper. Whatever their "disenchantment," the directors of both companies must have seen less and less need for maintaining the villages, especially in the depths of the Depression. Certainly other mill owners in the county were watching the "experiments" at Drayton and Whitney very closely and making their own plans.

World War II brought a temporary halt to mill village sales. Just as some owners had warned, the sudden revitalization of the economy produced a labor shortage in Southern textiles. Providing a house once again became a key way of enticing workers who might otherwise accept employment somewhere else. Not only did no company announce further sales after war began to look inevitable in 1940; Drayton, which had already sold

Clyde Camby, a serviceman from Pacolet Mills, home on leave. He is standing outside a home on Walker Street. (Courtesy, Don Camby)

Mrs. Allen's kindergarten class in the Spartan Mills village leaves the community building in 1941. (Courtesy, Ed and Mary Ann Hall)

over half its lots, froze further sales and began hasty construction on a handful of new houses on the village's east side. Beaumont, Startex, Glendale, and other mills began adding houses where they could on their old properties.

But other dynamics unleashed by the war proved the mill village's undoing after V-Day. The most obvious factor was the rise of textile workers' real wages. Under the wage scale set by the War Labor Board in Washington, textile workers suddenly found themselves making nearly the same money that their peers—construction workers, railwaymen—had been earning for years. The loss of governmental controls after the war affected textile workers in other ways—for instance, in leaving their unions more open to company attacks—but left the new wage scale in place. Textile workers were no longer a segregated class, the poorest of the (white) working poor; more and more, they had a bit of money to spend, or save, or invest. They were no longer solely dependent on the mill for housing, or on the company store for food and clothing. Many began to move off the villages voluntarily, buying homes in town or small farms in the country.

Second, the rise in wages was equaled by the rise in automobile ownership. Cars had been virtually unknown on Southern mill villages in 1930. A supervisor might have one, or there might be none at all. Over the course of the decade more and more made their appearance, and during the 1940s automobile ownership among working-class

Girls pose in front of the "picture show" in Enoree about 1942. This movie starred Henry Fonda and Olivia de Havilland. Margie Gentry is on the left. (Courtesy, Anthony Tucker)

School children in Glendale (Courtesy, Paula Suttles)

men and women in Spartanburg County became the rule rather than the exception. With cars, of course, came geographic mobility. Workers living off the mill village—several miles into the country, even—could commute to work in faster time than it took some village workers to walk.

Third, during the war Spartanburg's mills had been forced to look beyond their villages for help, either to the towns or to the farms. By the mid-1940s, a substantial percentage of workers at each of Spartanburg's mills not only were not living on the mill village, they had never lived on it. This development was the deciding factor. If mill owners could ensure a sufficient labor pool without the financial burden and administrative headaches of owning a mill village, then the mill village had become an anachronism. It had to go. And so, between 1950 and 1959, all but two of Spartanburg's mill villages, one after another, followed the pattern that Drayton and Whitney had set in the late 1930s.

The effect of mill village sales on the workers and their families was two-edged. Whether companies sold their homes directly or hired an outside real estate firm to do the job for them, the policy was always the same: offer each house first to its inhabitants, then to anyone who wanted it. To some, this was an opportunity; to others, a nightmare. As an Arcadia resident recalled, "A lot were worried: a lot had the money to buy theirs, but a lot of them didn't." Villagers who could not or

When Tucapau Became Startex

The following is an excerpt from an article that appeared in the Spartanburg Herald-Journal, *January 1, 1940, describing the transition that took place in Tucapau when the Montgomery family purchased the mill and village in 1936 from its New England owners. A representative of the new company is presumably the author.*

The visitor in 1935 saw lines of unpainted houses, in need of roofs, chimney repairs, flooring, latticing and screens. No efforts were made to have tidy yards, and across the bridge in the lower part of the village the main road was unpaved. He might see women carrying buckets of water since many of the houses were unpiped, and the existing piping was so small that houses away from the standpipe got little or no water. The community building had been deserted through lack of funds and the visitor got this reply from one villager: "What do we do when we're not working? Nothing to do but just roam the hills…"

Today the visitor who saw the plant in 1935 will be startled by the changes made in four years by Startex Mills. The machinery has been thoroughly modernized and the plant is one of the few air-conditioned mills in the South…The vast improvement in the village is so great that a visitor will wonder if he is in the same place. Newly roofed houses fronted with well-kept, flower-bordered lawns line the long street shaded with trees. The paved street continues across the bridge now for a mile to the old Greenville highway. No buckets are in evidence because a new pipeline runs through the entire village, and every house is piped for a water supply that is so adequate that bright green winter grass, which takes more watering, forms lawns of the majority of the homes.

Such a great change for the better could only be accomplished through the cooperation of everyone in the community. Few "roam the hills" any longer, and all join in the new interest in the village and community activities.

The earliest mill houses in Tucapau deteriorated rapidly under a series of corporate owners from New England. This photograph of South Main Street was taken in 1923. (Courtesy, Junior West)

According to the Cotton Mill Association of South Carolina, mills lost an average of $23.53 per year on each mill house they owned in 1939.

Inman Weave Room No. 1, Christmas 1946 (Courtesy, Jim Everhardt)

would not buy their homes knew they ultimately faced eviction.

The small numbers of black workers with mill village homes were especially vulnerable. At Drayton and Whitney, for instance, the houses that had been reserved for black workers since the earliest days were sold to whites, each of whom had to sign a covenant promising that their new property "not be sold, rented, leased, or otherwise disposed of, to any person of African descent for a period of twenty-one years." Black families that had lived on the villages for a generation or more were forced to leave.

Sale of the villages accelerated the widening gulf between workplace and community. Identification between mill and village had always been the backbone of mill life. By 1949, however, not everyone who worked in a mill lived on its village: and at Drayton and Whitney, not everyone who lived on the village worked in the mill. As the older generations of villagers began to die out and retire, their homes were not necessarily taken, as before, by other mill workers. By 1949 Drayton and Whitney were well on their way to becoming general working-class suburbs of Spartanburg, with fewer and fewer ties to the mills that had built them half a century before. The rest of the textile South was not far behind.

Robert and Nora Bell Kirby both worked at Pacolet Mills in the 1940s. Robert worked in the shop and ran the "dope wagon." (Courtesy, Don Camby)

Tragedy struck Spartanburg on May 21, 1944, when five people drowned in "12 feet of muddy oil-coated water" in the mill pond at Spartan Mills. Three men and two women, all African Americans, were in a Plymouth sedan that inexplicably crashed through the fence surrounding the pond. They were Frank Ezell, Willie Johnson, Dovie Ray, Beatrice Dandy, and Janie Woodruff.

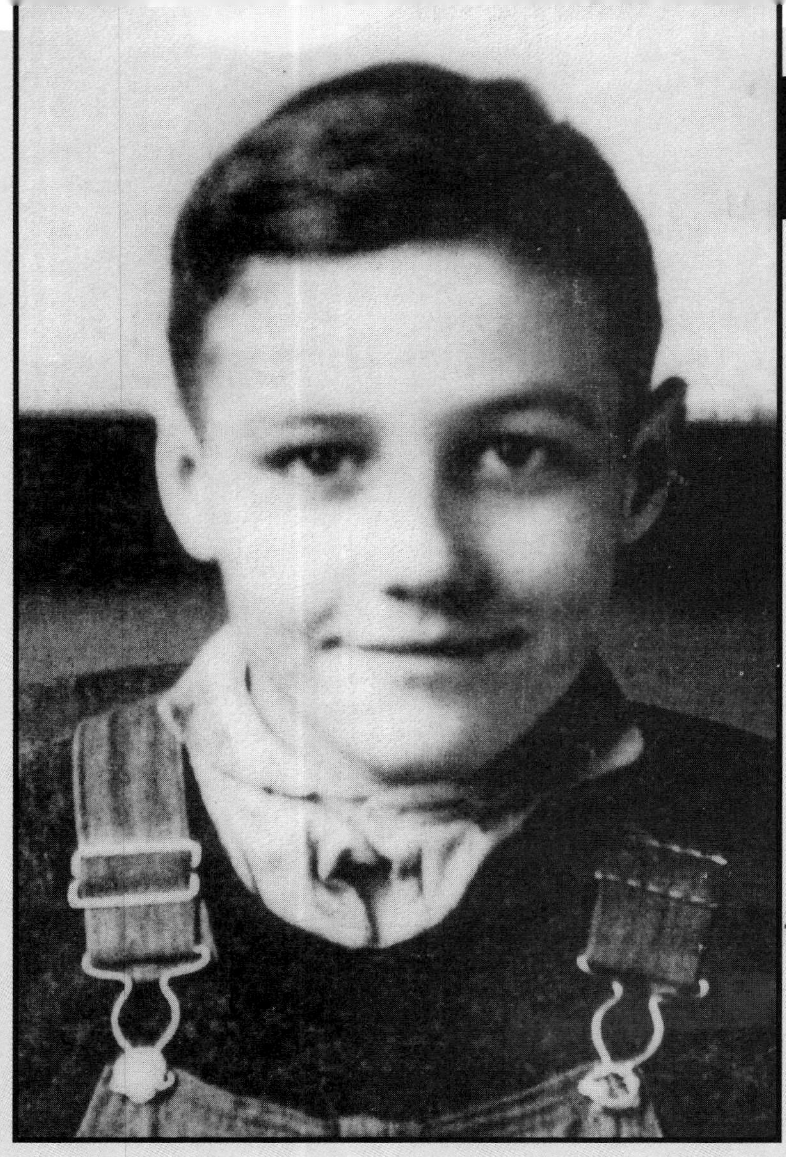

Boyd Israel

Boyd Israel was born in Whitney in 1925, the son of a card room employee and night watchman in the mill. As a teenager, he worked in both Whitney and Beaumont mills. Using the classroom experience from Whitney and Spartanburg city schools, he went on to get a master's in education degree at Columbia University in New York before returning to Whitney for a five-year stint in management at the newly named Pequot Mill. Later he received a doctorate in education and taught at Albany State College. Here he recalls two rituals from his childhood in Whitney.

Boyd Israel as a young boy in Whitney, about 1933 (Courtesy, Boyd Israel)

During the mid-'30s I lived in a village house across the bridge but close to the nearby Whitney Elementary School, which went from grades one to seven. As I remember, class sizes were small by today's standards and became slightly less in the higher grammar grades. Our principal, Mr. Peck, was a friendly transplanted New Yorker. He was friendly but firm in his contacts with me.

As I moved up the grades each year in the mid-'30s, I became aware of and began to participate in a strenuous mid-morning activity during recess. This practice involved the following steps: Upper-grade boys would exit the building and run quickly to the cyclone fence that split the northern front playground into two sections. A large oak tree stood tall beside the huge trunk portion of the tree. Each boy would climb the fence and reach for a lower limb to climb into. The last boy from class would do the same and become "it" until he was able to touch another boy. This continued until the end of recess was sounded. The participants became very agile in climbing from limb to limb at higher levels. Strong cords were used to bring smaller limbs together for greater mobility. Some boys were able to go untouched for the entire recess period.

I remember distinctly the day that was bound to come. A boy in my grade named Harold McAbee had gotten about as near the top as possible. He called out and came falling down beside me to the ground. The tree was emptied rapidly. Harold needed medical attention immediately from a visible gash on his head. He was taken rapidly the several hundred yards to the nearby village office of Dr. W. H. Chapman. The student recovered quickly, but school officials stopped all recess activities in the tree.

Another of the pleasant memories still firmly etched in my mind began as a young boy growing up in the Whitney village. A good many of the men with homes located on streets close to each other would purchase small young pigs to raise to full-grown size for slaughter and consumption during the winter cold weather.

The process followed a similar pattern each fall. Each family had a wooden pen and an adequate wire fence on company pasture land, usually away from the homes. The young pigs would be fed daily until ready for processing, usually after several cold, frosty nights. For the participating families "hog killing day" was a time of steady hard work but also one of genial fellowship by the men together with their chores and the women with a difference set of tasks. Close friendships were strengthened as the day wound down and each family assembled the different packages to be carried to the respective homes.

Curious young children were entertained and supervised by the older youths as the adults kept a watchful eye on them and kept track of the different activities needed to successfully complete the different kinds of meat for storage during the winter.

During the course of the busy day the atmosphere among and between each group of children, women, and men, reflected a rich blend of warmth, friendship, cordiality, and concern for each other. The women had prepared carefully in advance the tasty food needed for all ages based on previous fall festivities. In retrospect such activities are believed to have contributed in highly positive ways to the making of a more personally satisfying and enriching way of life for those in the Whitney community.

Whitney Mill and the village that flanked it, 1930s
(Courtesy, Herald-Journal Willis Collection, Spartanburg County Public Libraries)

Voices from the Village

Winslow Howard

Winslow Howard, born in 1915, came to Lyman with his family at age eight. His father was related to the mill supervisor and got a job as an overseer to help open the giant Pacific Mills complex. Winslow went to work in the mill in 1929 and soon after joined the National Guard. Here he tells about the day the General Strike and the "flying squadron" came to Lyman.

And so they decided that they were going to strike, every mill. In the meantime, I joined the National Guard, 16 years old—no, 17, 17. I was on the second shift at the time. I went to work at 3 o'clock until 11. I can remember that we had wind that they were going to cause trouble, this flying squadron. That's what they named themselves…Brother, they had a followin', I'll tell you. Well, I was up in the mill working the second shift at the time, and [the boss] Mr. W—, he was kin to Daddy, he come up, and he says, "Capt. A— wants you at the Armory Hall immediately." He went around telling everybody that belonged to get to the Armory Hall. So we went up there and put on our uniforms. They gave us rifles, ammunition, belts, and everything. Riot gear, we called it. Sent us down in front of the mill. Out in front of our mill there.

Winslow Howard (Courtesy, Mark Olencki)

And, boy, they were coming over the railroad bank. There were hundreds of them. That flying squadron was a squadron! They were just pouring over that bank. They was comin' to the mill. They was comin' up in it and shut it down. People were still in there running it, making cloth that we needed. We barely got from the Armory Hall down to there. We didn't march—we ran down there and made a ring around the big front door. Down at the bleachery gate…they had a big line of soldiers, guards down there, wearing the government uniforms. They came to the door there.

Sheriff [Sam] Henry was from Spartanburg—you probably heard of Sheriff Henry. A lot of 'em was about drunk, them old people, and they had their wives and babies like that. It was pitiful. I think about it, and I'm glad we didn't have more trouble. Some of them [National Guard] got hit in the head. Some of the boys hit 'em with clubs, picker sticks, and things that they took out with them. They were about that long [holds hands three feet apart]. Some of our boys were hit with 'em. They was gonna go up and stop our mill down. That mob, you wouldn't realize. I'd say there was 1,000 or more people around the mill door. I don't know how many around the bleachery door down there. But I imagine just as many. All our whole company was on guard with guns. Captain told us to fix bayonets—a bayonet about that long [holds hands a foot apart], fits on your rifle, and move 'em back. So we started that. But they overwhelmed us. They'd push in on us. We was jabbin' with the bayonet, hitting them on the hands, turning around right quick, doing what we was taught to do.

And Sam Henry came and he got in front of the mill. There was two drunks…One of them come up there and said, "We're gonna give you just one more chance," talking to old man [boss] W—. He wasn't scared of 'em. He was an old man. He wasn't scared of 'em. And Sam Henry come up with a deputy and went over to the door. He says, "The first man that touches that doorknob is dead!" I was as close to him as the TV over there. I heard him tell him. He had a pistol about that long [holds hands a foot apart]. He says, "I mean it, the first man, the first person, touches that door, is DEAD!"

And they go on, "Ah, he wouldn't shoot you. He's shootin' blanks!" But there wasn't any such thing. Our ammunition was the real stuff, and his was too. [They said] "Come on boys, let's go get 'em! We'll shut her down!"...

That was a heck of a time. And nobody went in. Cause Sam Henry, he done cocked it. He done pulled that hammer back. He said, "If you touch that door, I'll shoot you," and he would have. And then he told us, "Move on up a little more." We got to movin' up and we moved 'em out. Moved them across the street finally. The old sheriff, he was a good sheriff. Sheriff Sam Henry...We got them back down the hill and it was getting dark. So they began to vanish and go somewhere. The cars, you wouldn't believe it, the people coming over them banks. And so we went back to the Armory Hall, and they had some sandwiches made up, and they gave us sandwiches and a Coke. I believe somebody brought some coffee, I'm not sure. But who was in the mood to eat? None of us...

—Interview by Betsy Wakefield Teter

Sheriff Sam Henry
(Courtesy, Fred Parrish)

Remembering holding a gun outside the mill at Lyman
(Courtesy, Mark Olencki)

Voices from the Village

Annie Laura West

Annie Laura West was an employee at Saxon Mills and secretary of the union local when she was invited to participate in the "Summer School for Women Workers" at Bryn Mawr College in Pennsylvania in 1936. She was among four women from Spartanburg County who attended the college that summer, taking courses in social justice and labor organizing. The following is part of an essay she wrote that summer, "The Eight Months' Strike," a description of Saxon's 1935-36 labor disturbances, which included a shootout in front of the plant and a dynamiting.

SCRAP BOOK

SOUTHERN SUMMER SCHOOL 1936

Annie West's story was published here. (Courtesy, George W. Moore Jr.)

On July 30, 1935, the second shift of workers in the S Mills came out on strike. Of course we of the first shift were there outside the fence as a precaution, for sometimes trouble does start when a local union calls a strike. In this plant, over 85 percent of the workers were members of the union, and most of the new people who had been hired because they were non-union joined in a few days.

We, the organized workers, had been trying unsuccessfully to bargain collectively with management but had gained only a few things, while they continuously discriminated against our members. Some were laid off because of elimination of old machinery; others were stretched out past human endurance, so we decided we could not work that way…

I was on the Soliciting Committee during this long strike. The other organized workers in our state helped us; merchants and others, some of them lawyers, were very nice to us, and one local in Ohio sent us a check for $65. Some of our girls got jobs in FERA [Federal Emergency Relief Administration] and WPA [Works Progress Administration] sewing rooms; in fact one from each family was on some kind of work. Sometimes we would have to send for them all to come and help on the picket line.

The sheriff [Sam Henry] would sometimes come with 10 or 12 of his deputies and run over the picket line and place Negroes or scab guards in the mill to load cloth. On one occasion when this happened we had about 500 people on the picket line in about 30 minutes time. We very forcefully told the sheriff that we would take charge of those men when they came back out. He tried to run us out in the street, but we refused to go. Our shop committee called us into a group and asked us to let them come out and we voted with the exception of a few to let them come out and leave without molesting them. But we told them we would not let them get by any more…

We had many similar instances where machine guns were trained on leaders of our strike and a scab guard placed on the mill, but later on in this same week feeling was running very high. The men invited this scab to come outside the fence and fight it out. He would not, so some small boys started shooting him with sling shots. He was soon about 125 feet from the gate and opened fire on the crowd which contained about 200 men, women, boys and girls. The shots from this 45 free wheeler revolver went through a nearby garage and as far as 600 feet where they were found in the oil mill across the double track of the railroad. Some of the crowd opened fire with shotguns and the guard dropped. We thought he was dead but found he was only reloading. He used all his ammunition and then retreated after luckily not hitting anyone.

The ambulance was called, and he came from inside the mill fence. He was limping; his neck and the side of his face were very much riddled by shot and his eyes were glassy. We women begged the men not to finish him. In the hospital 60 or more shot were removed from his body.

Fifteen of our men were arrested including our shop committee and relief chairman, who had been fired off the job that the scab guard was on. These four men were 11 miles away cutting a tree for the CLU [Central Labor Union] Christmas program. They were released. But seven of these men were tried in magistrate's court and four of them freed while five were bound over to the next term of Sessions Court. When the trial came up 29 witnesses were placed on the stand for them, but in spite of that, three of them were sentenced to two and a half years in the State Penitentiary. We are trying to get them out for [the] men have wives and small children and we feel they are not guilty.

After an unsuccessful attempt to evict four families, their lights and water supply were cut off, but the State Department of Health made the company turn on the water. Then the UTW [United Textile Workers] sound truck came into our city and with some 400-500 people we paraded by the homes of the bosses and the mill president and scared them almost to death. I spoke over the sound truck several times—you could hear it 12 blocks away.

Later some dynamite bombs were exploded in the village by unknown persons and the union was blamed for it. When they did not know who did it, news stories with pictures about the union were published in the papers. But we were successful in winning a "working agreement," and now conditions are much better. We don't have a dictator any more. Here's to the UTW. Long may it live!

No one was ever arrested for the dynamite incident, which blew out windows in the teachers' cottage where Community Director Marjorie Potwin lived. Annie West was fired by Saxon Mills, along with several other union leaders, a few months after she returned from Bryn Mawr. The union at Saxon was inactive by 1941. Annie died in 1986.

Saxon textile workers and their family members parade through the mill village during the strike of 1935. (Courtesy, Caroliniana Library)

Laura Rodgers
(Courtesy, Mark Olencki)

VOICES FROM THE VILLAGE

Laura Rodgers

Laura Goudelock Rodgers came to Beaumont at age six with her mother, father, and siblings and moved into a four-room mill house in 1924. She married Isham Rodgers, founder of Ike's Corner Grille, in 1935. Here, she relates a tragic family story.

My daddy died in the mill in 1936. It was a rickety ladder is what I've always understood. He was fixing a generator and it was in the ceiling, and he had a wrench in his hand, and when the ladder gave way he fell and hit himself in the larynx with it, and he bled to death internally. My daddy was a good man, a very good man. I was in the mill that morning. I didn't work in the mill, but I was with my sister-in-law, and I told her, I said, "Gladys, when we get your side straightened up, let's walk over and see where Daddy's at."

Another spinner came over and said, "Laura, you better go home. Your dad's had an accident, and he's hurt pretty bad." They buried my dad at White Rose Cemetery at Pacolet. That's where he and Mama lived when they got married. That's where his folks lived. Uncle Walter Goudelock was postmaster down there.

There was six of us children living with Mama when that happened. After Daddy died, the mill wanted the house we was living on North Liberty Street and said they'd give her a six-room house on Beaumont Avenue. Mama had a cow and a garden, and she would have to put her cow over on Clinchfield where they had stalls because there wasn't no backyard down there, and the houses was too close together anyway, and she didn't want to go. [When she didn't leave the house], Mr. B—, the superintendent, fired Mama and my brother. Dawsey, my brother, couldn't get another job around here. They was upset about it, but Dawsey played the piano for a lot of them around here, and then he started playing for dances, so he made more money than the rest of us. Mama got this house over on Langdon Avenue. My brother helped her a lot, and then she had boarders. She had a garden. She had a cow. She kept a few chickens. She didn't get rich on it, but she got by on it. Isham and I lived with her. We had two children then.

Mr. [Walter] Montgomery bought the mill after that. He came through the mill one day and he asked Isham, "Are you happy?" or something like that.

And Isham told him, he said, "I'd be happier if we could get a house at Beaumont. We're expecting our third baby."

He said, "Mr. B— won't give you a house?" and Isham said no. He said, "What happened?"

And Isham said, "My mother-in-law wouldn't do what he wanted her to do."

And he said, "Well, I'll talk to him." So he came back through the mill a few days later and he asked Isham if he had talked to Mr. B— and Isham told him, no not yet, and he said, "Well, go ahead and talk to him. You'll get a house in Beaumont." And we did.

—Interview by Betsy Wakefield Teter

On the front porch in Beaumont, watching the mill come down (Courtesy, Mark Olencki)

Christine Gates
(Courtesy, Mark Olencki)

VOICES FROM THE VILLAGE

Christine Gates

Christine Gates began working for Reeves Brothers in Woodruff at 17 years old. Her father was an overseer in the card and spinning rooms. She retired from the mill in 1983. With the exception of a three-day stint at Enro Shirt Company, it is the only job she ever had. Here, she describes life in the mill during World War II.

I hadn't been married long when the Japanese bombed Pearl Harbor. We were living with my sister because we didn't have anywhere else to stay. We heard it on the radio or maybe read about it in the paper. I'm not really sure now...

We worked on Sunday during the war. We had to work seven days a week in order to make [cloth for] khaki pants and things like out of Army twill. We had Army twill that we made, different types of cloth, they more or less were for the Army at that time. The mill supplied them with the material...Working on seven days, you got overtime. They were a lot of people hoping the war would last because they got more money than they'd ever made, you know, and they was hoping that the war would last. They wasn't thinking about people over there risking their lives or anything. There was just some of them who just enjoyed getting that extra money. I didn't especially care for working on Sunday. I'd never worked on Sunday in my life, and it didn't seem right cause I'd always been taught to keep the Sabbath holy. I guess that was the start of all that working seven days a week at plants...

In the cloth room, I'd work many a times nine and ten hours a day because they'd be behind with some of the cloth...At that time I was divorced and had my son to raise, so I agreed and worked overtime. Most of that time I was standing on my feet even though we had stools. I was a fast worker. I was one of those nervous-type people, and when I worked on production downstairs you had to get so many boxes. They'd have maybe 240 bobbins a day, and you'd have to run 'em by the box...You got paid by the piece, but in the cloth room you got paid by the hour. It wasn't as strenuous on us...

In every department there were men and women. Even in the card room, there were even women down there...It was just hard work for everybody.

They brought more women in during the war, because they didn't have enough to fill the jobs with all the men being drafted. I believe that is about the time that Roosevelt made that law that they had to make a certain amount of hours and they couldn't work but a certain amount of hours per day. What was it they called it? The "good deal" or the "New Deal?" New Deal, yeah. He was the one that started that, and it seemed to help the textile people more than anything...Now, people used to work 12 hours a day. My mother worked in the mill when we were children and worked 12 hours a day and made fifty cents a day. Like I said, they put your children in there when they maybe get 10 or 11 years old. I had an uncle that went in the mill. He had bad eyes. I always felt he was much older than we were, but there wasn't but a couple of years' difference between him and my brother because he had been put in the mill to work.

[After the war] they kept them [the women] in there and hired the men back as they could. They had to give them a job back after the service. It was guaranteed that they get their jobs back. They had to get their jobs back, because they'd served the country, and they deserved it...I didn't think too much about what we were doing then [helping the war effort], but I knew that's what we were doing. We were making Army twill and the things they made the men's shirts and pants

[with] and the jackets they had. There was rationing, especially sugar and coffee. You didn't make much, but you could go to the store and for five dollars buy enough to feed your family for a week...You could buy eggs 10 or 15 cents a dozen back at that time. Like I said, anything you wanted you could buy for about five dollars a week. But we were making $2.49 a day...

I didn't feel like I knew anything else. I went to high school, and I took a business course, but I was always a nervous-type person, and the teacher stood over you, and I'd get even more nervous...My daddy wanted me to work in the mill. So I just continued to work in the mill and got used to it. I was happiest when I was working in the cloth room, because it was easier on you. It wasn't so hard on your hands and back...It used to be sort of respectable in the cotton mills years ago.

—Interview by JoAnn Mitchell Brasington

Outside the mill in Woodruff, now closed (Courtesy, Mark Olencki)

Spreading the Union Bug
Judson Brooks and the Una News Review

by JoAnn Mitchell Brasington

Judson Books, editor of the Una News-Review in 1936 (Courtesy, Lyn Sellars)

For 42 years, Judson L. Brooks was a voice of the cotton mill worker with his pro-union newspaper printed in Una.

A native of Brevard, North Carolina, Brooks and his family moved to the Upstate apparently around the turn of the 20th century. They, like so many others, came from the mountains looking for work in the cotton mills. They settled in an area of Spartanburg County that became known as Una, among people described as some of the most independent in the region. Una residents "didn't want to be pent up in a mill house. They wanted to be out where they could be foot-loose and raise their hogs, chickens and cows and farm their crops as they chose," local resident Warren Justice once said in a newspaper interview.

When Brooks went to fight in World War I, Spartanburg County textile mills were working at peak operation, wages were rising, and there was employment for many. A few years after Brooks' return, however, the Great Depression had arrived, bringing with it lower pay, fewer working hours, and stressful working conditions in the mills. In 1929, with dismal prospects for Spartanburg textile workers, Brooks and his brother Henry established a print shop and the *News Review*, a weekly pro-labor newspaper, in vacant rooms in Justice Grocery. Touting itself as "the working man's paper" and endorsed by the Central Labor Union, the South Carolina Federation of Textile Workers, and The Textile Council, the *News Review* cost $1 a year through the 1930s and '40s. By the mid-1930s, it was distributed to about 2,000 homes.

"He and his brother did all the writing," recalled Ellen Melton, postmistress of the Una Post Office and a friend of Brooks. "They knew labor front and backwards. And if Mr. Brooks had to go somewhere to get information, he'd walk. He didn't drive, and he didn't often ride anywhere either."

Without transportation, distribution of the *News Review* was a problem. Brooks and political rival Wilburn P. Justice, a Republican and owner of the grocery store where the *News Review* was located, both wanted to improve their area of the county. They worked together—Justice lobbying powerful Republicans and Brooks securing support from influential Democrats—to bring a post office to their neighborhood. The Una Post Office shared a building with Justice Grocery and the *News Review* press. The post office and community were named Una, by Brooks and Justice, in honor of Una Mae Abbott Justice, the first postmistress.

Those who knew Brooks speculate that he started the county's only pro-union newspaper because he wanted to help people at a time when they needed help the most. "He was a wonderful, religious, smart person who worked all his life helping people with labor problems," said Melton. "He wasn't a person to brag. He wanted to help everyone. He never had trouble with anyone and dedicated most of his time to his job."

While not welcomed by the entire community, the *News Review* nevertheless found mainstream advertising support in Community Cash Stores, Bobo Funeral Chapel, Heinitsh Drugs, Hub City Lunch, Hammond Brown Jennings, Floyd's Mortuary, Spartan Automotive Service, and numerous political candidates hoping to garner the working class vote with the endorsement of the labor movement, known as the "union seal" or the "union bug."

The *News Review* thrived during the early Depression. The newspaper printed weekly reports from union locals throughout the county—news of the neighborhoods of Clifton, Pacolet, and Tucapau, among others. The infamous stretchout system gave Brooks both local and national meat for his readers. Anticipation of the Strike of 1934 may have been the *News Review*'s finest hour. Articles on Roosevelt's National Recovery Act shared space with information on local meetings and pleas for support of striking workers throughout the area. Other subjects included child labor legislation, the minimum wage, and the purchase of tear gas by mills for local use during the Strike of 1934.

The *News Review*'s most outspoken rival was the management-sponsored *Textile Tribune*, also published weekly in Spartanburg by Robert DeYoung. In a 1936 *News Review* editorial, Brooks wrote, "The *Textile Tribune* refers to the *News Review* as a cesspool. This seems to be the attitude of the *Tribune* toward all cotton mill people and their friends."

After the defeat of the union movement in 1934, the *News Review* changed. By 1939 the masthead no longer listed union affiliations but read: Devoted to Progressive Farming and Wholesome Labor Conditions. Last published in 1971, the *News Review* moved away from local news in the 1940s and carried primarily national news and opinions relating to national labor issues. There were exceptions. In the early 1950s, Brooks ran a series of editorials criticizing people who were charging that the Rev. John B. Isom of Saxon Baptist Church (where Brooks was a lifetime member) was a Communist. By the 1960s, Brooks wrote only in generalities about unions with an occasional reference to an event long past. By this time, circulation was down to around 1,000.

In addition to publishing the *News Review*, Brooks devoted his time to improving the conditions in his community. He served on the local election commission from 1938 to 1977 and helped establish the local Water and Sewer District. Brooks held leadership positions at Saxon Baptist Church and eventually helped found Northside Baptist Church in Una.

Frances Tate went to work for the Brooks family in 1948. "Mr. Jud treated you like you was a human being," she said. "He treated everybody the same. It didn't matter if they were black or white. I cleaned their house, but they wouldn't introduce me as 'my maid.' They'd say, 'this is my friend Frances Tate.'"

Brooks died on October 30, 1981, at the age of 93. Services were held at Una Northside Baptist Church, and he was buried in Sunset Memorial Park. Along with a few civic affiliations, the small obituary states simply that he was "former owner of Una Print Shop" and "editor and publisher of *News Review*."

THE NEWS REVIEW

Endorsed By

CENTRAL LABOR UNION -:- S. C. FEDERATION TEXTILE WORKERS -:- THE TEXTILE COUNCIL

Volume Eight No. 25 — Una, South Carolina, Friday, June 18, 1937 — Five Cents Per Copy

LYMAN CASES SETTLED

HELPED WIN THE LYMAN VICTORY

Above is John C. Lanham, of Lanham & Lanham law firm. Upper right is John C. Williams, of Johnston & Williams law firm. Lower right is Furman Garrett, President Local 2191, Lyman. Others associated were the law firm of Gantt & Brown, of which we were unable to obtain a cut of either. These firms have proved themselves friends of the labor movement, handling many cases for workers without compensation, where they were unable to pay.

Company Pays Sixty-Three; Invites Twenty Back to Work

Great Victory For Labor, As Wagner Act, Coupled With Untiring Efforts of Union President and Labors Attorneys, 'Persuade' Pacific Mills They Had Better Behave

John Lanham's office on Tuesday morning, June 15, was a scene of general rejoicing and smiles, because sixty-three of the former employes of Pacific Mills, who for a long time had been discharged by the company, were receiving checks representing cash sums, and twenty others were glad to be receiving invitations by the mill, to return to their former jobs, with the benefit of two ten per cent wage increases.

Under the terms of the settlement, which brought to an end a struggle between Local Union 2191, United Textile Workers of America at Lyman and the Pacific Mill Company of more than two years duration, forty-four former employes were handed checks in the sum of three hundred and twenty dollars ($320.00) apiece, while nineteen

This issue details the settlement of discrimination cases filed by Lyman workers against Pacific Mills. (Courtesy, George W. Moore Jr.)

Tate still lives in the Brooks family home on Sibley Street in Una, a gift for the years that she cared for the Brooks family. The small, wood-frame house still holds photos of the Brooks family and memories long lost of the *News Review* and Judson Brooks. File cabinets from the *News Review* offices still sit on the front porch, blocked by old furniture and boxes of brittle magazines and papers.

Christian duty, political conviction, human nature—no one knows exactly why Brooks remained the sometimes unpopular voice of labor in a largely non-union county in the union-resistant South. But he did, and for 42 years the *News Review* spread a gospel of peaceful reform and concern for cotton mill people across the county.

Cotton Mill Music
The Front Porch as Stage

by John Thomas Fowler

Music around the mill hills of Spartanburg County was a rich and diversified blend of traditions during the heyday of textile life. Mountain, folk, blues, and church music made up a cross section of entertainment. As families migrated to the Upstate from near and faraway places, they brought with them their backgrounds, influences, and lifestyles. These folks shared their fiddle tunes, instruments, ballads, and church songs. From the mill houses of Inman Mills to the camp meetings at Glendale and Pacolet Mills, music became an important part of everyday life.

Before the 1920s Southern music was somewhat isolated to geographic regions. In other words, the type of music you were most familiar with depended widely on where you were born and grew up. Things changed very quickly in the early '20s. In 1923 an unknown fiddler, John Carson from Georgia, made the first commercial recording of fiddle music. To the astonishment of the record companies, the recording took off, and the country music recording industry was born. By the late 1920s radio stations sprang up to play and promote this old-time music. Textile workers in Spartanburg County would gather at the mill homes of family and friends to hear a little of WSM's *Grand Ole Opry* on Saturday nights. Many textile workers felt a close kinship with these tunes. "That music was our music," they would say.

Alvie Westmoreland Brackett was born in Inman Mills in 1919 and began working in the spinning room at age 15. She remembers when she was 10 years old, her mother played records on their graphanola. "Mother liked the Carter Family and Jimmy Rodgers—they sounded good," she recalled. Marvin Newman of Beaumont Mill fondly remembers when he was a young boy, "Friends and neighbors would come over on Saturday nights to listen to the *Grand Ole Opry*. We were the first to have a radio on the mill hill back then." For hard-working mill folks, owning a radio was not a necessity of everyday life. Many could not afford the luxury, yet they enjoyed the music. Another millhand, Bessie Holland Fowler, started working in 1931 filling batteries at the Glendale Mill. She recalled, "You didn't need a radio in those days. Everyone would play theirs so loud, you could sit out on the front porch and listen."

The front porch was an important feature for many a social gathering in the early years. As teenagers, Alvie Westmoreland and her other girlfriends would sit out on her porch and sing hymns, folk songs, and radio favorites for hours. "We were pretty good then," she warmly recalled. Some of those songs were *When They Ring the Golden Bell* and *O Beulah Land*. Along with porch gatherings, other places became important links to sharing and listening to music. Gene Franklin, a retired preacher, lived in the Clifton Mill community and attended the No. 1 Baptist Church as a boy. His father was "song director" at the church. Gene remembers in the late 1930s going to Noblitt's Store in Clifton. Friends would gather at the store and sing. "We'd sing hymns, hillbilly music, and old folk songs." He also loved to listen to the radio: "Roy Acuff and Jimmy Davis were who I liked," he recalled.

During World War II Spartanburg textile workers' children sang songs of patriotism. Hymns and sentimental songs were also favorites. Mill workers tuned their radios to local broadcasts of WSPA and WORD. Cliff Gray, better know as Farmer Gray, broadcast local farm reports, weather forecasts, and music and kept workers and their families informed of the ongoing war in Europe and Asia. Grover Golightly, another local radio announcer, hosted a very popular show called, "Cousin Bud and the Hillbilly Hit Parade." The show began in the early 1940s, airing in the afternoon. Locals, along with well-known musicians, would stop by and play on the live radio show. Arthur Smith, Hank Garland, and Roy Acuff were a few of the well-known guests that frequented the "Hit Parade." On weekends Golightly would sometimes broadcast from area mill villages. "I'd audition local musicians, and if they made the grade, I'd put them on the show," he fondly recalled. "That was how we paid them."

The Double X Boys, 1953. From left, Vernon Riddle, Cecil Prince, Jim Petty, and Wayne Lyles (Courtesy, Vernon Riddle)

Golighty, Gray, and other broadcasters worked to keep spirits up during this time of uncertainty. After the war "Cousin Bud" continued his live radio shows on WSPA. In the early 1950s WORD broadcast a live "hillbilly" radio show called "Dixie Jamboree." Although the show was aired only for a short time between 1952 and '53, many talented textile workers would line up to be auditioned for a short, featured spot on the show. Groups like the Double X Boys, Jack and

Roland, and Midnight Ramblers played "hillbilly music" on the Saturday night broadcast. Local mill folks tuned in from all over the county to hear the "Jamboree."

Dances were also popular in some of Spartanburg's mill villages. Although some families did not dance for religious reasons, all enjoyed the music and would go and listen. Vernon Riddle, a musician and past resident of Glendale, said that when he was a boy, his mother did not allow him to go to community dances. "She didn't think dancing was right, so she'd let me stand outside and listen to the fiddle music. It was good music," he remembered.

Thomas "Ted" Brackett of Inman Mills said his family were farmers by trade. Yet when he was a young man the lure of steady work, along with an easier life, led him to the cotton mills. At 10 years old in 1927 he began learning to play fiddle. His younger brother Leonard played the banjo. The brothers played at house parties and square dances in mill towns and farmhouses around the northern part of the county and were very popular. Ted smiled and remembered, "We'd play *Old Joe Clark, Cindy, Bile Them Cabbages Down,* things like that, and everybody would dance." The Brackett brothers continued to play together well up into their seventies. They were always the featured band at the annual Spartan Mill reunion dance held up until the late 1980s.

Religion also played a significant part of musical pleasure in cotton mill villages throughout the county. Early Sunday mornings, textile mill churchgoers sang out hymnal favorites at churches throughout the Upstate. Children were encouraged to participate in the music. Vernon Riddle was often asked to sing at Glendale Baptist Church when he was a boy. Being a little shy, he would always make up an excuse not to sing. It was his mother who promised him $10 if he would sing for the congregation. Vernon recalled, "That was good money in 1952, the first time I got paid for singing." As a teenager Gene Franklin played the "old-time pedal organ" during services. One of his favorite songs was *What a Friend We Have in Jesus.*

Many textile mill churches began with very little in the early years. Choir robes, hymnals, and even the existence of a piano would have been rare. When Bessie Holland was a teenager, she sang in the choir at Glendale Baptist Church. "We didn't have choir practice; we'd go and sing and that was it," she said. "Didn't have a piano then—we'd just follow the song director."

Textile churches played active roles in the community, and music was a focal point of many gatherings. On some mill hills in the early 1950s families would meet at fellow workers' homes and "put up loudspeakers" and have what was called "a singing and preaching." Vernon Riddle remembered, "We'd sing on Saturday nights, and it was loud! Families would sing, one person would sing. We would use guitars, or just sing without any instruments. It was like a camp meeting."

In Spartanburg County cotton mill folks loved their music. Whether it was homemade, church-made, or radio-produced, their music helped shape a part of who they are today. While some played music, others danced to the tunes. Many used music for worship or listened to it on battery-powered radios. Workers whistled simple melodies while walking down dusty mill streets. Mothers hummed sweet tunes rocking their babies to sleep. All this music defines our hard-working textile mill folks and their past.

Clipboards and Stopwatches
"Scientific Management" Comes to the Mills

by Toby Moore

The end of World War I prosperity heralded tough times for mill villages in Spartanburg County and across the South. Falling profits and stagnant or declining wage levels beset the industry even before the Great Depression: the gap between Northern mill wages and Southern wages, which had narrowed to 28 percent by the close of the war, ballooned to 65 percent by 1924. The industry responded by adopting a new and hotly contested strategy, one that would forever change the villages and the millhands who lived in them. Modern scholars call it scientific management or Taylorism, but angry mill workers at the time had another name for it: "the stretchout."

Scientific management, or Taylorism (named after a founding father, Frederick Taylor), attempted to break individual jobs down to a series of carefully timed motions. "Time and motion men," equipped with clipboards and stopwatches, swarmed into Southern cotton mills in the '20s and '30s, trailing millhands as they tended to their spinning frames and looms and dissecting each step of the production process. Supervisors used these studies to boost workloads, coordinating manual labor with the new, faster and more reliable textile equipment. In the place of the traditional rhythms of the turn-of-the-century cotton mill—where mothers tended babies as well as machines and frequent shutdowns relieved the tedium of long hours—modern mill managers envisioned an efficient and orderly assembly line in which workers were as interchangeable as the machines they tended. Mill managers argued that such measures were the only way they could survive without cutting wages even further.

Southern cotton mill workers, however, did not take such a sweeping change to their work lives lightly. The additional workload, they argued, was not matched by additional pay; as the Depression set in, wages were often actually cut as the number of machines assigned to each worker increased. In 1929, a series of strikes erupted. At Marion, North Carolina, "special deputies" opened fire on strikers at Baldwin Mills, killing six and wounding 25 more. In Elizabethton, Tennessee, 800 troops encamped to break a strike and quell the escalating violence. The notorious Communist-led strike at the Loray Mill in Gastonia, which led to the shooting of the Gastonia police chief and the revenge killing of a striker-balladeer, had its roots in the stretchout. In South Carolina, so-called "leaderless" strikes spread to involve nearly 80,000 workers; at Ware Shoals, 91 National Guardsmen, 12 constables, and two machine gunners were sent in to break a United Textile Workers strike.

In the early 1930s, men at Tucapau took a course in Practical Loom Fixing, sponsored by Furman College. The blackboard reads: "Poor Man's College." (Courtesy, George W. Moore Jr.)

The liberal magazine *The Nation*, long a critic of what it saw as the feudal structure of the Southern textile industry, sent a perceptive reporter named Paul Blanshard to cover the 1929 labor disturbances. Blanshard's dispatches depicted what he called the "strangest struggle that ever took place in the American factory." In Gastonia, "Communism and the stop-watch have brought to Southern cotton mills...a rebellion so charged with fear and bitterness that it resembles a civil war." In South Carolina, workers won strike after strike against the stretchout without much leadership of any kind, a telling omission that spoke to the millhands' perception that scientific management broke the rules that the mill owners themselves had written. "The presence of an outside labor leader would have challenged the philosophy of class paternalism which is the corner-stone of South Carolina life," Blanshard reported in May 1929.

Scientific management, whether it was more efficient production or more ruthless exploitation, represented far more than simply a change in production techniques. It represented a shift in the basic relationship between employer and worker. The Southern village system had been built on the use of personal relationships—real or claimed—in arbitrating supervision of the factory floor. The paternalism that had built the Southern textile industry relied not on written rules and personnel offices but on individual sets of obligations and rights between workers and management, workers and owners, and between workers themselves. Managers used favoritism, nepotism, social coercion, and reciprocal loyalties to keep the production lines running. Scientific management, however, sought to substitute for those sorts of relationships a more thoroughly modern contract between worker and management, one that rested on standardized workplace practices. By undercutting the family work unit and moving toward a system of substitutable employees, mill managers promoted a relationship that began at the factory door with the shift whistle and ended when the millhand left the building. Scientific management, in other words, was everything the mill village system was not.

In the end, the success of scientific management was ensured only by the reforms introduced in desperation by the New Deal and its Cotton Textile Code. Minimum wage laws, improved shop safety, the 40-hour work week, tentative steps toward the right to organize and the prosperity of the 1940s and 1950s paved the way for acceptance of what workers had once derided as the "stretchout."

The Negro Textile Leagues
Segregation and Athleticism
by Thomas K. Perry

Theirs was a legacy built in the shadows of white society ruled by Jim Crow laws, and it touched every part of life. Negro League teams endured both racial discrimination and "linthead" mockery. No mill team gained much notoriety, because town teams, both black and white, were considered more legitimate programs.

A significant rivalry developed between the Spartanburg and Greenville town teams in the early years, and by 1909 was a prominent feature of local baseball. The Hub City Gents showed off that year, winning 4-3 at home in front of 1,500 fans (July 5), and it wasn't even close the next day when the locals pounded out a 16-0 win. They closed out a perfect 8-0 season, sweeping Newberry 9-0 and 6-1 at Wofford Park (July 16), and beating Greenwood 4-3 later in the month.

The fortunes of the Negro Leagues paralleled those of the white teams. During the 1930s, the Spartanburg County Colored League was a solid organization of teams from the mills at Lyman, Arcadia, Converse, Clifton, Pacolet, and Drayton. Converse took the pennant in 1939 with a fine 11-3 record (5-1 in league play). Two years later, Arcadia was declared the first half winner while Pacolet claimed the second, but there was no mention in local newspapers of a playoff to determine a league champion.

Postwar years were also strong for both teams. July 4th festivities in 1946 featured the homestanding Pacolet Black Trojans against the Lyman Blues in the morning game, while the afternoon affair had the White Trojans play the Greenville Air Base. The shared field represented an unusual blurring of the segregated holiday celebrations, though no direct competition occurred between the black and white teams.

As textile baseball waned, there were still great performances. Pacolet's Willie Bailey hit for the cycle (single, double, triple and home run) in a win over the Lyman Blues (August 1, 1950). The next year, John Wesley Gossett of the Drayton Black Dragons provided his own brand of July 4th fireworks, tossing a 3-0 no-hit win over Jonesville's Black Tigers. Local fans were treated to barnstorming professional teams as well. The Philadelphia Stars and the Indianapolis Clowns of the Negro American League made a stop at Duncan Park (May 3, 1952), and added a bit more fun to the evening with an appearance by King

The Negro team in Pacolet, 1950 (Courtesy, Thomas Perry)

Tut, clown prince of black baseball.

At its beginning, the Negro teams of textile league baseball were only a generation removed from slavery; the close of the era nearly touched the passage of the first major civil rights legislation. Though segregated, fine teams and wonderful performances gave the Negro Leagues their rightful legitimacy. To have persevered, and to have been remembered, was but a measure of their greatness.

Blacklisting
Keeping Union Members Out

by Susan Willis Dunlap

Blacklisting was a common term in the 1930s that could be heard from mill village porches to Congressional hearings and was used to describe how some employers blocked the hiring of union-affiliated textile employees.

John Peel, president of the Greenville United Textile Workers, testified before the U.S. House of Representatives in 1936 on textile industry labor conditions. "We have hundreds of people in the States who are out of work and who have been out of work since the general strike (of 1934). They are blacklisted. But if you should ask me the question if I could produce a blacklist, I would answer by saying I cannot. But we know they have a blacklist. We have tried to get hold of it."

Workers certainly knew when they had been blacklisted. They often went from mill to mill and heard one explanation after another of why they could not be hired.

The National Labor Relations Board, formed under the Wagner Act of 1935, was the government's attempt to resolve the legal issues surrounding organized labor. In the First and Second Annual Reports of the NLRB, the Board reported their findings on allegations of blacklisting in several industries, including textiles. According to one of the NLRB's decisions, employers violated federal law "by blacklisting certain employees and letting it be understood they would not be reemployed." The Board ruled, "As the blacklist was based upon the union activities of the employees…the employer's action was the equivalent to a discriminatory refusal to reinstate."

Clyde Gilreath (Courtesy, Mary Willis)

Blacklisting took on a human face in the story of Upstate textile worker Clyde Gilreath (1906-1965), as told by members of his family. In his youth, Clyde was a talented basketball player, recruited by textile mills to play for their company teams. In 1925, the Piedmont Manufacturing Company won the Southern Textile League's "C" Division. Clyde held the trophy in the team photo.

Clyde courted Louise Porter in a Piedmont Manufacturing Company weave room. They married in 1927 and moved to Lyman, where Clyde made $22.50 a week in the new Pacific Mill. But after a disagreement with his Lyman boss, Louise said, "Clyde caught the P&N and left that day." Clyde moved his family back to Piedmont, where employment and another basketball team were waiting for him.

In Piedmont, Clyde was named the recording secretary for the local union. The union members met on Saturday mornings in the union hall over the grocery store. The employees of the Piedmont Manufacturing Company struck on May 28, 1934. The walkout included all 900 employees. In seven weeks the strike was settled, and all the employees were reinstated.

Clyde lost his job soon after he went back to work, and he never admitted to his family that he had been fired. According to Louise, the "company men" were carrying guns inside the mill because of the threat of flying squadrons. She explained, "Clyde was a weavin' and this other man takin' off cloth. And he (the other man) would have his gun, and he would lay the gun down when he'd go to take off a roll of cloth. With the looms vibratin', Clyde was afraid the gun would go off.

So he quit and come home."

Clyde's brother, Hovey (Mutt) Gilreath, had a different explanation. "When Clyde went back to work down there, why they got it in for him and they worked him out. They can't just come out and say, 'We're gonna fire you 'cause you done somethin' with the union.' They find so many things to fire you for, but the union was the main reason."

Now with four children, the Gilreath family lost their home in the mill village and moved to the country. Daughter Mary Lee remembered, "We would go pick cotton after we left the mill. We'd make a nickel in the afternoon. We'd have to pick enough cotton for Mama to buy school clothes for us. I can remember a cornfield that was close to the house where Mama would go over and literally steal corn. I mean, we were really living from hand to mouth."

Everyone helped the Gilreath family out as best they could. Hovey remembered that "We was workin' at Lyman…and every weekend…me and my wife, we'd buy a bunch of groceries and take 'em down there to 'em."

Clyde was unemployed for almost three years. "They blackballed him," Louise said. "One mill would send names to another mill and say, 'Don't hire this person.' Clyde went all the way down plumb as far as Columbia huntin' a job in the mill. He'd thumb and walk. He'd have blisters on his feet when he got home. He really tried to get a job back in the mill." Daughter Mary Lee was six years old when her father lost his job. "I think I really began to hate what it all stood for when Daddy was out of a job for so long. And I can remember him coming home at night and Mama saying, 'Well, any luck today?' And him looking so awful and disgusted, 'No luck today.' And day after day he would go out and try to find a job, and it was horrible."

Clyde paid a number of visits to S. M. Beattie, the owner of the Piedmont Manufacturing Company. Louise recalled, "Mr. Beattie told him, he says, 'I'm gonna try to get you a job, and I want you to go get it and stay on it, and stay out of the union.' He (Clyde) told him he would. So (Mr. Beattie) told him to go to Inman. Mr. Beattie paid his rent two months, 'til we could join him and get a house on the mill village."

At Inman Mills the better the job a worker had, the better his street address. The Gilreath family moved to the last house on "H" Street, the last street in the village. Louise recalled, "Clyde worked hard. He tied on warps. He'd perspire so they'd send him home to change clothes in the middle of the day." At Inman, Clyde seemed to have a change of heart about the union. Eventually, he was named director of personnel and moved his family to #8 A Street. In that position he authored company personnel brochures instructing workers why they should not listen to union offers. One of them, "A Conversation between Joe, the Employee, and the Labor Organizer" included this imaginary exchange:

> **The Organizer:** *Joe, how about joining the union.*
> **Joe:** *What for?*
> **The Organizer:** *We will see that you get a square deal.*
> **Joe:** *Nobody gets something for nothing. You want me to pay dues to you every month, to go into the plant and do my talking and thinking…You don't know anything about my job. You would just confuse and foul things up.*
> **The Organizer:** *You just don't understand what we are trying to do.*
> **Joe:** *I do understand, and that is the reason that you can take your union card and go somewhere else. You know, Mr. Organizer, you ought to get you a good job down here in one of our southern plants so that you could enjoy all the wonderful benefits that we do. You know, I don't believe that union is giving you a square deal.*

Cotton Mill Poetry
A Controversy Unleashed
by Doyle Boggs

Peter Richard Moody, Class of 1937, was one of Wofford College's outstanding graduates of his generation. He was a highly decorated Army Air Corps officer in World War II and then earned a Ph.D. in English at Cambridge University in England. He was a charter member of the faculty and later a vice dean at the Air Force Academy, retiring as a brigadier general in 1967, and then he moved on to Eastern Illinois University as its provost.

These later distinctions notwithstanding, Moody is still remembered in Spartanburg and by some textile historians as the author of a poem, "To a Cotton Mill Worker," that he wrote as a junior at Wofford. Before a 90-day controversy ran its course, his work had been read and commented upon by hundreds of people throughout Spartanburg County and the state of South Carolina. Moody still has a thick scrapbook that his mother assembled, as well as a detailed account compiled by his professor and mentor decades later. Still, in 2001, he looked back on the experience with a balanced perspective: "It seems to me that the story is more about how various people reacted to the poem than it is about me. After all, it was just a classroom exercise, and, on campus, we quickly put the incident behind us," he said.

Peter Moody was born in Dillon, South Carolina, on April 15, 1917. His father, a Wofford graduate, was a financial manager for a North Carolina textile firm, and the family lived in the Cooleemee mill village near Salisbury, North Carolina. Moody himself occasionally worked in the mill during vacations. Arriving in Spartanburg, the young freshman liked Wofford and its academic tradition. Moody was particularly drawn to a teacher of American literature and creative writing, Kenneth Coates, who had joined the faculty in 1928 at the age of 24 and liked to describe himself as a "flaming liberal."

Peter Moody was editor of Wofford College's student newspaper, The Old Gold & Black, in 1937. (Courtesy, Wofford College)

"One day, I walked into class and, without any comment, read a selection from Carl Sandburg and then Edwin Markham's 'The Man with the Hoe,'" Coates recalled years later. The latter poem was widely quoted at the time as a cry for better working conditions for agricultural laborers. He then told his students to write "about anything," imitating the style of either classroom reading. Responding to that assignment, Peter Moody penned, "To a Cotton Mill Worker, a la Sandburg." Coates wrote on the paper, "This is good! Show it to the editor of the *Journal*!" The professor shared the verse with some faculty colleagues and read it aloud to classes, but it attracted no special attention until a few days after it appeared in the April 1936 issue of the college literary magazine. Then, Coates speculated, a "townie" took a copy to his part-time job in a nearby mill office.

From there, "To a Cotton Mill Worker" spread from hand to hand and mouth to mouth. The editors of two weekly newspapers that competed for circulation among the mill workers were incensed. The management-subsidized *Textile Tribune* addressed Moody directly: "In your infamous slander, you include every cotton mill worker…You showed only too well how void your empty soul is of the traits of a manly man or a gentleman." Rumors spread that a mob would assemble, march to the campus, and "get Peter Moody." A night or two later, the young poet was studying in his dormitory when perhaps a dozen workers came onto the campus and tried to call him out. Eventually, after the authorities began arriving, the outsiders tired of the game and left, suggesting they might return. According to Moody, the confrontation did not really threaten violence, and he was not frightened. However, the Wofford administration decided to take precautions. The college augmented its night watchman with some more formidable security for a brief period. Dean A. Mason DuPré suggested to Moody that he take his best friend, Pickett Lumpkin, and go up to the mountains for a few days, which they did.

The Spartanburg *Herald* ignored the disturbance, as was its habit when racial or labor unrest seemed imminent. It did, however, chronicle business-as-usual news of meetings and choral performances at Wofford, suggesting that the campus was hardly under siege. Nevertheless, President Henry Nelson Snyder worked hard behind the scenes to defuse the situation. He exchanged polite letters with officers of the Central Labor Union, assuring them that the student writer would offer a public apology. Moody willingly signed a letter drafted for him by the administration. It was read at the next Monday night union meeting and appeared in several newspapers. The text read, in part, "Nothing was further from my mind than to condemn or typify the cotton mill man as I described this one. For having worked in a cotton mill and lived in mill villages, I know them well as a class—hardworking, honest people, and worthy of respect."

The crisis might have ended with Moody's act of contrition, but it was an election year. State Rep. William Fred Ponder, who had entered Wofford with the Class of 1937 and attended the college for a year, was running in the Democratic primary for Spartanburg County's one seat in the State Senate. He needed the textile vote. On April 22, on the floor of the House in Columbia, he spoke out: "Peter Moody is undoubtedly either insane or criminal…[Textile workers] resent the danger that said Moody offers to such a fine, Christian institution as he has been allowed to enter." Ponder offered a resolution directing the superintendent of the state mental hospital to dispatch a recognized psychiatrist to Spartanburg to "examine" Moody and publicly report the findings.

Ponder's resolution actually passed the House, and it remains to this day in the official record of the South Carolina General Assembly. The hospital's superintendent, Dr. E. L. Hoerger, a friend of the college who was well acquainted with the Moody family, duly showed up in Spartanburg. Though both Snyder and Coates advised Moody he did not have to submit to the evaluation, the student "thought it might be an interesting experience" and spent a pleasant half-hour with the doctor, responding to questions not very germane to the poem or his sanity. Subsequently learning that Hoerger was prepared to report in polite terms that "Moody has more sense than the members of the Legislature," the lawmakers quickly voted the whole matter to the table. They were too late, however, to avoid embarrassment. Regional, university, and national media filled the Moody family scrapbook with columns that cited Ponder's antics not only as a prime example of what passed for statesmanship in the South, but also as a genuine threat to free speech and academic freedom. On May 23, 1936, President Snyder finally felt compelled to enter a debate in the letters column of *Time* magazine with a rather lame defense of Wofford's management of the situation.

Peter Moody
(Courtesy, Wofford College)

Back home in Cooleemee, mill officials suspended Moody's father for several days, allegedly because they could not guarantee his safety among the workers. Cooler heads finally prevailed as the company's need for its manager's expertise became obvious. After completing ROTC summer camp in Alabama, the younger Moody came back to Wofford for his senior year. He edited the campus newspaper and was a finalist in the competition for a Rhodes scholarship. Coates long remembered Moody's commencement address, "full of deep affection for the college." Ponder lost his election but became a capable post-war South Carolina Commissioner of Labor. He died in 1971.

What is to be made, 65 years later, of this bizarre episode? In his official Wofford history, published in 1951, David Duncan Wallace suggested that the incident showed "without intention or anticipation, how the feelings of class prejudice may be suddenly or dangerously aroused." In his 1998 volume, *A Fabric of Defeat*, Bryant Simon opined, "Moody's savage picture wounded mill people…South Carolina lintheads indicated that they would no longer be passive victims of prejudice."

However, hidden in Coates' manuscript is an even more revealing theme. Throughout "New South" history, criticism of the region's peculiar institutions from any source was greeted by a highly personalized hostility to be feared or to be exploited. As this particular controversy raged, Coates claimed he went to Snyder's office for a talk. "I told the president that I had no idea that the poem would get twisted around in its meaning, as it had been. He replied, 'Well, you are a fool.' I could not and did not deny the allegation."

To a Cotton Mill Worker

(Free Verse a la Sandburg)

by Peter Moody

Your shoulders are humped and your head is bent; your dull dead eyes are spiritless and your mouth is just a hard straight line in a yellow face under the blue lights in the mill.

You are diseased and unhealthy looking, standing there in your faded overalls, with one suspender loose. Your voice is cracked and your throat and lungs are lined with cotton.

Every night the whistle blows and you plod home to swallow your bread and beans, comb the cotton from your straggly grey hair, wash your wrinkled face, and then lie down on your hard, unclean mattress until the whistle's blast calls you back to your machine in the mill.

You are narrow-minded and ignorant, you with your six years of schooling. And you are afraid, afraid of your bosses, afraid of being laid off.

You are desperately frightened by knowledge. Therefore you shun it and are content to stay a coward.

Recreation for you is in talking baseball and in seeing, on Saturday nights, some cheap Western movie full of guns and rope and horses and fights. And your pleasure is wasting your nickels in the drugstore slot machine.

Listen, lint-head—you are just another poor, illiterate cotton mill worker. You stand with a thousand others just like you for five days a week, eight hours a day, running and watching and nursing and tending a power loom, all for forty cents an hour.

What do you know about life?
What do you know about music?
What do you know about art or literature?
What do you know about love?
What could you know about anything?

You are dead!
You died on your sixteenth birthday when you went
to work in the cotton mill.

—Not all verses have been reproduced here.

Crowning a Queen
Textiles-Go-To-War
by Gary Henderson

A crowd estimated at 10,000 people gathered at Duncan Park Stadium on a warm June night in 1943 to crown Spartanburg County's "Cotton Textile Queen" and welcome home native son James F. Byrnes, the director of the country's war mobilization. The "Textiles-Go-To-War" event was staged to build *esprit de corps* of workers as they raced to keep production high for the military. Spartanburg County's contributions to the nation's war buildup were many because of the large number of textile mills that produced goods for the government. *Life* magazine editors recognized the importance of the mills, running a story and a series of pictures about the big show the week following the event.

"Lovely Nellie Maude Lanford" wearing her crown during the Textile Goes to War program. (Courtesy, Walton Beeson)

Nellie Maude Lanford, an 18-year-old textile worker's daughter representing Mills Mill, won the cotton lace crown and a $100 war bond and was one of several young women whose photographs appeared in the national magazine. Another of those was Mary Lane, who still believes a stumble on that stage might have cost her the title. "I turned the wrong way and about walked off the stage," Lane said as she laughed about the incident, 58 years later. Even so, being second best in the "Textiles-Go-To-War" contest gave Lane enough memories to last her a lifetime.

"It was a big thing just to get into town," said Lane, who grew up in a family of nine children in a mill house in Saxon. "There I was on stage." For years, Lane carried a neatly folded copy of the *Life* magazine story in her wallet. Lane, who took a job in the Saxon Mills spinning room so her family could live in a company house, was one of 25 women who represented each of Spartanburg's textile plants in the pageant.

"Mr. Law came in there and told me they were going to put me in the contest for Saxon," Lane said of mill owner John Adger Law. "I grew up very sheltered.I remember my aunt sneaked me into town one time to the movies to see 'The Snake Pit.'" But after she was picked to seek the crown for Saxon, Lane went to town in style. "They picked me up in this big car, drove me to

Mary Lane was runner-up. (Courtesy, the Spartanburg Herald-Journal)

town, and bought me a nice dress to wear in the pageant," said Lane, who can't recall the shop's name but said it was a "fancy" place. "I'd never even been in that store." Lane left the mill and her family moved from the mill village a couple of years later, after her stepfather died.

The war's signature could be seen all over Spartanburg in 1943. Newspaper headlines heralded news about the latest battles, textile mills changed their looms to turn out military materials, and Camp Croft soldiers from the base east of Spartanburg crowded downtown streets. "My stepfather would go out to the base and pick up a few of the soldiers and bring them home for Sunday dinner," Lane said. "But he wouldn't let them date his daughters. He fed them and took them home."

Still the camp's soldiers played a big role in Lane's life the night of the beauty pageant. The evening ended with the Cotton Ball Dance at the USO Center on North Church Street. Lane said lots of famous people were there. She remembers two: World War II pinup Betty Grable and comedian Zero Mostel, who went through basic training at Croft.

Lane's sister, Lottie Ravan, 73, remembered going with her big sister to the pageant and dance. "I always thought Mary was pretty," said Ravan, who was 14 at the time. "It was a big thing to me, something that did not happen to regular people, and all. I was very proud." Ravan said it seemed like the whole town turned out to see the pageant. "I was so excited because the Vox Pop Radio Show was broadcasting from out there," Ravan said. "But what I really remember is Mary's pretty clothes they bought for her." Ravan said she and her sister didn't dance at the USO, because her stepfather didn't allow it, but they had a good time anyway.

As with many Americans, Memorial Day is important to Lane. Her first husband died from an illness related to his time on the warfront. "To live through all that back then, and to see what they went through," Lane said, her voice trailing off. "People rallied toward them then."

"I tell you what I will remember about all this," Lane said, holding the photographs taken at West's Studios in Spartanburg. "It was a big deal, and it changed a young girl's life."

—Adapted from an article in the Spartanburg Herald-Journal. *Reprinted by permission.*

Women's Softball
A Fast-Pitch Life
by Karen L. Nutt

In the late 1940s, Spartanburg's textile mill softball league attracted the attention of scouts from across the country in search of players ready to play in a league of their own.

Female players, that is.

Shirley Wilson, a star catcher and clean-up batter for the Beaumont Lassies and the Drayton Darlings, was called up to play in the All-American Girls Baseball League. That league, which inspired the movie "A League of Their Own," had been created a few years earlier to fill a void in professional baseball when many baseball players were called up to serve during World War II. Wilson, then 14, had to turn down an offer to play because her parents were concerned that she was too young to be away from home for a long period of time.

At home, the women's fast-pitch games between the mill teams were the talk of the town. Top games sometimes drew hundreds of fans, many of whom would climb trees surrounding the stadium when the stands at Berry Field (located along the Asheville Highway north of Spartanburg) were packed. "Berry Field was full every night when we played," recalled former Beaumont pitcher Betty Poteat. "It was a very popular sport. There were not a lot of cultural activities to do in Spartanburg then."

Paying 15 cents a ticket, spillover crowds at Berry Field were common when Mrs. America, Fredda Acker, was on the mound—not to throw the ceremonial first ball, but to throw a lot of balls as she was a pitcher for C. M. Guest of Anderson.

One of the biggest local rivalries involved Beaumont and Drayton. "We had a thing like Clemson-South Carolina. It was very competitive," said Poteat, the only female player inducted into the South Carolina Fast-Pitch Hall of Fame (as of this writing). One of Poteat's favorite memories of her softball-playing days is when her Beaumont team beat Drayton for the state championship in 1947. Before then, Drayton had won a string of championships. Poteat was 13

The Drayton Darlings participated in the Southeastern Regional Tournament in the late 1940s, taking a side trip to Silver Springs, Florida. (Courtesy, Shirley Holmes)

at the time while most of the players were closer to 20.

At 15, Poteat accepted an offer to play professional women's fast-pitch. The offer came from the Chicago Match Queens, which had sent a scout to the Spartanburg mill fields. The scout convinced the principal of Poteat's school to call her out of class one day to discuss a possible career on the mound. Leaving her family, living with the team's owner and his wife, and leaving the Hub City to play in the Windy City netted her "more money in a week than my parents made in a month."

Lucrative as it was, it meant playing every day in the summer, starting school late in the fall and quitting school early in the spring, all of which were monumental sacrifices for a teen-ager. "You can't keep traipsing up there every summer, but I wouldn't take anything for the experience," Poteat said of her single season in big league. When she returned to Spartanburg, however, she was prohibited from playing in the mill league for two years in order to regain her amateur status.

Betty Brown Holder recalled the more unpleasant aspects—wool uniforms and metal cleats—and the scraps and scrapes that sometimes accompanied the games. "You could be rough. We carried a lot of skinned-up knees and elbows," Holder said. Wilson remembered when the Women's Army Corps (WACs) from Columbia came to Berry Field and a fight broke out after a WAC base runner knocked Drayton's first baseman off the bag and repeated the move against Drayton's second baseman. After getting up, the second baseman hit the WAC base runner.

"They started swinging," Wilson said. "I started to run, but the umpire held me straight up in the air and said, 'You don't need to be that.' He knew I was just a little kid." Nevertheless, the coaches tried to instill good character and sportsmanship, according to Holder. "They took it pretty serious. They didn't want you slouching on the field. If you appeared not to have your heart in it or you didn't want to perform, they didn't let you play," Holder said.

Many of the mills had their own fields for the practices, often held two or three times a week. Poteat spent extra time off the field practicing her game-winning pitches at home, using an old tire as a target.

—Adapted from an article in the Spartanburg Herald-Journal. *Reprinted by permission.*

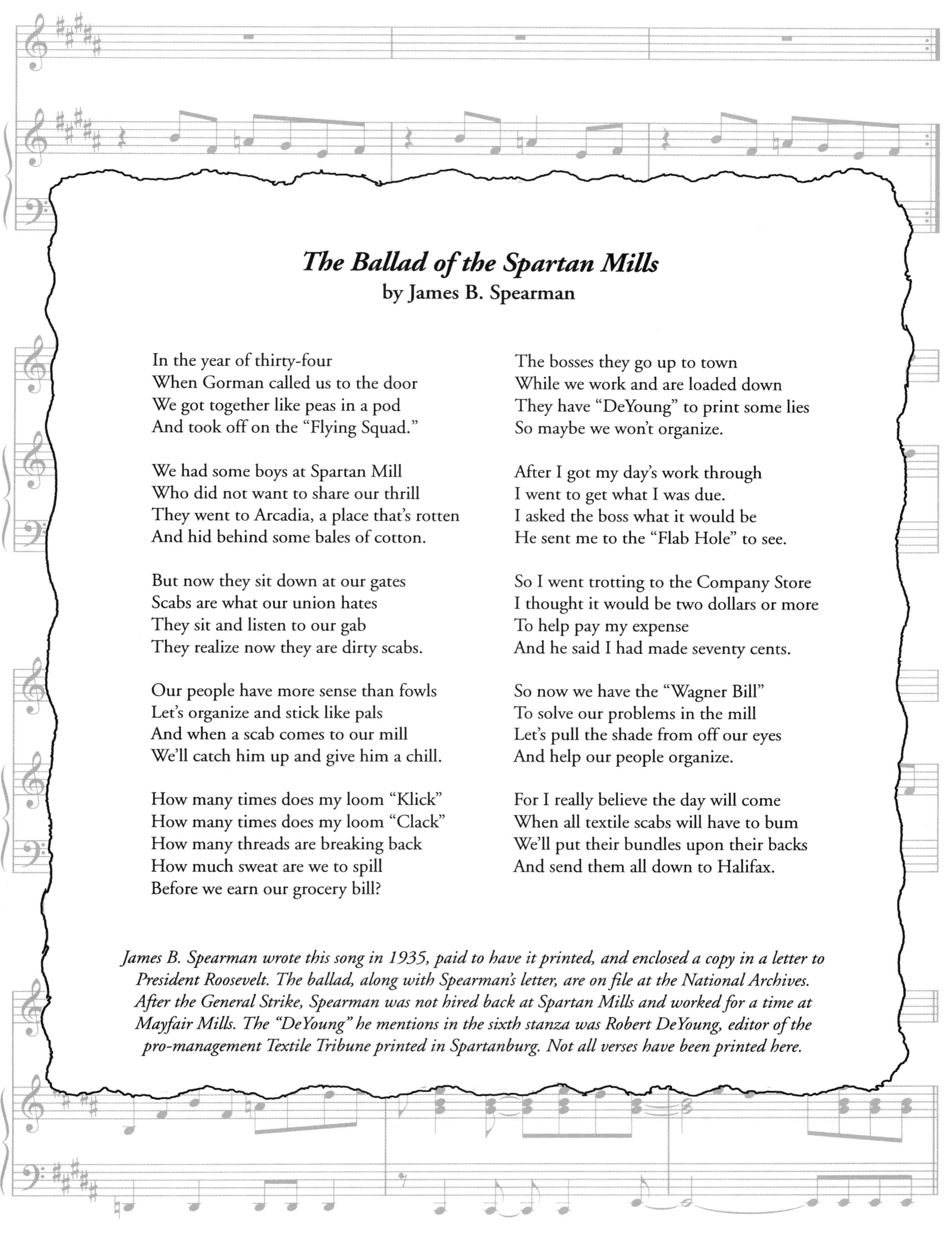

The Ballad of the Spartan Mills

by James B. Spearman

In the year of thirty-four
When Gorman called us to the door
We got together like peas in a pod
And took off on the "Flying Squad."

We had some boys at Spartan Mill
Who did not want to share our thrill
They went to Arcadia, a place that's rotten
And hid behind some bales of cotton.

But now they sit down at our gates
Scabs are what our union hates
They sit and listen to our gab
They realize now they are dirty scabs.

Our people have more sense than fowls
Let's organize and stick like pals
And when a scab comes to our mill
We'll catch him up and give him a chill.

How many times does my loom "Klick"
How many times does my loom "Clack"
How many threads are breaking back
How much sweat are we to spill
Before we earn our grocery bill?

The bosses they go up to town
While we work and are loaded down
They have "DeYoung" to print some lies
So maybe we won't organize.

After I got my day's work through
I went to get what I was due.
I asked the boss what it would be
He sent me to the "Flab Hole" to see.

So I went trotting to the Company Store
I thought it would be two dollars or more
To help pay my expense
And he said I had made seventy cents.

So now we have the "Wagner Bill"
To solve our problems in the mill
Let's pull the shade from off our eyes
And help our people organize.

For I really believe the day will come
When all textile scabs will have to bum
We'll put their bundles upon their backs
And send them all down to Halifax.

James B. Spearman wrote this song in 1935, paid to have it printed, and enclosed a copy in a letter to President Roosevelt. The ballad, along with Spearman's letter, are on file at the National Archives. After the General Strike, Spearman was not hired back at Spartan Mills and worked for a time at Mayfair Mills. The "DeYoung" he mentions in the sixth stanza was Robert DeYoung, editor of the pro-management Textile Tribune printed in Spartanburg. Not all verses have been printed here.

Cloth inspection, Inman Mills, in the 1960s.
(Courtesy, Inman Mills)

5 Textile Town Settles In 1950 to 1974

by David L. Carlton

THE POST-WORLD WAR II YEARS for Textile Town were in many ways much like those that went before. The textile industry entered the postwar era as Spartanburg County's dominant employer; in 1949, 20,000 workers, of whom over a quarter were women, ran the machinery. Despite the movement to sell mill-owned housing, and the increasing importance of commuting workers in the industry, nearly 28,000 Spartans continued to live in mill villages. The textile workforce fluctuated from 17,000 to 20,000 (depending on who was gathering and defining the numbers) throughout the quarter-century after 1950. For many thousands of Spartans, the life of work continued to be enveloped in what were by now traditional routines: doffing, filling batteries, tying ends, and fixing smashups in rooms that were noisy and dusty, but also full of familiar faces, people sharing a common experience and with common assumptions about how best to worship God and fill one's role in family and community. Children now spent more time in school than before, and ever-larger numbers availed themselves of new opportunities for technical training and college educations; but generation continued to follow generation into the mills, relying on them to help finance marriages and children, church donations and (increasingly) car and house payments. If early mill villagers were, as some historians have put it, "like a family," their employers as late as the early 1970s boasted that work in their plants was a "family tradition."

But behind the superficial continuity, mill people—workers, managers, and owners—were undergoing wrenching transitions. Economically, the mills lurched between prosperity and stagnation; though benefiting at first from pent-up wartime demand and American dominance in the world, the old problems of the 1920s and 1930s came back to haunt them, along with new difficulties arising from the new, open, global economic order. By the 1960s prosperity was returning, though perhaps for the last time in the history of the industry, and significant new mill construction was undertaken for the first time since the 1920s.

But the textile industry was now operating in a changed environment. A county dominated by small family-owned firms had seen some mills

The one-story John H. Montgomery Plant opened in Chesnee in 1965. (Courtesy, Walter and Betty Montgomery)

> ***"I heard that durned ol' cotton mill every night. Every night it'd make a racket all night long just like you was down there in it. I told Preacher I can't stand that. Let's get out and buy us some land and build us a house, and we did."***
>
> **—Lunette Owens**

The Furious Fifties

Local newspaper columnist Bob Craig describes "the Furious Fifties" in Spartanburg in this column that appeared in the Spartanburg **Herald-Journal** ***January 1, 1960. The textile mill culture was gradually being replaced by a more cosmopolitan attitude.***

Whenever a town acquires a Toddle House and a Howard Johnson's, you can stop calling it a town and call it a city.

The Furious Fifties will be remembered as the decade when Spartanburg did just that. In 1950 it lay asleep on the threshold of a new decade. In the ten years that followed, it awoke, yawned a bit, fretted, then blossomed into a bustling industrial city at the very Crossroads of the New South.

In short, during the 1950's we arrived.

Not since the Roaring Twenties have Spartans had so much to laugh about, talk about or brag about.

We will remember these ten years as the one during which we learned to use dial telephones, voting machines and barbecue grills.

This was the decade we got our own Sears Roebuck and Penney's. We got branch post offices, shopping centers, suburbs, commuters, traffic jams and green stamps.

The Furious Fifties will be remembered as the decade when we rebuilt everything in sight: hospitals, government buildings, schools, churches and even sewer lines.

In 1950, only the folks in the city of Spartanburg had water, fire protection and higher taxes.

By the end of the decade, just about everybody in the county was enjoying or was about to enjoy all the advantages of city living: water, fire protection and higher taxes.

Ten years ago, most Yankees who had heard of us at all knew us as Spartan*Sburg*, that town just this side of Greenville.

Today, more and more of them are dropping the excess "S" in the name. And more and more, they are beginning to think of Greenville as "that town on the other side of Spartanburg."

There is hardly a spot in the county where you can stand without seeing a new school, a factory, a gum wrapper or a beer can...

engulfed into large integrated corporations headquartered in New York. The new mills no longer used cotton exclusively, but blended it with new fibers made from petroleum in giant new plants, including some placed in Spartanburg by international chemical combines. The new plants were windowless, air-conditioned bunkers without smokestacks, cleaner but less romantic than the old multistory brick mills with their brilliantly lit blue windows. Moreover, they lacked villages; sited on the fringes of small towns such as Chesnee or Inman, or along the new Interstate Highway system, they relied for their labor on commuting workers, while the old villages began to slip into decay. They increasingly had to compete for their labor as well, as industrial diversification and broadening educational opportunities brought workers attractive alternatives to life in the mills. For one group of Spartans—African Americans—the Civil Rights era brought new opportunities, as a combination of federal muscle and labor-hungry employers opened mill jobs previously denied them. White workers responded to the integration of mill work with surprising equanimity, partly because the link between textile employment and community life had been severed, partly because of the proliferating new opportunities of the 1960s. Most striking of these new opportunities were those resulting from a new international presence in the Spartanburg industrial scene—a presence that would become ever more important in succeeding years. But the new prosperity steadily reduced the hold of textiles on the working lives of Spartans; by the 1970s, for the first time in the industrial age, the industry no longer accounted for the majority of manufacturing employees in the county. Its declining importance, combined with rising global pressures, began to squeeze Spartanburg's mother industry in a tightening vise, sending early signals that Spartanburg's days as Textile Town were numbered.

AT FIRST, IT SEEMED, THE GRIM YEARS of the Great Depression were past. Not only had World War II sopped up extra capacity, but the immediate postwar years saw unexpected prosperity; Spartanburg's cotton mills poured out goods to help satisfy the demands of Americans long deprived—first by mass unemployment, then by rationing—of the consumer goods that they craved, and to meet the needs of a war-devastated world.

The beginning of the Cold War, especially of the Korean conflict, kept military demand high. Prosperity was not limited to owners, either; workers were kept fully employed, and wages, while still low both by national standards and by those of other industries, rapidly converged on those standards.

Yet by the 1950s it was clear that some of the problems that had made textiles a "sick industry" even before 1929 had not gone away, while new difficulties had arisen to supplement them. Textile men had long been concerned about competition from the Japanese, who had made serious inroads into East Asian export markets even before Japan launched its imperial expansion. In the 1930s, moreover, Japan became for the first time a threat to American domestic markets. Now, in the 1950s, that threat was redoubled. Their old plants ravaged by the Pacific War, Japanese manufacturers quickly rebuilt with modern equipment, and in the 1950s launched a carefully targeted strategy of capturing selected American markets for coarse goods.

Local manufacturers, too, modernized, but not at the same pace as the Japanese. More seriously, they now found themselves in the unaccustomed position of being beaten at their own game. Their old competitors in New England sliding into oblivion, they found themselves no longer the low-cost producers; the Japanese could pay much lower wages, and, relying primarily on India for their cotton, did not face the government-supported raw materials prices paid by American producers. While these difficulties did not bring a return to Depression-era conditions, they did bring stagnation; from time to time on early mornings, white workers would join the African-American laborers who traditionally gathered on the courthouse lawn waiting for casual employment.

What could the industry do? One popular solution was consolidation. Spartanburg County's textile industry had long remained a bastion of the independent, locally-and-family-controlled firm; as late as the 1940s such families as the Montgomerys, Cateses, Chapmans, Converses, and Laws continued to dominate both the local mills and the business leadership of the community. To be sure, these families had close ties to Northern textile

Volunteers prepare bags of fruit and candy for 1,100 Glendale employees at Christmastime. The two men on the right are Clarence Crocker and James Jack. (Courtesy, Clarence Crocker)

> *"I am talking about a dead industry, dead as a door nail, dead as the passenger train, and dead as the country doctor."*
>
> **—Sen. John D. Long, discussing foreign imports at a Democratic dinner in 1970**

interests, especially the New York commission houses that marketed their products; Arcadia (after 1934 Mayfair Mills) was effectively controlled by the Dent family of the New York house of Joshua W. Baily and Company, and the Montgomery interests were so intricately entangled with those of the Milliken family of New York as to make them indistinguishable from each other. But generally the textile industrialists and the Spartanburg elite were closely intertwined; the welfare of the one group was that of the other. The one major "foreign" presence in the county was the Lyman Printing and Finishing Company, built in the 1920s by Pacific Mills, which in the 1950s came under the control of the New York converter M. Lowenstein and Company. Until the mid-1960s this huge plant was the largest employer in the county, and its enormous roof-top fluorescent "Pacific Mills" sign was a memorable nighttime landmark for travelers along the "Dual Lane Highway" between Spartanburg and Greenville.

The World War II era saw a wave of consolidations in the textile industry, as firms scrambled to position themselves to bid on government contracts and, later, diversify their offerings and gain control over marketing. In 1946 three Greer mills—Greer, Victor, and Apalache—were swept up in a massive merger with the commission house of J. P. Stevens and Company. Another New York commission house, Reeves Brothers, had long controlled the Mills Mill in Woodruff and the Fairforest Finishing Company west of the city along the North Tyger River; in 1945 it purchased Saxon and Chesnee Mills from John A. Law. More importantly, in the late 1940s the Montgomery and Milliken interests disentangled their holdings—the Millikens taking the Pacolet complex and Drayton, and the Montgomerys Spartan, Beaumont, and Startex, the former Tucapau.

Thus by the early 1950s "northern" interests had a firm foothold in Spartanburg's textile industry. But the industry retained its local flavor, incorporating the "outsiders" into the community and using the new connections for its own advantage. In 1947 a younger member of the Dent family, Frederick B. Dent, moved to Spartanburg to run the Mayfair operation, and soon became fully part of the local elite, as well as a leading figure in American business and public life. Reeves Brothers ultimately transferred its administrative headquarters to Spartanburg, where it remains today.

Moreover, the new, vertically integrated companies opened up new opportunities for a generation of postwar Spartans to penetrate the North. The old single-plant firms had required minimal management, typically located in a small, house-like building in front of the mill. Now, however, cotton-mill clerks with high-school educations were riding a postwar escalator into management positions. The rough-hewn, newly-minted Spartanburg textile executive on his first trip to New York became the butt of jokes as far away as Charleston–and some of the best jokes were those the Spartans told on each other. But for the shrewd and energetic, the new firms opened up opportunities and stimulation inconceivable just a few years before; the children of loom fixers and moonshiners were dealing with Jewish and Armenian marketers on Worth Street and Sixth Avenue, and spending their Saturday nights in Greenwich Village night clubs. Spartanburg was beginning to open to the world.

But the major new addition to Spartanburg's industrial community was Roger Milliken. The grandson of Seth Milliken, who had played such a pivotal role in providing entree into the industry for John H. Montgomery and other Spartanburg industrialists, Roger Milliken had become president

Walter Montgomery Sr., top left, with some of his employees: Harry Blankenship, Jean Tillotson, Maggie Satterfield, Hal Flynn, Dewey Flynn and Josh Woodruff. (Courtesy, Louise Foster)

An aerial view of Pacific Mills, 1965 (Courtesy, the Spartanburg Herald-Journal)

of the family firm, the commission house of Deering Milliken and Company, in 1947. In 1954 Milliken moved to Spartanburg—a move that likely ranks as the single most important in the history of Spartanburg's textile industry, if not indeed of Spartanburg itself. As head of what would become one of the world's largest privately-held manufacturing companies, and already possessed of an ample fortune, Milliken could have become one of the legendary business figures of the latter half of the 20th century had he made his career in a different, more cutting-edge industry—and had he not been such an intensely private man. To the cognoscenti of American business, he soon became legendary enough. A staunchly conservative Republican with a Yale education, he served as financial angel to the conservative intellectual movement of the 1950s, notably by bankrolling William F. Buckley's *National Review*. In the 1960s he became, in a controversial phrase, the "Daddy Warbucks" of the South Carolina Republican Party and a pivotal supporter of Barry Goldwater's 1964 presidential candidacy. In later years he became an ardent (some would say obsessive) opponent of economic globalization.

While politically he differed little from other textile industrialists, his impact on Spartanburg and its textile industry was epochal. Milliken was a true visionary who recognized that the southern textile industry could not long endure if it continued to rely on its traditional formula for success: bulk production of standardized goods using traditional methods and technology, but at lower cost than competitors from other regions. Those competitors had effectively been vanquished; as the Japanese challenge had shown, the "southern strategy" now had to face effective competition from abroad. Since Deering Milliken and Company was originally a marketing firm, Milliken was well aware that the advantage of American firms lay in their ability to create specialty fabrics tailored to the specific needs of their customers and to respond quickly and flexibly to changing demand.

He also recognized that key to such a strategy was a focus on research. The textile industry, and the southern textile industry in particular, had historically paid little mind to research, being

By 1961, Southern textile plants were responsible for about 89 percent of the nation's total textile production. The South had over 18 million spindles in 1965 compared to only 756,000 in New England.

—Author Timothy Minchin

Aerial view of the Milliken Research Center (Courtesy, Milliken and Company)

content to adopt the fruits of work done in other industries such as machinery and chemicals. Milliken, though, had the sophistication, the will, the marketing power, and above all the deep pockets necessary to launch a major private research initiative. The Deering Milliken Research Trust was launched in 1945; a small operation at first, it was housed alternately in Pendleton, South Carolina, and Stamford, Connecticut; by 1958 it boasted a staff of some 80 scientists and technicians and held 130 patents for improved textile machinery and innovative fibers, yarns, and fabrics. It was in 1958 that Roger Milliken moved the operation, renamed Deering Milliken Research Corporation, to its enormous campus on the northern outskirts of Spartanburg at the junction of U. S. Highway 29 (soon to be Interstate 85) and the North Pine Street expressway, then under construction. Spartanburg had never before seen the likes of the complex, with its minimalist architecture and elegant grounds; nor had it ever before dealt with the influx of scientists

Milliken Innovations 1950s & 1960s

Visa: A patented polyester fabric that doesn't wrinkle and releases soil easily in the wash. Used in table linens, industrial uniforms, and carpet. Initially made with radiation, later with a resin treatment.

Carpet tiles: Squares of carpet blended with a vinyl mixture on their base so that the weight of the fabric keeps it in place without glue.

Agilon: A stretch yarn used in hosiery and carpet.

Millitron dyeing machine: A computer-controlled machine that allowed patterns to be printed with exact color patterns and endless variation.

and executives it brought. There was some local resentment at the Ivy-League tone of the people with whom Milliken surrounded himself, and the tendency of some to isolate themselves from the rest of the community. But Milliken himself set a modest tone, settling in an unpretentious house in Converse Heights, and began quietly and privately to help reshape the economy and the life of the city. Among his first moves was to instigate the creation of the Greenville-Spartanburg Airport, which on its opening in 1962 brought direct jet service to New York.

CONSOLIDATION AND A NEW STRESS on research—these were two ways of dealing with the impasse the textile industry was finding itself in. Another strategy was to modernize the plant. Wartime and postwar profits permitted many firms to go on a buying spree to replace machinery that had not been updated before the Depression—a purchasing binge that spelled prosperity both for Spartanburg's small machine and metal-working shops and for the big Draper loom foundry on South Pine Street, which added 600 jobs in 1960. The new and improved machines allowed for higher speeds, lower labor costs, and better quality control. Electronic monitoring of machinery began to appear in the late 1950s, signaling broken ends that formerly had to be spotted by eye and eliminating the loud "bang-off" on an automatic loom. Materials flows were mechanized. Cotton mills met the competition of new synthetic fibers such as polyester by beginning to produce blended fabrics, combining cotton with specially designed "staple" synthetics in the opening room.

More strikingly, air conditioning was widely introduced for the first time. Not that it was new; Southern cotton mills were among the first adopters of this technology, and the very term "air conditioning" was coined by the early 20th century Charlotte mill engineer Stuart W. Cramer. The spread of air conditioning was slow, though; since southern mill buildings were designed to maximize ventilation and natural light, mechanical climate control was costly and, it seemed, dispensable. After World War II, however, equipment had greatly improved in efficiency, and wartime and postwar profits allowed previously strapped mills to make the investment. Moreover, improvements in electric lighting now allowed mills to dispense with natural light altogether. While air conditioning made working conditions more comfortable, and conditions for working fiber more controllable, efficient operation dictated that the great banks of tinted windows that had signified "cotton mill" from time immemorial be bricked up. What mill workers gained in comfort, the nighttime landscape lost in romance, as the mills now hid their great glowing lanterns under bushels.

Most importantly, new mill construction, such as Spartan Mills' John H. Montgomery Mill outside Chesnee, Arkwright's Cateswood Plant, or Inman Mills' Saybrook and Ramey plants, abandoned the old multistory design that had

Employees of Arkwright Mills pose for a photograph celebrating a safety award about 1960. (Courtesy, Mac Cates)

Inman Mills became the first cotton mill in the country to be fully air conditioned when seven large central units were activated on July 9, 1951. Riverdale Mills followed on August 31. The *Woodruff News* wrote: "These installations provide conclusive evidence that Textile plants in this area and in the southeast are moving ahead at a pace equaled nowhere else in the world, and measure to the ever increasing value of the south as the 'industrial area' of the future."

served the textile industry since its early days in New England. The older mills had been designed to be run from a central power source—first a water wheel or turbine, later a steam engine. With machinery driven by a complicated system of pulleys, shafts, and belts, it was most efficient to build multistory mills, with each department occupying a floor and materials moved from floor to floor with elevators. By the 1920s electric power had arrived in the Piedmont, carried by North Carolina tobacco mogul James B. Duke's massive power grid and making it possible to run each machine on its own motor; mechanized materials handling made it more efficient to move goods across one large floor than between four. Accordingly, beginning in 1945 with (appropriately) a Deering Milliken plant in Pendleton, the standard style of textile plant construction was single-story and (for efficient climate control) windowless. The new style of construction was not limited to textiles; similar low-slung boxes, often built by Greenville's phenomenal Daniel Construction Company, began to line the Interstate 85 corridor as other industries began to locate in Spartanburg.

Modernization was not fully met with equanimity by mill workers. New equipment always unsettled established work patterns, and could frequently become occasions for conflict; as late as 1964 doffers at Spartan Mills walked out in protest of a "stretchout." Nonetheless, the temperature of labor conflict, which had at times been red-hot in the 1930s, cooled considerably in the postwar years. The Congress of Industrial Organizations (CIO) launched a broad-gauged Southern Organizing Drive in 1946; dubbed "Operation Dixie" by the media, it ran into a stone wall of opposition from employers and apathy from workers. In Spartanburg—the South

Most mills, like Arkwright shown here, bricked up their windows with the advent of air conditioning. (Courtesy, the Spartanburg Herald-Journal)

Carolina headquarters for the Drive—only one plant, Glendale, was organized, and management successfully stymied a collective bargaining agreement.

Rising wages and full employment in the immediate postwar years were factors in the failure of unionization. The resulting ability to buy houses, cars, and major appliances that they could never afford before made workers feel that they were enjoying their share of the fruits of industry prosperity; one historian characterized worker attitudes with the question, "What do we need a union for?" As new, frequently more attractive industrial jobs began to appear in the 1950s, and especially the 1960s, workers were increasingly able to deal with their discontent by leaving the industry altogether.

But workers had over the previous two decades also gotten ample lessons in the futility of fighting management, which adapted to the federal oversight of labor relations established by the Wagner Act in 1936 with increasingly well-honed legal strategies as well as obdurate opposition. The shaky competitive position of the postwar industry steeled their antiunionism; the exemplary figure was, again, Roger Milliken, who in 1956 closed a plant in Darlington, South Carolina, rather than deal with a union—resulting in a legal case that remained unsettled until the Supreme Court ruled on it (against Milliken) in 1980. In 1954 industry anti-unionists joined forces with urban and rural businessmen and political leaders eager to attract outside industry to enact a state law (commonly called a "Right-to-Work" law) prohibiting collective bargaining contracts that required employees join a union as a condition of employment. This law made it difficult to maintain a viable union and even more difficult to organize one.

Top, In 1959, Inman Mills built its Saybrook Plant (shown behind the original mill), the first mill built in Spartanburg County in 24 years. (Courtesy, Inman Mills)

Left, Roger Milliken in the late 1950s (Courtesy of Spartan Communication Corporation)

The local industrial elite, in the meantime, maintained, and even strengthened, its vigilance against union activity. In 1959 Roger Milliken,

> **If one were a member of a mill village, he had no more chance of being invited to dinner by "respectable" white families than had a Negro, nor could he entertain much greater expectations of marrying into a "respectable" white family, going to the same college where their children go, belonging to their clubs.**
>
> ***—Author John Morland***

Frederick Dent, Walter Montgomery, James Chapman, Mac Cates, and others created the Spartanburg Development Association, designed, among other things, to "promote good human resource practices throughout the county for the purpose of strengthening employer employee relationships and the elimination of conditions that may retard economic growth"; "assist in establishing and supporting community educational programs having as their objective the creation of a healthy economic atmosphere in our community" (and "alert the citizenry to any influence national or local that might be disruptive to these objectives"); and "create a cohesive relationship between citizens and employers for the promotion of sound relationships, thus establishing a foundation for the maintenance and continuance of a healthy economic climate." The SDA was an example of new and more sophisticated employer approaches to labor relations: first handling potential conflicts with workers before they got out of hand (by the 1960s it was an article of faith with managers that successful unionization of a plant was evidence of management failure), and second, recruiting the community to help keep disruptive influences at bay.

The City of Spartanburg tied the subsidies it offered relocating industries to a no-union pledge; according to labor economist F. Ray Marshall, one local employer under pressure from his organized northern affiliates refused to deal with a union because he would "wind up on the chain gang" if he did. Maintaining a "union-free environment" became an objective pursued with sophistication by innovative lawyers, who, like Milliken's, could turn the legal framework estab-

This 1956 photograph shows a laborer in Glendale preparing to clean the cotton fiber. (Courtesy, Clarence Crocker)

A group of men gather for a break outside Saxon Mill. (Courtesy, Betty Livingston)

lished by the Wagner Act to their own advantage.

Some enclaves of union strength persisted from earlier times, notably at the six plants of the Clifton Manufacturing Company. Other union toeholds appeared in newly arriving nontextile industries such as metal-working and apparel; when New York-based Jonathan Logan, Inc. began making dresses in the county in 1959, it imported its contract with the International Ladies Garment Workers' Union. The AFL-CIO targeted Spartanburg for an organizing drive in 1961, which was generally counted a success because it strengthened the ILGWU, which by 1965 could boast 1,000 local members. Nonetheless, union membership—let alone union bargaining strength—remained paltry in the county; generally mill owners regained full control of the workplace after the tumult of the 1930s.

While the mills sought to fend off sharpened global competition by modernizing their plants and tightening their control over the workplace, they also sought to cut costs by withdrawing from their traditional roles in the life of their surrounding communities. The dissolution of the mill village, begun before World War II, accelerated afterward, and was largely complete in Spartanburg County by the mid-1950s. Typically in Spartanburg the houses were sold to the workers; in neighborhood after neighborhood crawl spaces were bricked in, siding went up, and hedges were chopped down, as the new homeowners celebrated their new opportunity to assert their individuality. On the

Boys and girls of preschool age play the same games, often in the same group. Among the favorite games are hide-and-seek, tag, froggie-in-the-millpond, drop-the-handkerchief, London Bridge and ain't-no-bears-out-tonight. Parents make swings for their young children, using long pieces of twisted cloth as strands. As they grow older boys prefer rough games like capture-the-flag, cops-and-robbers, fox-and-hound, hickory man, throw-the-tin-can, baseball, and football. They are sometimes taken fishing and hunting with their fathers and older brothers.

—Author John Morland

The Enoree school got new playground equipment in 1950. Led by Donnie Duncan, they took turns for their first trip down the slide. (Courtesy, Anthony Tucker)

downside, however, mills ceased to perform home maintenance as they had traditionally done; as workers moved out or died, the new occupants were frequently transients, and over time the villages began to deteriorate. In 1969 the last company stores in the county, run by Spartan Mills at Spartan, Beaumont, and Startex, finally closed their doors. Services that had traditionally been provided by the company—police, fire, water and sewer lines, schools—were turned over to the county, to special service districts such as the Una Water District, or to newly incorporated towns, their "doughnut" limits carefully drawn to exclude the mill from their tax bases. In 1950 the numerous small single-school districts of Spartanburg County, including mill schools, were merged into 12 districts, and two years later into the present seven; mill village schools were now part of larger systems, if they survived at all, and as high schools expanded, mill children increasingly rode school buses out of the villages. To some degree the mills continued to watch over their former wards, now neighbors, sponsoring scout troops, offering free periodic dental checkups, and running summer recreation programs. But the traditionally close intertwining of mill and community began slowly to unravel.

Perhaps the best indicator of that unraveling was the decline of mill baseball. Baseball in the villages had been more than recreation; it had been at the very core of community identity. Teams were semi-professional, with players paid by the mills, though usually with a fig leaf of "legitimate" employment on the outside crew. Teams were organized in leagues; the Spartanburg County League had been in existence since 1913, and by 1950 had been joined by the Textile Industrial League. Black millhands, though few in number before the 1960s, played in segregated textile leagues, and the Spartanburg County Colored League flourished from the 1930s into the 1950s. Mill teams along the western border of the county participated in other leagues. According to Thomas Perry, the game's loving historian, over 30 teams from Spartanburg played

Most of the millhands wear working clothes. For men these consist of overalls or blue jeans with blue cotton work shirts, or khaki shirt and trousers, and low-cut shoes which are no longer good enough for Sunday wear. Women wear cotton print dresses and aprons, and oxfords with low sturdy heels, or inexpensive barefoot sandals in the summer.

—Author John Morland

In the late 1960s African Americans made up 29 percent of the Upstate population but held just 8 percent of the textile jobs. In 1960, the median annual income for black families in each of the 14 counties of the Upstate was below the federal poverty level of $3,000.

—Author Timothy Minchin

textile league ball in 1950.

The 1950s, though, saw sharply declining attendance, as village life became less cohesive, and as alternative forms of entertainment and recreation, such as television and car travel, became more widely available. Young men who had in the past automatically played for the village team as literally the "only game in town" now had other choices as well. In the 1950s games once attended by thousands were now attended by several hundred; a game that had once paid its way began to lose money. Starting with Inman in 1951, mills began to abandon the practice of paying players; by 1960 the remaining mill teams were strictly amateur, increasingly reliant on high school talent. Spartanburg textile teams continued to play into the 1960s, but, with the player pool, fan base, and mill support eroding, they finally withered away, leaving derelict ball parks on the fringes of formerly vibrant communities.

To be sure, the end of the mill village was hardly the end of Spartanburg's blue-collar community. Rather, Spartanburg's mill workers were identifying increasingly with a larger geographic community defined by culture. One such source of identity came from music. Music had been a critical part of mill-village life from the beginning, as migrants from nearby farms and mountain hollows brought their songs and instruments to their new homes. The first "hillbilly" recording ever made featured "Fiddlin' John" Carson, a sometime Atlanta millhand; the Piedmont textile belt nurtured such popular artists such as Charlie Poole, Dave McCarn, and the Dixon Brothers, who in the grim 1920s and 1930s offered their fans both diversion and sharp-tongued social commentary.

The reach of mill musicians was broadened with the rise of the broadcast media. Radio broadcasts had begun in Spartanburg in 1930, and live performances on WSPA provided important early exposure for such locally-rooted string-band greats as Don Reno and Cleveland County, North Carolina's Earl Scruggs. Local listeners also had access to clear-channel stations in Charlotte and other major Southern cities, including Nashville's WSM and its Grand Ole Opry. As Nashville began to meld the varied styles of southern rural music into "country" after World War II, mill pickers such as Arcadia's Buck Trent made their way there to pursue their careers and make the home folks proud with their recordings and broadcast appearances. Country music simultaneously played to nostalgia for a rural past and assured its fans that they were not alone in wrestling with the dislocations of modern life; it was working people's music, and its prominence on the airwaves told them that they and their heritage were special. As workers improved their lives in the postwar years, the polished "Nashville Sound" became an emblem of their new respectability.

Another special bond, for white male working-class southerners at least, was offered by the rise of a spectator sport that was peculiarly theirs: stock-car racing. While much has been written about the role played by ex-bootleggers such as Junior Johnson in bringing the stocks to prominence, many of the great racers, from the beginning down to the late Dale Earnhardt, were rooted in the mill villages. Spartanburg's own contribution to stock-car legend was Whitney Mill's David Pearson. Pearson, who began his career on dirt tracks at age 17, joined the NASCAR circuit in

Spartanburg Herald, Thursday, Aug. 4, 1960—Page 13

CLOSE OUT SALE
½ Price
—on—
- Men, Women and Children's Clothing
- Hardware
- Household Goods
- Paints, etc.

Everything But Groceries And Drugs Included In This Sale

SALE CONTINUES THROUGH AUG. 31st

TERMS: CASH – NO REFUNDS
ALL SALES FINAL

CLIFTON MILL STORE NO. 2
Clifton, S. C.

Company stores were going out of business in the 1960s. (Courtesy, Spartanburg Herald-Journal)

Nicknames in the Village

I was born in Pacolet Mills in 1945 and lived there until the spring of my senior year of high school, at which time my family moved to Pacolet Station. In talking with the friends I grew up with, we have remarked many times on the large number of people who lived there who had nicknames. This was just part of life in the mill village. Many Pacolet residents did not even know the real names of those with nicknames.

Thanks to the following for their contributions to this list (which is not complete by any means): Hurricane, Suitcase, Zip, Pacolet Blackie, Doolie, Pecker Head, and Gary.

—Don Camby

These 60 were chosen from Don's list of more than 200 nicknames of residents of Pacolet Mills:

Alley Oop
Antelope
Billy Goat
Blue Jack
Boonie
Broke Arm
Bucket
Bugs
Cabbage Head
China Boy
Coal Train
Contact
Cotton Eye
Crisco
Dee Cooter
Drooper
Funky Foot
Gizzard
Goob
Gourd Head
Gumbo
Hambone
Hoe Cake
Hubba
Ish
Jay Pig
Juice
Knuck
Lead Bottom
Lick
Little Shoes
Mongoose
Mush
Onion
Pea Vine
Pidge
Powder Mouth
Pud
Puud
Red Buck
Roach
Rosebud
Rough Bucket
Satisfied
Shatter
Skee Bo
Snibby
Soldier
Spread Natter
Teaspoon
Therm
Tot
Truth
Two Barrel
Vitamin
Weenie Mutt
White Eye
Worm Oil
Yokum
Zoomy

1960; before he retired in 1986, he became the second stock-car driver to pass $1 million in earnings, and remains second in wins on the prestigious Winston Cup circuit only to the great Richard Petty. Not all Spartanburg racing fans were with Pearson (Petty's followers were numerous and vociferous), but the stock-car circuit and its good ol' boy drivers were the shared passions of factory lunchrooms throughout the county. Fewer and fewer were the work crews whose whole lives intertwined with each other on shop floor and in village; but NASCAR, and country music, provided the idiom by which they recognized each other and bonded together.

As the era of the mill village came to a close, Spartanburg's textile industry became increasingly a "normal" industry—a site of production, rather than a dense web of relationships reaching out to its surrounding community. In one respect, however, it persisted in a distinctly southern peculiarity: its racial segregation. From the earliest days of the post-Civil War mills it was commonly understood that mill production jobs were for whites and whites alone. African-American workers were allowed to take menial positions as cleaners, or do hot, heavy work in the all-male opening room, but by custom they were forbidden to work alongside (let alone over) white women and girls at the spinning frames and looms. Industrialists were not fully happy with this custom, as they eagerly eyed an untapped pool of cheap labor; but white workers, and public opinion, were adamant that the color line not be crossed, and experiments undertaken by, among others, John H. Montgomery and Seth Milliken with all-black workforces failed. The State of South Carolina wrote custom into law in 1915, requiring segregation in cotton-mill employment. Mill managers accommodated, and generally kept African-American workers at the margins; in 1950 African-Americans (virtually all male) made up less than five percent of the textile workforce in the county.

Such was the state of affairs when the civil rights movement began to gather steam around the beginning of World War II. Equal employment opportunity was among the Movement's earliest goals; the first federal agency with a civil rights mandate was the Fair Employment Practices Committee, established by Franklin D. Roosevelt under pressure from the black union leader A. Philip Randolph in 1941. Desegregating the textile industry was an especially attractive means of broadening economic opportunity for black workers. Because textiles was an industry in which most production jobs set low skill requirements, but provided important on-the-job training in factory discipline, they were ideal entry-level industrial jobs for people traditionally discriminated against both in their ability to gain industrial experience and to gain education. Because the industry was centered in the South, mill employment was geographically close to the largest concentrations of African-American population in the nation, at a time when a wave of farm mechanization, and in places such as Spartanburg the shift from cotton to less labor-intensive crops such as peaches, was wiping out many of the agricultural jobs traditionally held by blacks.

Finally, African-American leaders gained important new leverage over the mills during and after World War II. As the traditionally white-supremacist Democratic Party came increasingly to rely on black votes to maintain its control of national power, and as America's new role as a world superpower forced national leaders to face up to the hypocrisy of racial discrimination in a nation ostensibly defending freedom and equality, the federal government began—slowly and haltingly—to bring more of its weight to bear in support of equal employment opportunity. In the meantime, government, especially military, production, became vital to the mills during World War II, Korea, and the Cold War, suggesting that a firm federal policy of nondiscrimination backed by the threat of withholding contracts might force

The exterior of the company store in Fairmont (Courtesy, Spartanburg Regional Museum)

Second shift workers in the Inman Weave Room No. 2 in 1951 (Courtesy, Jim Everhardt)

> **Overall, in a one-year period, TEAM's aides found employment in the textile industry for more than 500 black residents of South Carolina. More than 80 complaints were filed with federal agencies.**
>
> ***—Author Timothy Minchin***

the desegregation of the mills.

On the other hand, resistance remained strong among both white workers and their employers. Posters advertising Ku Klux Klan rallies ("The white public is invited") were common sights around mill villages in the 1950s and 1960s, and the Grand Dragon of the South Carolina Klan lived in the blue-collar neighborhood of Hayne, near Saxon and Arcadia. Employers, for their part, were scarcely less resistant, especially if desegregation resulted in new federal government restrictions on their freedom to manage their workforce. In the 1950s continuing southern Democratic power in Congress, and a Republican administration reluctant to push for change, kept civil rights forces at bay. The millhands' old political champion, Olin D. Johnston, was now a powerful Senate committee chairman, and a bulwark of Jim Crow on Capitol Hill, along with his colleague J. Strom Thurmond and a virtually united phalanx of Southern political leaders.

By the 1960s, pressures began to mount against the industry's traditional racial practices. In 1961 newly elected President John F. Kennedy issued Executive Order 10925, strengthening requirements for nondiscrimination in employment by businesses doing business with the federal government, and established a Committee on Equal Employment Opportunity chaired by Vice President Lyndon B. Johnson. The National Association for the Advancement of Colored People (NAACP) quickly seized upon the opportunity in July 1961 by filing complaints against a number of Spartanburg textile plants. Unfortunately, this early effort came to naught, as the Committee dismissed all but one of the complaints, having ascertained from the Army that none of the facilities in question were currently working on government contracts. But Spartanburg's business and political elite reacted vociferously to Kennedy's moves. In June 1961, Arthur Erwin, the Executive Secretary of the Spartanburg Development Association, attended a hearing called by the Committee on

African-American workers, like this forklift operator at Mayfair Mills, typically stayed outside the mill until the late 1960s. This man is bringing bales of cloth inside to be inspected. (Courtesy, Mayfair Mills)

The only jobs available to African Americans in the 1940s and 1950s were outside the mill. These women worked in the hotel at Startex. From left: Rosa Smith, Leola Jones, and Daisy Barr. (Courtesy, Junior West)

Equal Employment Opportunity and challenged its legality, calling it "a tremendous bureaucracy in the making which will hold the power of life and death over much of the industrial might of this country."

The major turning point came with passage of the Civil Rights Act of 1964. Title VII of the act outlawed all forms of employment discrimination on the basis of "race, color, religion, sex, or national origin," and set up a permanent Equal Employment Opportunity Commission (EEOC). The EEOC was to receive complaints of discrimination and attempt settlement through mediation; if mediation failed, it could authorize complainants to bring action in the courts. With important new weapons at their command, African-American leaders began to make some headway against the textile color bar; by 1967 black employment in Spartanburg's mills had risen to 9 percent of the total. Yet, in a county where African-Americans made up over 20 percent of the population, and where textiles still accounted for over 60 percent of total industrial employment, the advance was modest, to say the least.

In that year the federal government and a consortium of private activist organizations launched a one-year program called Textiles: Employment and Advancement for Minorities (TEAM); headquartered in Greenville, it specifically targeted firms in South Carolina. According to historian Timothy Minchin, TEAM's staff quickly discovered that the Spartanburg textile community was a tough nut to crack. Power in Spartanburg had always been closely held, and the local black community (one of whose leaders joined the TEAM staff) had a long tradition of working with white leaders behind closed doors rather than engaging in open activism. As a result, TEAM was prevented from working openly in the county by the insistence of whites on maintaining full control of the pace of desegregation. To be sure, some progress was made in introducing Spartanburg blacks to textile employment, but mills insisted on keeping their efforts quiet, TEAM's leader complained, "so the whites won't find out what they are doing for Negroes." In one case a mill arranged to provide bus transportation for black workers—on condition that they continue to ride the bus even after they could afford to buy cars.

Quiet as the process was, however, integration of the mills proceeded; by the mid-1970s the proportion of black workers in the county's mills was around 15 percent. Moreover, in view of the frequently violent heritage of race relations about the mills, it was carried through with strikingly little worker resistance. Mounting federal government pressure, and the industry's heavy dependence on military contracts during the Vietnam War, gave textile executives cover in introducing black workers. White workers, for their part, were aware

Elizabeth ("Libba") McGee of Spartanburg was named the National Cotton Council's Maid of Cotton in 1950. That year, she traveled 40,000 miles on a goodwill and fashion tour. This photograph was taken in 1954. (Courtesy, Spartanburg Regional Museum)

of the fragility of the industry and acquiesced in the interest of retaining their own jobs. Managers continued to resist, particularly when the issue was one of nondiscrimination in advancement to more highly skilled and supervisory positions. But as the 1960s progressed, they found themselves under increasing economic as well as legal pressure to extend employment opportunities to blacks, as a spreading labor shortage began to squeeze textile manufacturers; in 1966 one unidentified local manager told *The New York Times* that "the color line has been broken down, perhaps forever....The pressure has brought integration in the mills and brought it fast."

Indeed, by the middle of the 1960s Spartanburg's textile industry, which only a few years before was barely able to maintain employment of its existing workforce, was faced with the tightest labor market it had ever seen. The deficit was due in part to the industry's own prosperity, fueled in part by increased military spending with the American troop buildup in Vietnam. Mill employment in Spartanburg County showed a modest rise from under 20,000 in 1958 to a near-postwar high of 21,500 in 1966, as the county saw its last-ever burst of new mill construction. But the major source of pressure on the textile workforce was coming from an unprecedented new source: industrial diversification.

Traditionally, Spartanburg's industrial experience had been dominated by the textile industry, and shaped by its peculiar heritage, and at the beginning of the post-World War II era, that

Whitney Mill, known as Pequot Mill in the 1950s, made the nation's first round bed sheeting and had actress Lucille Ball as the first customer.

—Author Michael Leonard

dominance seemed destined to continue. Unlike other textile centers such as Greenville, where the textile economy was now dominated by absentee corporations such as J. P. Stevens and Dan River, Spartanburg's old family-firm elite still held firm control of most local mills, and indeed was augmented by the arrival of Frederick Dent and Roger Milliken. One consequence of that dominance, rumor had it, was a lack of interest in allowing the entry of alternative employers, especially ones that might disrupt the local wage level or introduce unions. When Camp Croft, the large World War II training camp southeast of the city, came up for sale in the late 1940s, it was purchased by the Spartanburg County Foundation, controlled by local textile leaders, with one explicit aim being to keep the property out of the hands of a single large purchaser; as late as the 1970s

Cloth inspector at Glendale (Courtesy, Clarence Crocker)

a common story, spread by word of mouth among blue-collar Spartans, held that the "large purchaser" they wished to block was the Ford Motor Company. Among the objectives of the Spartanburg Development Association was to "aid the community in the selective establishment and retention of sound employers who can be welcomed as a good corporate citizen in our community, thus eliminating, as far as possible, employers who would exploit our citizens or in any other manner retard our growth."

Nonetheless, it was clear by the 1950s that Spartanburg had to temper its reliance on its traditional breadwinner industry. Not only were the mills stagnating, but (especially gallingly) it became clear that the Hub City's old struggle with Greenville for primacy in the Upstate region was now lost. By 1960 Greenville was one-third larger than its former rival, its postwar prosperity anchored in a more diversified economy. Belatedly, Spartanburg began to play catch-up. In 1953 the General Baking Company built a $1 million plant off Hearon Circle north of the city. The first major industrial catch, though, was the Kohler Company of Wisconsin, which brought a $5 million investment and 500 jobs to a large complex near Camp Croft. Despite its "smokestack America" origins, Kohler fit well into the Spartanburg business community's anti-union mindset; the company came in part to escape the United Auto Workers, who had at the time embroiled its Kohler, Wisconsin, plant in what would be one of the longest and fiercest conflicts in American labor history. (However, Kohler's Spartanburg plant was eventually organized by the International Brotherhood of Pottery and Allied Workers in 1972.)

The next major industry to locate in the county, the Butte Knitting Mills, was more problematic, for it was quickly organized by the International Ladies' Garment Workers Union, which had a long-standing relationship with its New York-based parent, Jonathan Logan. Nonetheless, it otherwise fit into traditional patterns; it was of an industry, apparel, which was allied to textiles, and employed a workforce of similar skill and not dissimilar wage level. Under its vigorous head Andrew Teszler, the son of a Hungarian refugee, Butte was built around the new technology of double-knit fabrics, which became a fashion phenomenon in the 1960s. By the mid-1960s

Right, Weave room, Inman Mills, in the 1960s. (Courtesy, Inman Mills)

Spartan Shield

September 5, 1962

ON MARCH 21, 1927, a group of Spartan's overseers, second hands and section men posed for the photograph above. In the intervening 35 years, most of them have either died or retired, but two of them are still active at the mill, with several others still following other lines of work. One man in the photo cannot be identified, but the others are:

Front row, left to right: Raymond Lancaster, John Lancaster, Luke Lancaster, Hobert Fine, J. T. Turner, Carl Taylor and A. R. Ochiltree. Second row, from the left: J. S. Underwood, W. M. Foster, Henry Calvert, Matt Dye, Otis Waldrop, T. C. Bullington, Hob Henderson, John Thomas, Will Waters and Herschel Treadway. Third row: Bob Shields, Ralph Cummings, Paul Greer, Jim Hudson, Leander Foster, Howard Rhinehart, Eulis Gowan, Clyde Knippe, Seymour Hancock and Burnett Roberts. Fourth row: Weldon Burnett, Roscoe Sawyer, Clarence Taylor, John Hudson, Horace Hancock, Jim Smith, Brady Gray, Monroe Rice and C. E. Edwards. Back row: Dennie Campbell, Gilbert Pye, Clarence Greer, Bruce Easterly, Jeff Adair, Alfred Burnett, Joseph Travis, Wilford Roland, the unidentified man, and Rome Gowan.

Of this group, only Weldon Burnett and "Luke" Lancaster are still active at Spartan; and Roscoe Sawyer works at Beaumont.

Below, This 1962 issue of the Spartan Shield spotlighted a group of 1927 overseers. The newspaper, published by the management of Spartan Mills, regularly spotlighted the activities of mill workers. (Courtesy, Kenneth Burnett)

south; the Draper Corporation, which had operated a loom foundry on South Pine Street from the late 1920s, sharply expanded in the early 1960s. More significantly, European textile machinery builders looking for a toehold in the U. S. market began establishing presences. Notable among these were Rieter AG and Sulzer AG, both of Switzerland, who established sales and assembly operations next to each other on I-85 not far from Milliken. In time their machinery, and that of other European producers, would largely displace American equipment and would drive the technological revolution that would sweep through the industry in later years. Of even longer lasting importance, though, was their role in establishing a foreign, specifically German-Swiss, presence in Spartanburg, a presence that by 1975 had burgeoned into an international business community attracting worldwide notice.

Most of the machinery manufacturers were small-scale, and the foreign-based companies left the high-tech end of their production in their home countries, limiting their Spartanburg operations to assembly and service. The greater impact on local jobs came from the synthetic fiber industry. Synthetics, especially polyester, were coming into their own in the 1960s, and surging demand forced a major expansion in plant capacity, most of which was located in and around the Piedmont textile belt where the customers resided. Synthetic fiber production was a sophisticated chemical process perfected by researchers at giant chemical and petroleum companies; it was manufactured in massive quantities in enormous, expensive facilities. Because the industry was so capital-intensive, and because a few giant producers effectively controlled the market, the synthetic fiber industry was far different from the highly competitive, low margin

Butte had passed the Lyman print works to become the largest employer in the county with over 3,000 workers. More importantly, it worked with Deering Milliken's research complex just up I-85, and with incoming machine manufacturers such as Sulzer Brothers, to develop knitting machine technology, providing the technological and entrepreneurial anchor for the double-knit boom of the 1960s and 1970s.

The 1960s proved a pivotal decade for diversification. Some of the new arrivals, such as Union Camp, Firestone Steel Products Company, and the Beverage-Air Corporation, were not textile related; however, many of the most important developments occurred in industries auxiliary to Spartanburg's traditional base. The collapse of the New England textile industry led the textile machinery industry to finally follow its customers

Fairmont Mill manufactured Hula Hoops briefly in the late 1950s. This fad didn't last long, and the plant was closed by 1962. It was destroyed by fire in 1977.

Butte Knit employees, members of the International Ladies Garment Workers Union, occasionally went out on wildcat strikes in the 1960s. (Courtesy, the Spartanburg Herald-Journal)

In an event that became known as the Midnight Massacre, Deering Milliken laid off 600 management and administrative employees at its headquarters in Spartanburg in the late 1960s.

textile industry; in particular, it could pay its workers considerably higher wages.

Thus it was a signal event in the industrial life of Spartanburg when synthetic fiber plants came to the county. Reeves Brothers began experimenting with polypropylene fiber manufacturing at its Saxon plant in the early 1960s; in 1964 it sold the mill to a chemical firm, Alamo Polymer, a joint enterprise of Phillips Petroleum and National Distillers. In 1965 it announced a major expansion, turning the former cotton mill into a major fiber-making facility and dramatically increasing

Manufacturing plant of Swiss-owned Zima and Kusters on Business I-85 (Courtesy, Mark Olencki)

employment. But the major chemical development lay along I-85 near the Pacolet River, where the Hercules Powder Company built a chemical plant to make polyester feedstocks and was followed by what became the massive Hoechst Fibers (now KoSa) polyester complex.

All these new developments built upon Spartanburg's textile legacy. Traditionally the mills had attracted little other industrial activity, buying their machinery and raw materials from outside the county and shipping their product to the nation and the world; now, the suppliers were coming to Spartanburg, drawn by its concentrated markets, by the transportation facilities centered there (the railroads, and now the interstates and—Roger Milliken's initiative—the regional airport)—and, far from least, a workforce that had long outgrown its rural roots and had become fully at home with the factory. The wave of new opportunities of the 1960s began to transform workers' lives in a manner comparable to the changes wrought by the first upsurge of the textile industry in the late nineteenth century. The newer jobs frequently paid higher wages and offered greater opportunities for advancement than the textile economy had traditionally afforded; employers such as Hoechst skimmed the cream of the workforce within a radius of 60 miles, drawing its workers from surrounding counties in both Carolinas. Demand rose for new skills, which were provided through South Carolina's Technical Education system, including the Technical Education Center at Spartanburg. In the meantime, a strong industrial base, population growth, and rising incomes allowed increasing numbers of Spartans to leave manufacturing altogether and seize other opportunities in service industries. As late as the 1950s observers of mill life worried that the horizons of those growing up in the villages remained constricted—that, seeing no opportunities for a different life, children cut their educations short to take mill jobs and assume their traditional roles as supports for their families. The 1960s, though, saw increasing numbers graduating from high school, taking training at TEC, and seeking higher education, at among other places, the new Spartanburg campus of the University of South Carolina. Public education sharply improved; with the opening of Hoechst, Spartanburg District Three, which embraced the Pacolet Valley mills and which had traditionally been among the poorer districts of the county, instantly became one of the wealthiest districts in the state.

As life got better for Spartanburg's workers, life became more complicated for the traditional employers. Industry-wide wage increases punctuated the 1960s, as mills were forced to bid against increasingly attractive job alternatives. By the mid-1960s, Spartanburg's labor shortage was making national news; gas station attendants and housemaids were reportedly abandoning their old jobs for new ones, and there were even reports of employers raiding each other for workers. Old-line companies such as Inman Mills and Mayfair Mills began to advertise for workers; radio commercials touted the "family atmosphere" of the factory floors, complete with testimonials from generations who had followed their forebears into the card, weave, and spinning rooms. By the late 1970s, mill personnel offices would begin posting signs in English and Spanish.

The 1960s, then, broke yet another link between Spartanburg and the textile industry. Despite appeals to nostalgia for the "family-like" atmosphere of the mill community, civil rights lawyers were already challenging the mills' tradi-

The number of textile jobs in Spartanburg County peaked in 1973 with 21,742 people employed in the industry. The high-water mark for apparel jobs was 3,462 in 1971.

> *"We have explored every possible avenue, including a different product mix and a revamping of machinery, but could not come up with any reasonable alternative that offered any promise for reestablishing the plant as a profitable operation."*
>
> **—Clifton Manufacturing President Stanley Converse on the closing of his No. 1 mill in 1969**

Stanley Converse, president of Clifton Manufacturing (Courtesy, the Spartanburg Herald-Journal)

tional practice of using families to recruit labor, rightly noting that it discriminated against blacks. Moreover, workers themselves realized that the newer opportunities were not only frequently better paying, they were also more stable. For the problem of textile imports had not gone away; indeed, the import problem continued to get worse throughout the 1960s, exacerbated by the rising labor costs resulting from broader American economic prosperity. Especially hard hit were the old-line, family-owned mills that continued to form the bedrock of Spartanburg's textile economy, complained James Chapman Jr. to *The New York Times* in 1969, after a 50 percent increase in imports forced him to cut back the hours of Inman Mills' workers. Such mills, traditionally selling to merchants and other manufacturers, rather than final consumers, lacked brand recognition, producing bulk greige goods of the sort increasingly being displaced by foreign production. At the same time, though, Chapman and other Spartanburg clothmakers were caught in a labor squeeze; "If we go below five days a week, we're going to lose our labor to other industries," he told the *Times.*

Thus, even as new, state-of-the-art mills were being built, and were struggling to piece together a workforce, some of the oldest mills began to go under. Glendale, where the waters of Lawson's Fork Creek had driven spinning frames since the days of slavery, shut down for good in 1961; virtually all of its workers quickly found employment at other mills. The major loss, though, came with the long decline of the Clifton Manufacturing Company. The oldest textile firm in the county, it remained controlled by the Converse family into the 1960s, its three mills hugging the banks of the Pacolet River or perched on the hills overlooking the gorge down which the horrendous floodwaters of 1903 had surged. By the 1960s the mills were museum pieces, continuing to make do in a world for which they had not been built; former workers later told historian G. C. Waldrep III of seeing President Stanley Converse combing junk piles for spare parts for his machinery. In 1965 the company passed into the hands of Dan River Mills of Danville, Virginia, which in 1969 and 1971 shut the mills down—taking with them the principal TWUA local remaining in the county.

At the time, Clifton's demise was commonly attributed to ruinous competition from imports; the truth, of course, was more complicated. However affectionately the historic structures were regarded by the generations who remembered toiling inside them, however venerable the communities they had underwritten, the mills were industrial dinosaurs. Squeezed between steep banks and a swift-flowing river, designed around shafts and belts driven by water, with antiquated layouts and little room for expansion, the days had long been numbered for the Cliftons. Now even their workers turned their backs, as those still able to adapt (many could not) began to drive up the river to Hoechst, or to other, increasingly far-flung employers. Clifton's fate was a sign of the times; it was also a premonition of what was to come.

> *The super came down there and I was on the second shift, and he was leavin' one evening, and I met him in the supply room. We done heard that it was going to close. I asked him, "Are they gonna close down?"*
>
> *He wouldn't look at me. In a few minutes he raised his head up and says "Paul, yeah."*
>
> *I said, "Would you advise a man to go get him a job?"*
>
> *He looked at me and said, "I believe if I could find me one, if I was you, I'd get it."*
>
> *I left the next week. I got a job at Valley Falls. They put the notice up the week I left. There was a lot of people that had worked there years and years, and they hadn't worked nowhere but there. I worked five years at Valley Falls. I didn't think they'd close that one down, but they did.*
>
> **—Clifton loom fixer Paul Jones about the closing of Clifton No. 1 in 1969**

Voices from the Village

Alfred Dawkins
Alaree Dawkins

In the early 1960s, Alfred and Alaree Dawkins became the first African Americans to work inside Reeves Brothers' Chesnee Mill, integrating a factory that historically had been all white. Alfred was promoted from an outside maintenance job, and Alaree previously held a domestic job. Here they tell the story of how they were hired and what life was like for them inside the mill.

Alfred and Alaree Dawkins (Courtesy, Mark Olencki)

Alfred: I went there in maintenance. Pick up trash, I guess you could say, pick up trash, and clean bathrooms, pick up yarn and carry it to the waste house in bales. I think [when] I went there it was paying $1.25 an hour probably for a 40-hour week, sometimes 45 hours, 48 hours a week. All over 40 hours was overtime. It was okay. I think I was trained for about three or four days before I took the job by myself. It was okay. It wasn't real hard work. It was just steady.

My father, he worked there before me. But he was not in the direct plant, but he [was] in what they called the opening room. That's where the cotton bales would be opened...There were some [blacks] in the receiving. That's where bales of cotton come in on the railway. The blacks shipped the cloth back out; a lot of it went out by truck.

I really didn't apply [for my first job inside the plant]. They put me up there. I think it was because I was black. We had, if I remember, seems like it was 95 percent of our orders were out by the government. And they had to put blacks in keep to those orders. So I think that's why I moved up. They just asked me would I like to move to a better job. And told 'em, "Yeah, if the money's right." They told me if I did, they gonna hire some black ladies, and if I did, they would like for me to have my wife to come down and talk to her, then me and her both could come in later on a different shift.

I was a-doing doffin' then. What doffin' is, is taking thread off a machine. It would go from cotton to thread on a bobbin, and I would take that bobbin off, replace it and start it back up, fill the bobbin back up again. That was my job. It was okay considering [it was] during the '60s. We had a few words—me and a few of the whites—but we settled it peacefully. Like some would speak, forgetting that I was standing there, black stuff like this...I had to let them know that I was standing there because I didn't back down. I wasn't Uncle Tom and I wasn't going to be an Uncle Tom for the few dollars that I was making.

Alaree: My job came about through my husband because he was the first black man to go inside the plant on a good paying job. So they were going to have to hire some black women so they asked him to have me come down and apply for a job...I hadn't applied for another job in a factory because black women didn't work in the factory. So, I am the first. I was a spooler hand, they called it.

The lady that trained me, she was very nice. But I didn't learn until later that they asked every one of the women in this department to train me and nobody would. But this lady, I'll never forget her. I think they put me with the best [employee] they had. So they said I get six weeks to train for this job. They gave me two weeks. And put me by myself. But I did it, and I was good at it!

I think maybe [the women didn't want to train me] because, you know, they had not been around us that much, and they really didn't know what to expect. And I had worked for them all my life, you know. And I think that was it. I don't think it was that they hated me. You know, as a person. I just think that I was black, you know. And that's it…

The lady [who trained me] told me that they said that I wouldn't make it. You know: "She won't. She won't do this. She won't do this. She won't." I can do anything anybody else can do! If those ladies could do it, I could do it! That's what I said when I went in the plant.

[Later] I applied for a better job, and my boss would always tell me that he needed me on the spooler because I was such a good hand. But I thought, "I don't care how good a hand I am. If an opening come, and I apply, I should get it because I'm next in line." But he kept passing me over. And something [sabotage] was happening to the work I was doing. So my little boss man, D—, he was so nice, he would be with me all night. He worked third shift to see if I was doing anything to cause the problem with my work. And D— would say, "Well, 'Ree, I can't see a thing that you doing." And he would follow me for nights.

So then it just kinda got out of hand, and me and him just we would just fuss every morning and go on. Then Alfred had to have a few words with him, and then after a year, I quit. And so I went to work at this other little place, so in about a couple of weeks he sent for me to come to his house, my boss man did. So I went down there, and he said, "'Ree, I want you t' come back to work."

I said, "Did you find out what was happening to my work?"

He said, "Oh, yeah, we found out."

And I said, "Somebody was doing it, wasn't it?"

And he told me, "Yes they was." See, they were trying to get rid of me, but they found out who it was. But he never would tell me, and I guess it was best he didn't. But then when I went back to work, back on the same job…he kept bringing in people, new people and put them on the job that I wanted. So that went on for a while, and I got tired. So I told my father, I said, "Daddy, the next time you see the plant manager, you call me 'cause I'm goin' to come out there and talk to him."

I was on third shift, and Daddy called me one day about 12 o'clock, and I went down there. They took me in this little office, and we talked, and I told them, "You send for my boss man because I don't want you to think that I'm lying. I will say it with him standing right here." And they told me, "We don't need him, 'Ree." That's what they said.

That night, my boss man told me to go on that better job that I wanted. It was in the twister room. I went in the twister room. And then on, they just started moving me on to the better jobs. To the creelers, which was a good job. A little more money. And then I went from the creelers to the cloth room, and that was as good as you were going to get.

I worked at Reeves Brothers until 1995 when they closed. I think it was a good job. We made, you know, good money. We made what everybody else made…It put you in the middle class, I guess. It was good because we were making as much as the whites.

—Interview by Gladys Coker

Outside the shuttered mill in Chesnee (Courtesy, Mark Olencki)

Mac Cates
(Courtesy, Mark Olencki)

Voices from the Village

Mac Cates

MacFarlane L. Cates Jr. was the third generation of his family to serve as president of Arkwright Mills. During his lifespan, he saw Arkwright go from a vibrant mill community to a crime-plagued neighborhood with no connection to the mill, now demolished more than 20 years. Here he remembers his boyhood, spent on the streets of Arkwright.

Now as far as my first recollection of the plant, in the Depression, we moved down here and lived in a house that at one point was used as an office, and we stayed in that house, not the big house that you see right here [across the street from the Arkwright Mill office], but a smaller house, actually. We moved down here when I was about four years old. And I can remember we were living on the corner of Glendalyn and Clifton in Converse Heights. I remember my father coming home and saying the bank was busted, and that was the start of, I guess, the Great Depression. And, I wanted to go downtown and go help pick up the money. I had a little bank that I was putting pennies in, but I couldn't get it open until it was full, you know. That wasn't the case. So we moved to the mill village. I guess that's my first recollection of the mill itself when I was not quite four when we moved down here.

My father ran the company, and we moved down here to be with things. And, I guess it was some of the most enjoyable part of my life that I can recall because it was a fun time. I had lots of friends here at the time. And to see the contrast if you come down through the crack houses or whatever's going on up there in what's left of the mill village. Back then no one locked their doors. Everyone knew one another, and it was a perfect place to not even lock your house up. So it's a real difference in time what goes on in the old village now...

The mill village had indoor plumbing. They were, of course, small mill houses. There were much larger houses up from the Baptist Church up there—Arkwright Baptist Church—there were larger houses, but they still had privies. So it was a real contrast...It must have been back in the '20s somewhere when they got indoor plumbing 'cause the privies were still around, but they weren't being used, because they had indoor plumbing. But the big houses that were just outside of the mill village, some of these fine houses that were up, not even a quarter of a mile away, they still were using privies.

We still had mules, had a mule barn. And where the community center is up here, they actually grew some cotton up there, right on into the late '40s. There was about two or three acres at that community center, and they'd get maybe half a bale or a bale out of it. We weren't growing the cotton. I don't know who did it, but we had mules down here as a child, and of course they would deliver ice, and I guess nobody, very few people had refrigerators as we know them today. They had iceboxes. In the summertime a great remembrance is that everybody went barefooted as soon as they could, but the horse-drawn ice wagon would come through, and you'd leave a card on your porch to tell him how many pounds you'd want of ice, and the guy down on the street would chop the ice up to get the right block shape, or what have you. And, you'd be barefooted, and there's a running board on the back of that horse-drawn wagon, and you'd jump up on that running board, but your feet would get real hot coming 'cross that road. You'd get on there and when he chipped up the ice, he'd give the children a chip of the ice that had come off. So you had the great contrast between the cold ice and your hot feet. That was sort of fun...

One of the things that got us through the Depression (that's an interesting sidelight), the building that's across the street out there, that's a smokehouse. During the Depression we were smoking hams and shipping them all up and down the Atlantic Seaboard, really. How that happened, Otis Clemmons was the store manager, and he came here from Smithfield's, the people that do Smithfield Hams. He had this know-how. So we started smoking hams, and that, really, probably kept the company alive during the Depression. So anyhow, the sad thing is after Otis got pneumonia, that was back when people couldn't handle too well, he died. So our know-how of smoking hams went with him...

Before he died, the people that furnished the raw pigs—the raw hams so to speak—their son, Jack, gave me a baby pig. So I had a baby pig back up in the back yard of my house up there. Well, we had chickens. We raised chickens, too. And, if you've ever robbed a chicken's nest to get eggs, you know that you don't rob a chicken's nest barefooted but maybe once. That stuff will stick to the arch of your foot like peanut butter to the roof of your mouth...But my pig, Elvira, went off to the fair, and met some boyfriend there, and when she did we had a litter of pigs. But the whole idea was that once we got the pigs, the mill business started doing better. And when we had the litter of pigs, we really started going great guns. And there's an old Irish superstition that pigs bring good luck, and with that, my mother started collecting pigs. We had China pigs and wooden pigs, and there's some damn pig in here [looks around his office], no that's a turtle. It's my father that's back around there. He was known as Cooter to my mother because he was so slow. She called him Cooter because that's another word for turtle...

They used to have the bootleggers. They would come down through the mill village and back up over the railroad to escape the cops. One night, they had one come through here, and he was going so fast he went airborne over the railroad track and landed in a ditch right in front of our house. My father and Mark Shook, they went down—all of us went down—to see what was going on, and I don't know how many fruit jars of moonshine, most of them broken, but some of them together. And so the cops got here and apprehended him out here on the trestle. And my father was trying to see if he could keep a few quarts for having performed the helping in the capture, but he didn't get any.

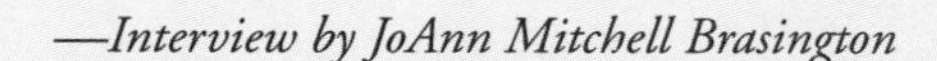

—Interview by JoAnn Mitchell Brasington

At the site of the now demolished Arkwright Mill (Courtesy, Mark Olencki)

Voices from the Village

Rosalie Tucker

In the early 1960s, the International Ladies Garment Workers unionized the fledgling Butte Knit operation in Spartanburg. Two Spartanburg women, Frances Caldwell and Rosalie Tucker, were key to that effort. Here, Mrs. Tucker, born in 1923, remembers how it happened and what it was like to work at Butte, which grew to more than 2,000 employees before shutting down in 1985. She begins by describing short stints at factories outside Spartanburg:

I went to work at Enro Shirt Co. in Louisville, and they said, "You have to join the union," and I said, "That's all right," because my husband was a union member. [He was a carpenter.] That's why he was traveling; the union was sending him. It happened in Detroit, Michigan, when I went to work there at a factory making uniforms. They said, "You got to join the union." I said, "All right." So I learned quite a lot up there about the union when they didn't know anything here about it.

When I went to Butte Knit, when I first went over, I went there with a friend of mine from a plant up here in Inman where I was working. She said, "I'm going to Butte Knit. You want to go too?" I said, yeah. She went in as a supervisor and I went in with her.

Rosalie Tucker
(Courtesy, Mark Olencki)

The union, they was organizing Butte Knit, so I went outside one day, and the man said, "Hey, you want a union card?"

I said, "Yeah, give me one." I didn't know what they was doing, you know. Down here, you know, it surprised me. Anyway, I took one, and I said, "In fact, give me a handful," and they did. I went back in the sewing room and started handing them out, and do you know I give everyone in that sewing room a card? I said, "When you sign it, give it back to me." I signed up I don't know how many people. Just about all the sewing room.

I just told 'em it was a good thing. I said you have a representative that represents you. You have a pay increase. I knew this from 'cause I had that up in Michigan and Kentucky. I said, "They'll negotiate your pay increase, they'll negotiate your paid holidays." I said, "You're going to have to do something real bad and something happen to you before they can fire you." You had job protection. I just explained it to them. They said, "It sounds good to me."

That Joe Ferguson [ILGWU labor organizer] said, "You're the best organizer I ever seen."

I said, "Well, the only thing is that I already knew what I was doing." I wasn't afraid they was going to fire me because I knew it was a law that they can't fire you for trying to organize a union unless you do something real bad. If you go in there and jump up and down and raise sand and stuff, they'd fire you.

The union got in mainly because of the piece rates. They paid so much money for each dozen you done. If they set it at 100 pieces for eight hours, you had to make over 100 to make any money. If you done more than they set the piece rate at, they'd up the rate, how many more you were supposed to get. That's what the main thing was in there. It was easy to organize. We had an election. It was in the cafeteria, and you go in and you vote yes or no on these little cards, and they dropped them into a little jar. The majority of yes won. It took about three or four days. They had to count them, the cards. The union come right in.

We had a sit-down strike in there one time. All them girls. Frances Caldwell was the president of the union then. She had talked to me, and them girls had been in an uproar about their piece rates and writing grievances. They had a stack that high and was getting upset. And when they get like that, you get a crowd of women—they just want to go on and walk out of the building and forget their jobs, but you can't do that. You got to have a reason to do it. So all over the plant they talked and planned and said we're going to have a sit-down strike in the cafeteria at lunchtime. We're not going back to our machines, and so, you know, they all stuck to that thing. They all sat in there. I said, "I'm going wait and see." I told Frances—she and I had dinner together—I said I'm going to sit here and I'm going to see what happens. I said, "I just want to see." And then when they didn't move, I said, "Oh yeah, they're gonna do it."

One black lady went back in there, and the supervisor said, "They are all in there, and you go in there with them." They had the business agent to go over there and personnel and talk to 'em. Wasn't anything I could say. I wasn't gonna say nothing. They did this on their own. They told them they'd get things straightened out, and they could go on back to work. Well, they got them all back to work.

It just would flare up like that. And one time they all got mad, and they left, and they was going to the union hall, striking, and the guards tried to lock the gate up there to not let 'em out. And they said, he better open them gates and let them out, and they all went to that union hall up there, and they come up from the plant to talk to them, and they got to fighting with one man up there. And it was something else. We finally got some of that straightened out, you know. But every time they'd miss with their rates they'd get mad. It's not only one rate, but a lot of girls getting the same rate. They was good about doing that. They'd get to making a little money, you know, they'd put their rates up where they wouldn't have to pay for them. Which wasn't right, wasn't right.

I retired at 62. The next year they closed down. I knew they was gonna close. I just got out just in time.

—Interview by Betsy Wakefield Teter

In the late 1960s six representatives of Union Local 581 from Butte Knit traveled to Charleston to march for better working conditions for hospital employees there. They wound up in jail for four days. From top left: Alvin Gregory, Dwayne Anderson, and Ben Hall. From Bottom left: Brenda Hoover, Frances Caldwell, and Rosalie Tucker (Courtesy, Rosalie Tucker)

Voices from the Village

Clarence Wilburn

Clarence Wilburn, born February 25, 1923, took a job as a laborer outside of Drayton Mill in 1936, at a time when jobs inside the mill were reserved for white people. He served in the military in World War II and returned to Spartanburg to work another "outside" job at Spartan Mills in 1946, finally moving inside the mill in the mid 1960s. Here, he reflects on the time before and after integration of the mills.

I started on the outside when I started in 1946, mostly when all the help was separate. Black was in one department, the outside department. Whites, they mostly had all the inside skilled jobs. That's the way it worked up until '66 or '67. I think that's when they first started integrating the plant. One or two went in at the time, slowly, but they had two different types of skilled work. Outside paid one thing, inside paid another. And that's the way it stayed just about until it ended. If you didn't move up on the inside job or take a better job, pay scales stayed just about the same. It didn't increase. Of course you got your pay raises, but they still never did catch up with the top paid job unless you transferred to one. In the recent last years, things got pretty good. Everybody moved up, you know. You come to be fixers, overseers, second hands, better jobs, better opportunities.

Clarence Wilburn (Courtesy, Mark Olencki)

[Before integration] most of them was on the outside, and all of them was men. They didn't have women. All the heavy work, mostly—all the grass cutting, cotton handling. When they started integrating, they started opening up a few jobs. You got the pay scales, but if you didn't move into one of those jobs, you still stayed on your basic scales, what you was on the outside. They never did catch up with those jobs on the inside. What I said it was, was "black scale" and "white scale," you know.

When I first started, it was 36 cents an hour. That's what the basic scale when I was coming back from service was. When I went in the service, they wasn't paying but 18 cents. When I went to Spartan Mills, that was in 1946, basic wage scale was 36 cents an hour. That was top money. You started from that and went on up to what it is today.

The labor union took over, and they had to put some [blacks] in there. That's how it basically got started. Because when I first went there, they didn't hardly let you walk inside the mill, much less talk about work in there. As the years went by, it got better and better. Pay salaries kept increasing.

I had all kinds of experiences. [I remember] drinking out of a water fountain when they first started that. George C— he was the first incident happened. He was going do what he had to do. He went in there and started drinking out of it. He said, "Well, you'll just [have to] lose this job." After that, they raised a whole lot of hell about it. At that time they had separate bathrooms, separate everything, separate water fountains. At the water fountain, they had whites over here, blacks over here.

You wasn't allowed to even transfer out of your department unless you asked your supervisor. That took place so long ago 'til I just can't halfway even remember it. I remember the people. Of course all the people I talk about are deceased. You were so glad when [integration] did happen—I never even think about the time of the early years.

Back in them days, you didn't have a choice. You did what you were told. If you didn't do what you were told, you didn't have a job. That's just the way it worked back then. You didn't have a union, you didn't have any kind of representation, you didn't have nothing, you was just there working. Whatever the supervisor told you to do, that's what you did. If you didn't do it, they fired you right on the spot. You didn't have no committee or nothing you could go to in the beginning. Whatever he said, that was it. He hired you, he fired you. It went on that way, clean on up…'til they created a union. I can't remember the year that union first come there. Things got better. Every year, things changed. It was a nice decent place to work there in the end. It was hell, though, to start with. I can tell you that. If you stayed there a week or two you was lucky. 'Cause everybody back in those days—textile mills, railroad, Draper, them about the three main highest-paying jobs there was back in them days. You was lucky if you got a cotton mill or either one of those places. Before I left there, though, everything was all fine and nice. I had all my supervisors, overseers, all of them was black.

I talk with a lot of young folks and tell them about different things and what happened back in them days and how things have come along. They think you're crazy. They don't think things have been that bad. When you had to go to work, you couldn't even ride a bus. And if you did, you didn't have but two or three seats, and that was at the back of the bus. And if there was two or three people on there, the bus driver wouldn't even stop. He would keep on drivin'. You just had to walk. I had to walk to Spartan Mills several times. Unless you had some kind of old car. And back in them days, wages was so cheap, you couldn't even hardly afford no car then. These young people this day and time, you talk to them, like my own kids, I talk to them, they don't believe me. They think time's been just like it is all the days. But, thank God, things changed. It took a long time comin' but finally…It still ain't what it's supposed to be, but it's a whole lot better than it have been.

—Interview by Danny Shelton

Outside Spartan Mills before demolition began (Courtesy, Mark Olencki)

Heat, Noise, and Lint
Working Conditions in the 1950s
by Mike Hembree

A frog in the weaving room?

It hopped onto the foot of Kathryn Mabry Holden, who was working at the Clifton Manufacturing Company No. 2 plant. It came through a window, she guessed, because the basement of the brick mill was below the water level of the Pacolet River, which rolled along adjacent to the plant.

"I felt something real cold touch my foot, and there was a great big frog that had hopped on me," Holden remembered. "You talk about screaming. You could hear me over the looms. The poor thing had so much lint on his legs he couldn't hardly hop. I caught him in a big piece of cloth and let him out the window."

The huge brick mills of Spartanburg County, many built in the last half of the 19th century, survived into the middle years of the 20th as monolithic reminders of a different time. Aging centerpieces of the villages they spawned, they continued to produce textile products in conditions that typically were far from modern.

The factories were busy on three shifts in the 1940s and '50s, contributing to the World War II effort with cloth used by the military, then providing jobs for returning soldiers. The men and women who worked in the old mills during those years remember conditions that usually were less than ideal. They talk of an odd dichotomy—they remembered the times as good ones and the work as a shared enterprise with friends and neighbors, but they also remember the dirt, the noise, the grime, the flying lint, oppressive workloads—and particularly the heat. "I wouldn't be on the job 15 minutes until it looked like somebody had thrown a bucket of water on me," said Holden, who worked almost 15 years in the No. 2 plant at Clifton in the 1940s and '50s. "That's the way it was. You had to wear your hair up. The perspiration would roll off everybody. You wore an apron to work because of all the perspiration and oil. I've gone out of that weave room with my apron wet practically all over."

LaBama Lyda Phillips in the weave room at Glendale (Courtesy, Clarence Crocker)

The late Jesse Franklin Cleveland, a great nephew of John B. Cleveland, one of the Clifton mill company's original directors, went to work at the Clifton No. 1 plant after World War II. He started as a floor sweeper and eventually became a company vice president. In an interview, he once remembered putting his lunch pail "up on a wire that was hanging down off a pipe because sometimes there'd be some pretty good size rats in those places."

Down river at the Pacolet mill, Rosell Suttles Pressley, who started work in the spinning room in 1951, said the mill was "real hot, even in the winter." She remembers dirt and cotton flying through the air and the non-stop noise of the machinery. And the "bossmen," as they typically were called: "I had some good ones and some bad ones," she said. "Some were pretty rough. I had one who was hateful to everybody. He stayed on people, and you already were doing all you could do."

Holden said supervisors sometimes upset workers by increasing production demand. "I've seen people go after a bossman like a hen after a June bug," she said.

The pace of work in the mills differed from location to location, job to job—even side to side, as the long rows of machines were called. Breaks were few and far between and often came only when a worker was "caught up," and his or her assigned machines running smoothly. Often, workers say, they would ask a co-worker to cover for them while they stopped to visit the restroom (the "waterhouse" in mill jargon) or to smoke a cigarette. Lunch breaks—if they could be wedged in—were brief.

Pressley, whose mother had started to work at Pacolet at the age of nine, said she usually didn't attempt to eat during working hours, especially on her first job at the mill. "I was slow at first; maybe that's why I didn't get to eat," she said. "I guess I was scared, too. You had to keep the job running. There was no time to eat."

Holden said she usually took a sandwich and a pickle—maybe even a rabbit leg—for her meal at work. "You sat on a box at the end of the alley or where a bunch of rolls of cloth were stacked up or on the steps and ate," she said. "You just put your food up beside somebody's toolbox until you got time to eat." Soft drinks, candy and cookies could be bought from the so-called dope wagon, a cart that was pushed through all the mill departments during each shift.

Eldred Quinn started work at Drayton Mill in 1939. He described Drayton as a "fancy" weaving mill, one in which the looms carried different blends of thread and weaved more complicated cloth than most other mills in the county. At its height Drayton employed 1,200 people.

"It was hot in some of the departments, but I remember the lighting and ventilation being good," Quinn said. He said he usually didn't have time for a lunch break but often ate a sandwich while walking up and down the alley, tending his job. "A loom fixer might pinch-hit for a weaver so you could go to the bathroom and back," he said. "But the looms didn't stop. They wanted maximum production out of them."

Quinn, whose grandfather drove a wagon into the North Carolina mountains to recruit workers for the Clifton mills, worked several years at Drayton and decided to pursue a weaving job at nearby Whitney Mill before leaving for the United States Navy and service in World War II. "It was a lot harder there (at Whitney)," Quinn said. "I had 48 looms, and I never got all of them running all night long. They were running broadcloth, and it would work you to death. I worked one night and never went back."

The work often was long and hard—and the workplace loud. Holden said the weaving room at Clifton No. 2 was so loud that "you couldn't talk to anybody unless you were real close to them." It wasn't wise to get too close to some folks. Many chewed tobacco or used snuff, and the refuse often wound up on the floor or between machines at some plants. "And the sweepers had to get all that up," Holden said. Rosell Pressley remembers spittoons being used for a while at the end of machine rows at Pacolet.

Dewey Pressley, who would meet his future wife, Rosell, in the spinning room at Pacolet, was a doffer, a job that involved replacing full bobbins with empty ones on spinning frames. Speed was important as workers typically removed a bobbin with the left hand as they put on another with the right.

"I started working as a floor sweeper and learned to doff on my own time," said Pressley, describing a frequent path to better paying jobs in the textile mills. "You couldn't get a break. There was too much to do, and you had to keep the job running. It made corns on your fingers."

The Last Strike
Clifton's Bitter Showdown

by G.C. Waldrep III

To most workers and managers, the labor union movement in Spartanburg County's textile mills was just a memory by the time the nation emerged from World War II. Rising wages in a booming postwar economy, as well as gradually improving working and living conditions, made the desperation of the 1930s seem more and more distant. Where unions held on, as at Spartan Mills, they were weak.

The exception was Textile Workers Union of America Local 325, covering the three plants of the Clifton Manufacturing Company. In general CIO unions were more militant than their AFL counterparts (such as the Spartan union): they fought harder for higher wages and better working conditions, and they bore the brunt of manufacturers' attempts to destroy what was left of organized labor in South Carolina during the 1940s and 1950s. The local union at the Cliftons wasn't the only TWUA-CIO left in South Carolina, but it was one of the strongest, and by the late

Striking workers gather on the bridge into Clifton during a major strike at the plant during 1949-50. The strike, which pitted neighbor against neighbor, was the last major textile strike in Spartanburg County. (Courtesy, Mike Hembree)

1940s textile manufacturers across the Upstate had targeted it for extinction.

Clifton President Stanley Converse was reluctant to disturb the good relations he'd built with his workers since the turmoil of the 1930s, union or no union. But, pressured by his fellow manufacturers, it appears that Converse began engineering a final showdown with Local 325 in the spring of 1949. Radical changes in workload and job assignment were the rule that spring. Coupled with the company's inaction in the face of earlier union complaints, these changes raised workers' tempers to the boiling point. To them, the union contract was a dead letter if it didn't guarantee them some practical control over work in the mill.

Although a formal strike vote wouldn't be taken until December 8, the six months' strike at the Cliftons actually began at 7 a.m. on October 31, 1949, when loom fixers at the No. 3 mill (Converse) refused to accept new job assignments handed down by their supervisors. The strike started with a guerilla campaign in which the loom fixers would come to work on each shift, turn their pick clocks to prove they had reported for work, and then disappear. Sometimes the loom fixers would retreat to the No. 3 tower, sometimes to the men's dressing room; sometimes they would fan out into the rest of the mill, trying to build up worker morale for the all-out conflict they knew was coming. Without loom fixers, one by one the Converse plant's old-model looms fell silent. On the morning of November 8, general manager T. I. Stafford ordered No. 3 shut down and sent the workers home. The night before, loom fixers at No. 1 and No. 2 had voted to back up their colleagues at No. 3, and within a week Stafford and Converse had closed those plants too. The strike was on before the national union leadership of the TWUA knew anything about it.

Once they did, however, they understood that losing the Clifton strike would be the end of textile unionism in South Carolina. The TWUA had not wanted a showdown at the Cliftons; they would have preferred to spend their money on a campaign against one of the larger textile chains. But once the Clifton strike had begun, they had no choice. The union poured money into the Cliftons, where over a thousand workers—union and nonunion alike—were now idle.

The strike was peaceful until January 10, 1950, when Stanley Converse announced plans to reopen the mills with supervisors and however many workers would come back to work on his terms. From January 24, when Converse reopened the doors, until April 25, when the company and the union finally negotiated a new contract, the Cliftons were armed camps. The majority of Clifton's workers supported the union, but a small minority opted to cross the picket lines. "Most of the people felt so strong one way or the other," paymaster Betty Hughes Carr recalled in 1993. From the union side, Ruth Barber agreed: "You felt like where you were standing up for something to better things, and you felt like, well, the others were going against that." Anyone who crossed the union line was subject to abuse and harassment from desperate unionists, who felt they faced losing everything they had gained through the union if the strike failed.

The union won: that is, the company finally agreed to renegotiate a contract, and though it wasn't a good one by nationwide standards, both it and Local 325 endured until the final closing of the Clifton plants in 1971. But the victory was a mixed one. As communities, the three Clifton villages never really recovered. Bitterness from the strike endured long after the mills had reopened, as did divisions over whether the fight had been worth it. "I don't think most of them's ever got over it," former striker Elbert Stapleton said in 1993. Georgia Seals, the wife of a strike committeeman and a prominent union supporter herself, agreed: "It was a sore spot, to a certain extent, for a long, long time. It still is, a little." Some believed, in retrospect, that the union had accomplished very little, that wages and working conditions in the Clifton mills would have improved with or without a union. Others believed that Local 325 was responsible for those improvements, not only at the Cliftons but all across upstate South Carolina, where other manufacturers were forced to match union workers' gains or else risk unionization of their own plants.

Either way, there was a cost to standing together as a union, as the men and women of Clifton learned. They were the only ones, as a Saxon worker had put it at the height of the union movement in the 1930s, to "keep the faith."

Walter S. Montgomery Sr.
At Spartan's Helm for 67 Years

by Lisa Caston Richie

Walter S. Montgomery Sr. left a legacy of leadership and philanthropy that will remain long after the last loom is removed from his former textile kingdom. Born October 18, 1900, he worked briefly with Montgomery & Crawford Hardware Co. before joining Spartan Mills, the company his grandfather founded, in 1922. After gaining experience in carding and weaving, he was named treasurer in 1926 before taking over in 1933 after the deaths of his father and brother. He would serve as president for 43 years, and then as chairman of the board for another 24, continuing to build Spartan Mills through the early 1990s. Presiding over a company that grew to 5,000 employees, his policy became one of growth and acquisition, with a focus on plant modernization, safety, employee relations, and diversified product lines.

During his tenure, he chaired industry organizations like the South Carolina Textile Manufacturers Association and the American Textile Manufacturers Institute, consistently remaining a key ally to Roger Milliken in the campaign to restrict foreign imports of textile and apparel. He was also a major benefactor for the Spartanburg County Foundation, the Boy Scouts, Spartanburg Regional Medical Center, the United Way, the Chamber of Commerce, Converse College, and other local colleges.

His first project as president was the purchase and transformation of Tucapau Mill to Startex Mill in 1936, with numerous others to follow. The Startex project included the installation of modern heating, ventilating, humidifying, and air conditioning systems for the plant, and a complete remodeling of the village homes. His crews had to rebuild, paint, and roof each house, along with adding indoor restrooms. He endured—and prevailed in—a protracted clash with

Walter Montgomery Sr., shown here aboard an aircraft carrier during his post-World War II tour of Europe, was an avid photographer and moviemaker. (Courtesy, Walter and Betty Montgomery)

Tucapau Local 2070, a bitter conflict that included the firing of dozens of union members. Simultaneously, he emerged victorious in smaller battles with unions at Spartan Mills and Gaffney Manufacturing.

The financial investments Montgomery made in the 1930s paid off when his employees helped him reach unprecedented production goals that led to national recognition. In 1941, Montgomery filed a necessity certificate with the War Department proposing machinery needed to devote all materials from Beaumont Manufacturing for Army and Navy use. The government allocated almost $800,000 to equip the mill for the production of fabric that would be used for tents, rafts, ammunition belts, medical equipment, and other miscellaneous needs of the armed forces. Both Spartan and Whitney Mills also successfully operated for wartime production throughout this period.

After the war, weakened foreign industry created a boom in business for the domestic textile industry. This couldn't last forever, though, and Montgomery was highly critical of the Marshall Plan, the U. S. initiative to assist defeated countries in rebuilding. According to Montgomery, the very program that was meant to help the world would ultimately harm most American industries. He felt that by giving new, modern machinery to our defeated foes, the United States put its own companies at a disadvantage.

Despite his fears, Spartan Mills flourished under Montgomery's leadership. From 1941 to 1950 alone, the company's value increased fourfold. After 1950, he purchased the Dixie Shirt Company, challenging Spartan Mills to enter the new area of specialty production. During the post-war boom, Montgomery acquired additional plants, including Powell Mill, Whitney Mill, and Niagra Mills, and created a finishing business that included support branches. He built some of the most modern and expensive U.S. textile plants of their time: the John H. Montgomery Plant in Chesnee and the Rosemont Plant in Jonesville. In the early 1950s, he also initiated the company's first comprehensive safety program, an issue that he would continue to emphasize throughout his career; in 1990, Spartan Mills was named the safest textile company in the United States in its size category by the American Textile Manufacturers Institute. Montgomery also began a warehousing operation, Montgomery Industries, and was well known for his propensity to speculate on the cotton futures market.

Montgomery, or "Mr. Walter" as his employees affectionately called him, wasn't only concerned with business. He was a sportsman, a humorist, and a humanitarian. It was not uncommon to see him out in the mills among his employees. He was even accustomed to taking off his coat in an effort to make them feel comfortable around him, and he often called them by their first names and remembered to ask about their families. He valued—and honored—loyalty.

The late Vernon Foster, a 40-year employee of Spartan Mills, liked to tell the story of Montgomery's benevolence to his father, a long-time loom fixer at Spartan No. 1. A supervisor discharged his aging father one day, claiming he was too old and weak to run new, heavier looms that had just been installed. "Mr. Walter heard about it," Foster said, "and arranged for a reinstatement in No. 1 mill, at a time when transfers between the two mills were simply unprecedented."

Montgomery also had a great interest in the community, its people, and their welfare. He played a monumental role in the growth, development, and success of Converse College, and was the chief organizer of the Spartanburg County Foundation. Because of these achievements and so many others, he was inducted into the South Carolina Business Hall of Fame in 1986 and received the state's highest honor, the Order of the Palmetto in 1994.

Montgomery claimed to have been a linthead since the day he was born, and he was determined to enrich the lives of those who shared his community. "I've always said that attitude and perseverance are the most important assets a person can have," he once said. "The key to success is to keep working at something you believe in and enjoy. It's hard for me to believe that anyone can be unhappy and be successful."

Montgomery received his education in public schools, at the Hastoc School for Boys and the Virginia Military Institute. He was married to Rose Bailey Cornelson of Clinton, and they were the parents of Walter S. Montgomery Jr. and Rose Johnston. He died April 26, 1996, at the age of 95 at Spartanburg Regional Medical Center, one of the organizations he had supported for years. His last words were reported to be: "I've got to get my boots on." Even then, he was thinking about work. Whether he is remembered as a textile giant or a quiet humanitarian, Walter Montgomery greatly influenced the development of Spartanburg.

A Fallout Shelter for 1,600
Milliken Prepares for Nuclear War
by Tanya Bordeaux Hamm

While the world was worrying about the possibility of World War III, one Spartanburg textile company was taking extreme precautions to protect its employees and their families from the potential dangers of the Cold War.

In the early 1960s, as industries across the country were preparing fallout protection plans, the Milliken Research Corporation, then known as Deering Milliken Service Corporation, or "Createx," constructed a gigantic shelter and developed a detailed plan in case of nuclear war. Executives at the company determined that, because of the proximity of Air Force bases in Greenville and Atlanta, nuclear fallout could reach Spartanburg two to four hours after an attack.

The shelter, designed to hold Milliken's 500 employees and 1,100 of their family members, was constructed in the basement of the Prototype Mill, an experimental cotton-spinning facility adjacent to Milliken's research offices. The Prototype Mill was newly constructed at that time, with a basement approximately 130 feet wide by 400 feet long. The basement, almost completely underground, was centrally air conditioned and ventilated. The total budget for fallout preparations was reported at $86,882, including a shelter menu of items such as applesauce, saltines, peanut butter, Tang, Pream, and "multipurpose" meats and vegetables that didn't require cooking.

The fallout protection plan also included a guide for employees and their dependents to follow in the event of an attack. Published in December 1962, the handbook gives detailed information on shelter organization, administration, and operation, including a schedule of activities if the shelter was used. It encouraged families to place the following items in the trunks of their cars in case they must go to the shelter: a blanket for each family member, medicines, infant supplies, bars of soap and a towel for each family member, several changes of underclothing, items for personal hygiene, a first-aid kit, a Bible and hymnal, flashlight, and small games, toys, and books for entertainment.

According to the guide, daily activities in the shelter would begin at 7 a.m. and would include work duty, rest periods, religious services, recreation, and nursery school for children ages three to six. The shelter was furnished with power supply and radio equipment, and plans were in place to publish a daily one-page newspaper for occupants.

Col. Marshall H. Strickler, Milliken's Civil Defense coordinator, prepared the fallout protection plan, which he presented before Congress in June 1963. Strickler, who retired from Milliken in 1982 and moved to Clemson, said he presented the plan before Congress during its fourth week of hearings on the Kennedy administration's $175 million Fallout Shelter Incentive Payment Plan, designed to spur shelter construction in public and non-profit institutions. He later presented a method for developing a similar plan to local industry leaders during the Piedmont Industry Defense Institute, held March 5, 1964, at the Cleveland Hotel in Spartanburg.

"This was in a time when the Cold War

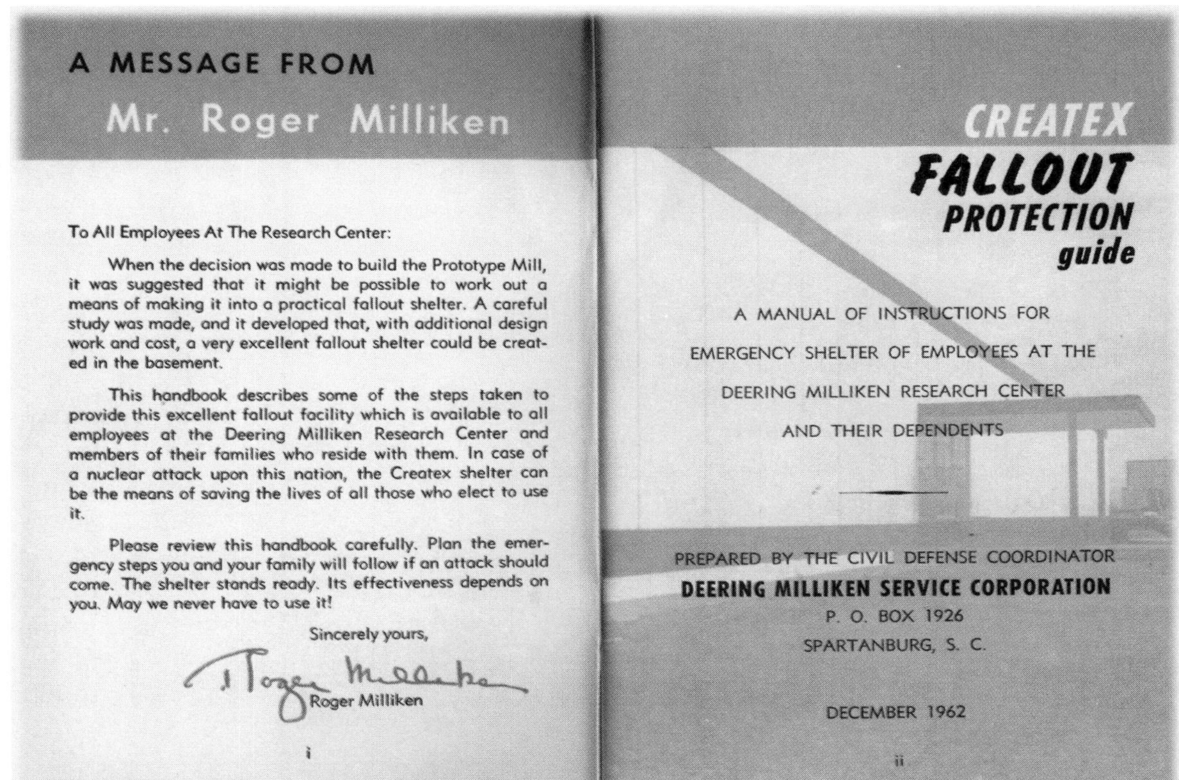

A MESSAGE FROM

Mr. Roger Milliken

To All Employees At The Research Center:

When the decision was made to build the Prototype Mill, it was suggested that it might be possible to work out a means of making it into a practical fallout shelter. A careful study was made, and it developed that, with additional design work and cost, a very excellent fallout shelter could be created in the basement.

This handbook describes some of the steps taken to provide this excellent fallout facility which is available to all employees at the Deering Milliken Research Center and members of their families who reside with them. In case of a nuclear attack upon this nation, the Createx shelter can be the means of saving the lives of all those who elect to use it.

Please review this handbook carefully. Plan the emergency steps you and your family will follow if an attack should come. The shelter stands ready. Its effectiveness depends on you. May we never have to use it!

Sincerely yours,

Roger Milliken

i

CREATEX

FALLOUT PROTECTION guide

A MANUAL OF INSTRUCTIONS FOR EMERGENCY SHELTER OF EMPLOYEES AT THE DEERING MILLIKEN RESEARCH CENTER AND THEIR DEPENDENTS

PREPARED BY THE CIVIL DEFENSE COORDINATOR
DEERING MILLIKEN SERVICE CORPORATION
P. O. BOX 1926
SPARTANBURG, S. C.

DECEMBER 1962

ii

A message from Roger Milliken in a booklet for the company fallout shelter (Courtesy, Col. Marshall H. Strickler) was pretty warm, and people were putting up fallout shelters across the country," he said. "Thankfully, there it stayed, unused, from 1962 until the 1970s."

According to Richard Dillard, director of public affairs for Milliken, the company abandoned the shelter sometime in the 1970s and donated all of the useable supplies to the Red Cross and other community agencies. The shelter space has since been used for other purposes, he said.

Making History
The Battle to Desegregate

by Betsy Wakefield Teter

With cotton mill jobs out of reach for most African Americans in 1961, a small group of black residents quietly made Spartanburg County the first battleground in the fight to integrate the textile plants of the South. Though unsuccessful in their efforts, they blazed a trail of federal intervention that ultimately opened the doors to blacks across the region by the late 1960s.

Buoyed by a new federal law that prevented discrimination by U. S. defense contractors, about two dozen black residents began visiting the personnel departments in Spartanburg County mills in early summer 1961. In some cases, they filled out job applications; in others, they left their names and addresses when applications were denied. Assisted by the NAACP, they filed the nation's first discrimination complaints against southern textile companies and brought federal scrutiny to most of the major mills in Spartanburg County: Beaumont Mills, Spartan Mills, Arkwright Mills, Saxon Mills, Clifton Manufacturing, Niagra Mills, and Deering Milliken's Drayton Mills. They also brought national attention to the issue when a story about their charges appeared in *The New York Times.*

Geneva Hines, a 43-year-old resident of Southern Avenue, was one of the complainants. She wanted a job as a battery filler at Saxon Mills. "At Saxon Mill, the lady only took my name and address," she said in her complaint, now housed at the National Archives in College Park, Maryland. "I expected to be able to fill out an application blank. I believe that no application was available to me because I am a Negro, and the mills here do not want to hire Negroes as bat fillers."

Geneva Hines's visit to Saxon Mill was set in motion by President John F. Kennedy, who signed an executive order creating the President's Committee of Equal Employment Opportunity on April 7, 1961. That law, a steppingstone to the Civil Rights Act, prevented companies or institutions with U. S. government contracts to discriminate on the basis of race. Kennedy placed Vice President Lyndon Johnson in charge of the federal initiative.

Henry Prysock, right, attended by Carl McClure, offloads cotton in Glendale. Such "outside" jobs were the only mill jobs available to African Americans in the 1950s. (Courtesy, Clarence Crocker)

The NAACP was already chomping at the bit when the law went into effect. Clarence Mitchell, director of the Washington, D. C. bureau of the NAACP, wrote Johnson that very day to inform him of state laws in South Carolina that prevented blacks from working inside the mills. The South Carolina segregation laws were so strict in 1961 that it would be a violation for black and white workers to pass each other on the staircase. Mitchell pointed out in his letter that new federal law, in effect, now prohibited South Carolina textile companies from doing business with the U. S. government. "We therefore call upon you as Chairman of the Committee to take all necessary steps to enforce compliance with the Executive Order in the State of South Carolina," he wrote to Johnson.

Meanwhile, down in Spartanburg County, a plan

was set in motion to test the new directive.

On May 25, Bobby Twitty, a 21-year-old black truck driver for Spartan Mills' Niagra Plant on Williams Street, applied for a position inside the mill as an operator of a cloth rolling machine. "Even though I have never had an opportunity to operate a cloth rolling machine, I have watched the men operate them," he said in his complaint. "And with a High School education, I am confident that I can operate the machine. When I applied for the job as operator of cloth rolling machine, Mr. E- said he would talk to me about it, but to date he has not. Also, about 10 days after I asked for the job, he hired a white man for the job."

The NAACP logged 38 such complaints from Spartanburg County residents in the month of June. The federal committee decided that 28 of them were within its jurisdiction. But by early August, the committee had determined that most of the Spartanburg County mills had ended their Defense Department contracts. Only Saxon Mills, owned by Reeves Brothers, had a current contract with the government, making cotton uniform twill for the Army. A brief article appeared on the front page of the Spartanburg *Herald-Journal* announcing that several unnamed local mills had been cleared of discrimination charges.

The NAACP, meanwhile, was enraged. The months wore on with little success in their campaign. In an address before a labor union group in New York City, Herbert Hill, the NAACP's labor secretary, charged that Southern textile companies had gone on a "contract sit-down strike" against the Defense Department and other U. S. procurement agencies to avoid integration. "An unpublicized conference was held April 21 of major Southern cotton textile companies in Charlotte, North Carolina, and there industry representatives discussed ways to circumvent the President's anti-bias order," he charged. The Defense Department had not received one textile bid since the effective date of Kennedy's order, he claimed. "In Virginia, North Carolina and South Carolina, Georgia and Alabama," he declared, "there is not a single Negro employed as a weaver, spinner, loom fixer or in the carding rooms. Negroes are employed only as sweepers, janitors and cleaners."

While most of the mills in Spartanburg County were let off the hook, the government, however, did keep the heat on Saxon Mills, conducting a full investigation, which lasted through the summer of 1963. Unfortunately for the African Americans, the government concluded that Saxon had broken no laws. Of the four who filed complaints against Saxon Mill, none had textile experience. And by 1963, Saxon was experiencing a severe economic downturn, reducing its number of workers from 365 to 145. Other local mills were facing similar declines in business. The 220 white workers who had been laid off from Saxon had more experience than the black applicants and therefore were first in line for the jobs, the committee concluded. Although the NAACP had asked that a textile training program be set up for black applicants, "there is now a large surplus of qualified textile workers [in Spartanburg County], and no need for such training," wrote Hobart Taylor, executive vice chairman of the President's Committee, in his final report.

It would be five more years before African Americans worked beside whites inside most local textile mills. But the small group of local blacks made history. Their complaints were the first of 80 against southern textile companies logged by the NAACP between 1961-1964.

Funny Uniforms and College Stars
The Southern Textile Basketball Tournament

by Thomas K. Perry

In the autumn of 1905, L. P. Hollis, YMCA secretary at Greenville's Monaghan Mill, traveled to Springfield College in Massachusetts speak with Dr. James Naismith, creator of basketball. Equipped with Naismith's book and a ball purchased in New York, he brought the new game south.

Hollis was the catalyst in organizing the Southern Textile Basketball Tournament in 1921, an event dedicated to developing the competitive talents of children in the cotton mill communities. Mill kids took to the new game—quirky rules, funny uniforms, and all.

The first year it was a men's-only gathering, but in 1922 the ladies played right alongside. The Monaghan Mill lasses defeated Greer's Apalache Mill 98-2 and claimed the first women's championship, led by the scoring of Oveida Henderson and Minnie Heath (50 and 48 points).

The basketball team from Spartan Mills in the mid-1920s. (Courtesy, Lisa Caston)

Crowds of several hundred gathered around the ropes that cordoned off the court—such popularity was not easily kept secret. By 1929 game results were telegraphed to radio station WLW in Chicago and broadcast worldwide.

More teams entered the league each year, and soon the Textile Hall floor was divided into adjoining courts to accommodate the expanded schedule. Such an arrangement posed a unique problem, since players were often confused when a referee's whistle sounded from the other court. And there were other moments of confusion. In the 1931 Class B men's semifinals, Dunean (Greenville) and Converse (Spartanburg) mills battled to a tie in both regulation and the first overtime, when the scorer discovered an error putting Dunean a point ahead. Converse protested their loss, and the game was replayed the next evening. The Spartanburg boys lost 56-27.

"The Greatest Athletic Meet in the World" was good for professional baseball players shaping up for spring training. Bill Voiselle (New York Giants and Ninety Six Mill) and Art Fowler (Cincinnati Reds and Converse Mill) hustled their way straight into a new season on the diamond.

The 1940s introduced the basketball world to Earl Wooten, the greatest player ever to grace the tournament. He was MVP in 1947, averaging 36.3 points per game for Pelzer's Class A men and was a major reason the tournament drew more than 20,000 fans that year. A 12-time All-Tournament selection, Wooten thrilled crowds with matchless skills for 21 years and was afforded legendary status long before he retired from the game.

Mills began recruiting from the college ranks in the 1950s. Ellerbe "Big Daddy" Neal (Wofford) and Jim Slaughter (USC) were only two among many stars that graced the competition, and big-time talent made way for the Open Division experiment in 1956, named because any industrial team could participate. Neal scored 49 in the Open finals that year, but his American Enka team lost to Dunean 119-105 in one of the greatest championships ever played.

The Open Division became a permanent fixture in 1959. All-Americans Jack Salee (Dayton University) and Tommy Kearns (UNC) joined Piedmont Mill's eventual champions, and big-name college guys made the event a regular season-ending destination.

But it was the 1965 Open Division semi-final that provided both the greatest game and individual match-up in tournament history. Mikro (Charlotte) had Billy Cunningham, UNC All-American and ACC Player of the Year, and the Greenville Old Pros countered with Fred Hetzel, Davidson College All-American and Southern Conference Player of the Year. Billy C scored 39 as Mikro pulled away for a 112-78 win.

Another member of that team was Robert Sorrell, the first African-American player to participate in the tournament. He more than held his own with better-known teammates like Cunningham, but his biggest contribution was paving the way for players like the incomparable Henry Logan. In 1966, the Western Carolina University great starred with Sunshine Cleaners, and led the Columbia, South Carolina, team to the Open title.

If there was a saving grace to the tournament from the 1970s to the 1990s, it was in promoting the area's fine high school talent to the college ranks. In 1977, that included Larry Nance, Horace Wyatt, and Terry Kinard (future Clemson greats); and Zam Fredrick and Willie Scott (soon-to-be USC stars). The 1993-94 games featured Mauldin High's Kevin Garnett, with performances offering glimpses of his future NBA prowess.

Yet, there were more players than fans in attendance, and the biggest draw was often the Old Timer's Reunion. Even Pearl Moore, the nation's leading scorer at Francis Marion College who led Spartanburg's Four Star Sports to the women's title in 1986, elicited little fan interest. Any mentions in the media during the later years were brief game accounts or perhaps an article on days gone by.

Created first for the children of the mill village, the Southern Textile Basketball Tournament later became a showcase for some of the best college players in the nation and, finally, a haven for college coaches scouting high school talent. The good doctors Hollis and Naismith would be proud.

Working Class Hero
Beaumont Woman's Supreme Victory

by Ross K. Baker

On April 24, 1963, a case was argued before the U. S. Supreme Court that would impose upon governments at all levels a severe restriction on a fundamental First Amendment right—the free exercise of religion. The case was *Sherbert v. Verner*, and it is one of those cases justly referred to as "landmark." The fact that it was overturned a quarter of a century later does not render it any less significant because the principle it laid down continues to infuse the debate over "free exercise" with its peculiar and explosive energy.

The Sherbert in *Sherbert v. Verner* was Adell Hoppes Sherbert, a widow with five children who had worked at the Beaumont Division of Spartan Mills since 1924 when, as a 22-year-old, she took a job as a spool-tender. At that point, she was a veteran millhand, having begun work at the age of 11 at the mill in Cowpens. She worked continuously at Beaumont until July 28, 1959, when she was fired for refusing to work a Saturday shift. The mill recently had increased production and bumped the workweek from five days to six. She declined to work because she was a member of the Seventh-Day Adventist Church whose doctrines designate Saturday as the Sabbath because it was that day that Jesus, as a Jew, would have observed the day of rest.

Sherbert, a quiet, red-haired woman, had joined the church two years earlier after deciding that the message at Beaumont Methodist Church did not fit with her understanding of the Bible. She and a friend from Drayton would travel across town to the Seventh-Day services each week. The mill gave her three warnings that refusal to work would lead to her dismissal. At that time, she was taking home a weekly paycheck of $32.56. She sought employment at three other mills in the area, but could not find a job.

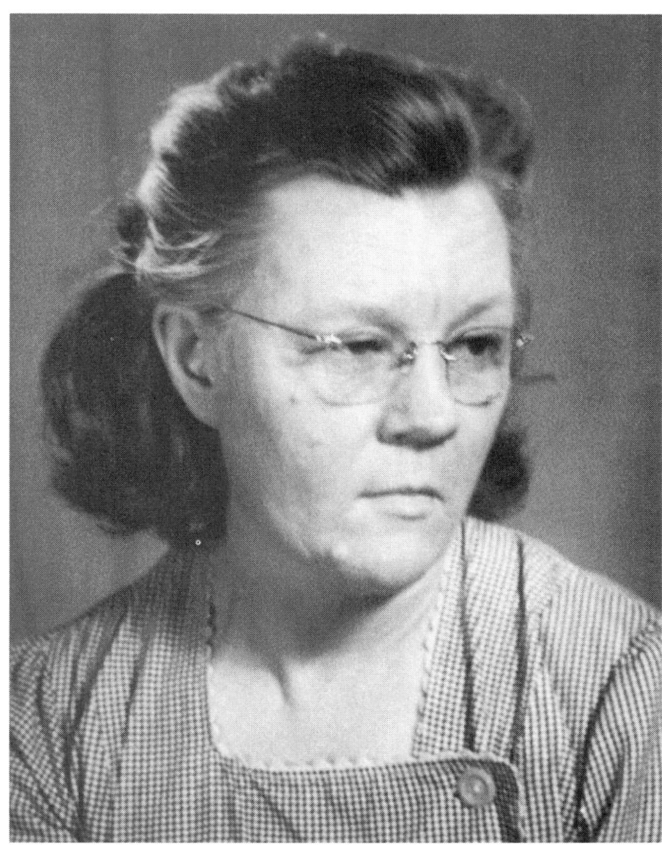

Adell Sherbert (Courtesy, Frank Sherbert)

Sherbert's application for unemployment compensation was denied by a claims examiner for the South Carolina Employment Security Commission. The examiner's denial of the claim was upheld by an appeals board, and on January 5, 1960, Sherbert took her case to the Court of Common Pleas for Spartanburg County for a judicial review of the Commission's decision, but here too she lost. In his decision, Judge J. Woodrow Lewis held that by refusing to work on Saturday she had made herself unavailable for work and therefore could not claim benefits.

She appealed once again to the Supreme Court of South Carolina, naming as the principal defendant, Charlie Verner of the South Carolina Employment Security Commission. Her appeal, in the case of *Sherbert v. Verner*, was rejected May 17, 1962.

Adell Sherbert then took the next step and applied to the U. S. Supreme Court. Forty years later, her eldest living son, Frank, suggested in an interview that pursuing the lawsuit had been encouraged by the Seventh-day Adventist Church as a test case. Filing briefs in her favor were the American Jewish Committee, the Anti-Defamation League of B'nai B'rith, and the American Civil Liberties Union.

On June 17, 1963, the court held that the denial of unemployment compensation to Sherbert restricted the free exercise of her religion.

Justice William Brennan, writing for the 7-2 majority, asked whether South Carolina had "some compelling state interest" in applying the unemployment law so as to interfere with Sherbert's religious practices. He answered it by saying, "Only the gravest abuses...give occasion for permissible limitation."

The Sherbert test for allowable state action would remain the constitutional standard for the next 25 years until rejected by the Rehnquist Court in the case of *Smith v. Employment Division.* But it remains the standard that has been embraced by Congress in its efforts to nullify the Smith ruling and restore "the Sherbert test" as the standard to which both the federal government and state and localities must adhere in enacting statutes that restrict the free exercise of religion.

For Adell Sherbert, the victory was not an unqualified triumph. Interviewed on the day of the Supreme Court decision, she told a reporter, "It makes me feel good," but quickly pointed out that she was still unemployed and added, "The textile mills work on Saturday...This is the only work I know. I've tried and tried to find work but it seems every place works on Saturday."

Aside from the ironic inability of Adell Sherbert to find a job that did not require Saturday hours, the case of *Sherbert v. Verner* seems to have left little residue of bitterness in the Spartanburg community. A spokesman for the Beaumont Division of Spartan Mills declined to comment at the time on the court's decision, and Adell Sherbert continued to live in the small house on Southern Street, just behind the mill, that had been sold to her by Beaumont in 1956. Another paradoxical touch is that the one job that Sherbert could get was as a babysitter. Among her most frequent charges were the grandchildren of Walter Montgomery Sr., the owner of the mill. She also worked at the nursery school at the Episcopal Church of the Advent where the Montgomery family worshipped.

When the Supreme Court rejected the standard laid down in *Sherbert v. Verner* in 1988, Congress reemployed the statute in 1993 when they passed the Religious Freedom Restoration Act. That law was found unconstitutional several years later in the Texas case of *City of Boerne v. Flores.* Congress struck back with the Religious Land-Use and Institutionalized Persons Act of 2000 that sought to restore the Sherbert standard. Currently, a number of cases testing the constitutionality of the legislation await the Supreme Court's attention.

As for Adell Sherbert, she died on December 12, 1989. The five-paragraph obituary in the Spartanburg *Herald-Journal* made no mention of the central role she played in one of the most important cases in American constitutional law.

Selling Arcadia
A Mill Wrestles with the Details

by Betsy Wakefield Teter

By 1950, most of the mills in Spartanburg County were moving quickly to sell the houses in their villages. But that wasn't always an easy process. The extensive personal archives of Greenville real estate agent Alester Furman, housed at Clemson University, show the obstacles that Mayfair Mills had to overcome before it could dispose of the 298 houses in Arcadia. Half the houses in the village had been built about 1905; the other half were constructed about 1920.

When Mayfair first contacted the Furman Co. in late 1950, Furman estimated that the sale of the houses could bring as much as $694,000 if they were financed through the Federal Housing Administration. If a conventional lender financed them, the average home price would be slightly lower, and Mayfair stood to make $600,000, Furman wrote. However, he cautioned, the FHA would require some renovation to the homes before they were sold. Representatives of the FHA toured Arcadia on February 6-7, 1951, and took notes of what they found. They were especially concerned with the older homes in the east end of the village. H. E. Bailey, state director of the FHA, wrote to Alester Furman on Feb. 8:

> **A close check of all the houses this week included discussions with the occupants of the deficiencies. In practically all cases it was learned that roof leaks varying from one to four per house existed. Interior foundation piers were in a number and in poor condition. Numerous installations of wood stiff-knees had been made in an effort to reduce floor tremor, and fully 90 percent of all window frames and window sashes were rotted to the point where repairs**

and replacements are necessary. Chimney tops generally are badly in need of repointing. It is only fair to purchasers that they be afforded a reasonable period of repair-free occupancy.

He also noted that 141 of the homes in Arcadia did not have bathtubs. Furman, in turn, wrote to Fred Dent, president of the mill: "In line with your expressed feeling that the mill preferred low sales prices to an extensive repair program, the inspection was made more with the view of keeping repairs to a minimum. We regret to report that the inspections results were very disappointing."

Despite the bad news, Mayfair chose to proceed with the FHA program and committed to an extensive renovation program in the village. In late spring, the mill sold 261 houses at an average sales price of $2,640. Proceeds of sale—before renovation costs were figured in—were $611,795. The Furman Co. took a total commission of $32,199 for its work.

Later that year, Fred Dent wrote Alester Furman requesting an appraisal of 16 acres the mill had chosen to donate to School District 6. He took the occasion to update Furman on the condition of Arcadia:

"The physical condition of our former cottages has changed considerably within the short period of time since the sale, and we are gratified to note the interest which people are showing in their houses. It is a most commendable attitude and, of course, enhances the overall success of our village sale a great deal."

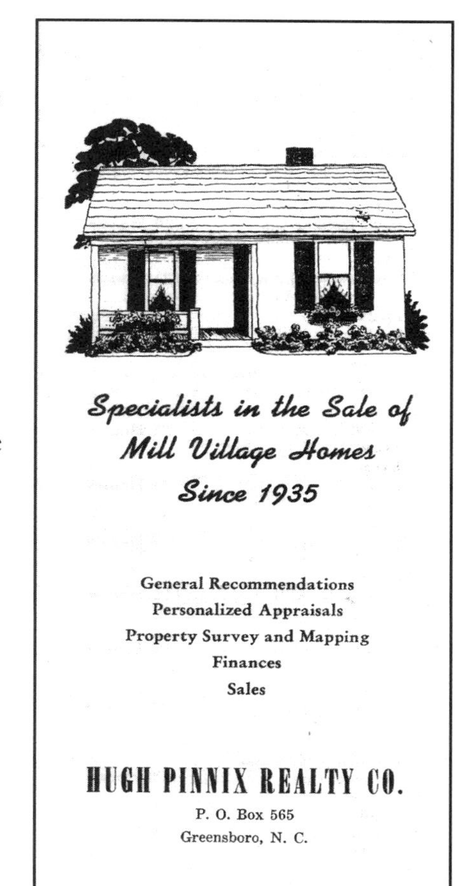

Brochure from a real estate company offering its services. (Courtesy, Converse College)

Butte Knit
Decline of a Double-knit Dynamo

by Betsy Wakefield Teter

Few communities in the world benefited from the double-knit fad as much as Spartanburg, the center of a sprawling homegrown enterprise called Butte Knitting Mills. But illustrating the old adage, "the bigger they are, the harder they fall," the collapse of Butte Knit was the largest single economic blow in Spartanburg's history. When the market for the once-hip fabric crashed in the late 1970s and early '80s, the double-knit dynamo disappeared as quickly as it had arrived.

The short story of Butte's meteoric rise and fall began in 1960 when New York dressmaker David Schwartz decided to cash in on a fashion craze sweeping Europe and dispatched Andrew Teszler to Spartanburg to organize the Butte Knit division of Jonathan Logan Inc. Butte successfully got out ahead of other U. S. manufacturers with an initial location at Interstates 85 and 26 (the "Crossroads of the New South"), creating a plant that was the only completely integrated knitted garment manufacturing plant in the country. Butte workers made the yarn, knitted it, cut and sewed the fabric into ladies' garments, then shipped it nationwide under the labels Act III and Butte Knit.

Double-knit, a strong, two-layer fabric made using a machine with a double set of needles, soared in popularity as women abandoned the conservative fashions of the 1950s and began wearing more casual pants suits, tunics, and jackets. Young women wore brightly colored stretch fabrics identified with hippie culture, and men's fashions followed, peaking with the leisure suit/disco craze of the middle '70s. One ad promised that double-knit fabric could be worn up to 67 days without washing or ironing and still maintain its crisp appearance.

All the while, Butte kept growing in Spartanburg with a goal "to present to the average American woman a new, young, exciting approach with knits in sportswear and separates category." From the beginning, Butte planned to expand the company as rapidly as conditions

Butte Industrial Park was a four-building complex on Sigsbee Road (Courtesy, the Spartanburg Herald-Journal)

warranted. From 1960 to 1976, Butte expanded 24 times, building 2.2 million square feet of factory space in the Spartanburg area and adding 2,400 employees. In addition to the plant on I-85, Butte built the Buckeye Forest spinning mill and dyehouse west of Spartanburg on U.S. Highway 29; a four-building complex called Butte Industrial Park off New Cut Road; and, near the end, a cotton yarn plant at I-85 and U.S. Highway 221. Simultaneously, Butte built or acquired 26 sewing factories from Pennsylvania to Florida to Arkansas. Sales of its women's wear soared 17-fold between 1966 and 1971.

The demands of meeting production led Butte employees in Spartanburg to unionize and establish a local chapter of the International Ladies Garment Workers of America. Periodically, the predominantly female workforce went out on wildcat strikes, though labor relations with New Yorker Andrew Teszler were generally characterized as good.

By the late 1970s, however, "leisure suit" clothing became a fashion laughingstock. The demand shifted to woven—rather than knitted—fabrics. What little market remained was quickly being swallowed up by imports. Production of Butte fabric dropped by two-thirds between 1976 to 1980. Butte leaders explored the idea of changing product lines. "But there was no way we could convert to a weaving plant and be competitive with those that already exist," said a disconsolate Richard Schwartz, son of the founder and company president.

On September 9, 1981, Butte announced two plant closings and the largest single layoff in Spartanburg County history—1,200 people. Four years later, the local sewing factories and shipping department were closed, too, ending the Butte era.

Double-knit survived as a niche fabric, used primarily in some baseball and football pants and jerseys. One of the Butte plants provided an entry into Spartanburg County for Japanese textile maker, TNS. Another was marketed unsuccessfully as "industrial condos." All still stand, mostly empty, as a reminder of the days when Spartanburg was the dean of double-knit.

Andrew Teszler
A Life Cut Short

by Tanya Bordeaux Hamm

Keen organizational skills and an innate ability to motivate people are qualities colleagues say led Andrew Teszler to a high-profile life as a young industrialist and philanthropist in the 1960s. Teszler, who established two textile operations in the Spartanburg area, was known for bringing the double knit market to the United States and for his generosity in the community, making donations that have lasted well beyond his short lifetime.

In January 1960, when he was 28, Teszler moved from New York to Spartanburg, where he opened Butte Knitting Mills for parent company Jonathan Logan off Interstate 85. Under his leadership, Butte grew into an enterprise with more than 3,000 employees. But in 1970, Teszler decided to start his own "vertical" knitting mill similar to Butte and left to found Olympia Industries, working first in a small office in the Hillcrest area. With a 400,000-square-foot plant in Woodruff, Olympia knit polyester fabrics and employed 900 people. It also had operations in Tuscaloosa, Alabama.

On May 6, 1971, at age 40, Teszler died suddenly of a heart attack while working late in his office. His new company had barely started up.

Teszler's death came as a tragic surprise to the Spartanburg community—he had no history of heart disease, although his mother had died of a heart attack. In a tribute, the Spartanburg *Herald-Journal* ran long lists of remarks from friends, co-workers, and community leaders who described him as brilliant, energetic, inspiring, hard-working, and concerned. Decades after his death, people remember him for these qualities as well as for his organizational skills, ability to get things done, and motivating management style.

Andrew Teszler (Courtesy, the Spartanburg Herald-Journal)

"He was basically a textile genius. He made the finest fabrics made in America, and was very respected by everyone in the industry and by each of his employees," said Ralph Iannazzone, who worked with Teszler at Butte, then at Olympia.

Iannazzone, who is retired from the clothing industry and lives in Florida, said he and Teszler considered one another brothers, and their families have kept in close touch through the years. "Not only did Andrew know textile machinery from A to Z, but he knew how to move people in a wonderful manner. Under him, we exceeded our own expectations," he said.

After his death, employees recalled Teszler literally giving an employee the coat off his back after the man said he admired it. When the five-year-old son of a veteran employee in the dress-inspecting department had leukemia, Teszler organized a blood drive for the boy, then flew him and his parents to St. Jude's Hospital in Memphis in his private jet plane. It is also said that he gave another employee $1,000 when his house was destroyed by fire and paid for one young man's college education.

Gifts Teszler made to several Spartanburg institutions before his death have made a significant impact on the community. Wofford College's "Sandor Teszler Library," named after Andrew's father, was made possible through his gifts, and donations Andrew made helped build the education building at Temple B'Nai Israel. He purchased the first intensive care heart unit and the first heart scanner for Spartanburg General Hospital, and spearheaded a campaign to raise $950,000 to open the Charles Lea Center in 1969. This prompted the center to recognize him by naming its school the Andrew Teszler Learning Adjustment School.

Born and raised in Yugoslavia, Teszler was a graduate of Leicester College of Technology and North Carolina State University. He learned the textile business from his father, Sandor, who operated a textile business in Yugoslavia before coming to the United States. In addition to running his mills, Andrew Teszler was a Wofford College trustee and served on Governor John C. West's Commission on Human Relations.

Teszler was honored for his accomplishments when he was named "Man of the Year" by Phi Psi Fraternity, North Carolina State University's School of Textiles fraternity, and awarded an honorary degree. The "Man of the Year" award recognized Teszler for his substantial contribution to the advancement of the textile industry as well as his contributions in the community.

"My father grew up in a factory, literally, so it was in his blood," said his son, David Teszler of Atlanta, who was 12 at his father's death. "It wasn't so much that he was a genius in textiles; he was a genius in business and in motivating people."

David Teszler said the thing that struck him most about his father was this ability to get the best out of people. He recalls one example of his father's motivational skills that came after he brought home a poor report card. He placed the report card—containing Ds and Fs—on his father's bed after school. The next morning, Andrew Teszler called a young David in as he was dressing for work.

"I remember he asked me what happened, then he gave me $20, quite a bit of money for a child," David Teszler said. "When I asked him why he gave it to me, he replied, 'because you told me the truth.' To this day, I don't lie. He just knew how to motivate you in that way."

An International Influx
Following Textiles to Spartanburg

by Laura Hendrix Corbin

When foreign imports produced with cheap labor began threatening the textile industry in South Carolina in the early 1960s, the industry ironically looked overseas in an effort to infuse new technology to keep itself viable. Nowhere was the search more intense than in Spartanburg.

Roger Milliken and other textile leaders—along with the late Richard E. Tukey, executive vice president of the Spartanburg Area Chamber of Commerce, and former Governor John C. West—led South Carolina to begin luring German and Swiss textile machinery operations to come here to be close to the local mills. Because U. S. textile plants were quickly filling with faster European-made machinery, they needed those companies' technicians nearby to make sure the equipment was properly installed and maintained.

Tukey and others also knew the long-term value of diversifying the state's economy through this "reverse investment," building on the foundation of the textile industry. Spartanburg became the leader in attracting international companies, and Tukey was recognized as the man who single-handedly molded the state's worldwide reputation as a premiere business location. Tukey, who led the Chamber from 1951 until his death in 1979, lured more than 60 foreign firms to build in South Carolina, most of them in the Upstate. Today, Spartanburg County is home to more than 100 international companies representing some 20 countries.

Tukey's old friends tell of how he put foreign businessmen up in his home, introduced them to banks and community leaders, and led them by the hand through setting up not only their U. S. business operations, but their U. S. homes (these legendary stories became known as "Tukey Tales"). Tukey's wife Jenny and other "Chamber wives" shepherded the foreign business

Swiss-owned Rieter AG was one of the first international textile machinery companies to set up operations in Spartanburg. (Courtesy, the Spartanburg Herald-Journal)

leaders' wives around the community, showing them the grocery stores and beauty shops, touring the local schools with them, helping them get acclimated to the community, making them welcome.

For decades, business leaders such as Milliken, Tukey, and Dr. Paul Foerster, the German retired head of the former Hoechst-Celanese Spartanburg facility, were "prime movers of economic development, shaping an industrial policy that is also a foreign policy," wrote Dr. Rosabeth Moss Kanter, Harvard Business School professor and author of *World Class: Thriving Locally in the Global Economy*. Her book, published in 1995, spotlighted the way Spartanburg had managed to make itself an international business center.

The development of the area as an international manufacturing hub happened in stages, starting with sales and service operations and later branching into manufacturing. One of the first European textile machinery sales operations in Spartanburg opened in 1959 when Brian Lyttle came from Ireland to start Hobourn Sales, a division of Hobourn Areo Components of Rochester, England. Lyttle purchased the company in 1964, changed its name to Brian Lyttle Inc., and represented several European machinery manufacturers.

Rieter AG, a Swiss textile machinery manufacturer, came the same year, starting as a sales and service operation, expanding later into manufacturing. Sulzer Ruti Inc., a subsidiary of the Swiss-based firm Sulzer Ruti Limited, opened an office in Spartanburg in 1963 to provide sales and service to its growing customer base in the U.S. textile industry. Company officials saw the potential for Spartanburg to become a hub of worldwide commerce and trade, and it thrived here.

The transition from sales and service to manufacturing began in earnest in the mid-1960s when German-based Menzel Inc. decided it was more practical to build machinery in Spartanburg. Now three times its original size, a diversified Menzel now derives less than 40 percent of its revenue from the textile industry.

Kurt Zimmerli, Swiss-German head of Zima/Küsters/EVAC, was one of those whom Tukey squired around. Zimmerli established Zima in Spartanburg in 1969, immediately becoming active in the community, helping found a bank and serving as chairman of the Spartanburg Area Chamber of Commerce's international committee for 20 years.

Among the other international textile-related firms arriving in Spartanburg in the 1960s and early 1970s were Staübli, Bruckner Machinery, Mahlo America, Zimmer Machinery America, Otto Zollinger Inc., Graf Metallic of America, SACM Textiles, Consultex, and Texpa America.

As the twenty-first century dawned, there were at least 35 international suppliers to the textile industry located in Spartanburg County. While most were German and Swiss, also represented were Austria, Belgium, and Japan. The four Italian companies located in the county were all textile related.

They paved the way for an influx of international companies of all kinds, leading Spartanburg to designate itself as having "the highest per capita diversified international investment of any area in the country." Before snagging the economic development plum of BMW Manufacturing Corp. in 1992, Spartanburg County already had lured the likes of Michelin Tire (France), BMG Entertainment (Germany), Alcoa Fujikura (Japan), Polydeck Screen (Germany), adidas American (Germany), Beresford Box (Canada), Trimite Powders (England), and Dare Foods (Canada).

These companies were looking for something more than a welcoming business climate and Southern hospitality. They wanted good hard workers, and they found them. The work ethic of the textile workers was rooted in an agrarian society; the adage "a good day's work for a good day's pay" was often quoted by heads of international firms based here. "The human 'product' is right in Spartanburg," said Hans Balmer of Symtech. "The labor force has a very good attitude of wanting to work, doing a good job, respecting what a company is doing for them, and being thankful for it, not taking it for granted."

Spartanburg also was the leader in establishing a technical education system and a workforce-training program—the Center for Accelerated Technology Training—that is second to none nationally. The area embraced the need to upgrade skills and, through the 1990s, continued to prepare workers who could meet the demands of an ever-changing industrial base.

By 2002, South Carolina had the distinction of being "one of America's most powerful magnets for foreign investment," boasting 500 internationally owned businesses representing 23 nations and employing more than 89,000 people. During the 1990s, more than $16.5 billion was invested in the state from international sources, creating more than 49,000 jobs.

Hoechst-Celanese
Making Fibers from Chemicals
by Betsy Wakefield Teter

Chamber of Commerce President Dick Tukey, left, and Gov. John West, center, welcome Paul Foerster and the giant German manufacturer Hoechst to Spartanburg. (Courtesy, Spartanburg Herald-Journal)

Eighty-five years after it first beckoned cotton mills to Spartanburg County, the Pacolet River enticed another kind of textile enterprise to its banks—a plant that began innocuously in 1965 as Hercules Powder Company and eventually became known as Hoechst-Celanese. At its height of operation, the German-owned company grew to employ more than 3,000 local people, primarily making man-made fiber and filament to supply Southeastern textile mills.

Ultimately, that one plant would produce 400 million pounds of polyester staple a year—an amount that dwarfed the size of Spartanburg County's cotton crop at its height in the late 1920s (when Spartanburg County led the state with 135,000 acres under cultivation, one-fourth of all land in the county).

The precursor to that enterprise, though, was Hercules. Based in Wilmington, Delaware, Hercules was the manufacturer of a petroleum-based chemical used in the production of polyester fiber. The company liked the Spartanburg site because it needed cooling water from a river for its operations, plus a place to discharge its treated effluent. The Pacolet, already home to six textile mills, was just the place. But when the Spartanburg Chamber of Commerce officials first welcomed Hercules, they had no idea of the scope of the manufacturing facility that would ultimately wind up occupying that site.

Before Hercules could even open the doors to its $12 million plant, Hoechst—the largest chemical company in Germany—decided to come to the United States to make its polyester staple, sold under the brand name "Trevira." Hoechst wanted to go head-to-head with such heavyweights as DuPont (whose polyester was known as "Dacron") and Fiber Industries, a British-American joint venture with a big plant under construction in nearby Greenville. By joining Hercules

German fiber manufacturer Hoechst merged with U.S. competitor Celanese in the 1980s. Together they employed about 3,000 at the Spartanburg facility. (Courtesy, Louise Foster)

in Spartanburg County, they could have one of their major suppliers on an adjoining piece of property.

Plus, almost every customer Hoechst targeted was in the vicinity. "In the mid-'60s, if you drew a circle of 150 miles around the intersection of Interstates 85 and 26, otherwise known as the 'Crossroads of the New South,' you had probably 80 percent of U. S. textile spindles," recalled Paul Foerster, Hoechst's original and long-time plant manager. Together Hoechst and Hercules formed a new company for Spartanburg called Hystron Fibers. The "Hy" in the name referred to the English translation of the German word Hoechst, which means "highest." The "stron" came from "strong," a description of the mythical figure Hercules.

Together, they began by turning out 25 million pounds of polyester staple a year, which was sold to mills like Mayfair, Spartan, and Milliken for the production of cotton-blended fabric. From the start, wages at the facility were among the highest in the area—about 25 percent higher than those paid by textile mills. "We had to do that or we would have had unions at our door right away," Foerster said.

Over the years there were numerous expansions. One of them was a plant to make filament, a tiny silk-like polyester thread; hundreds of miles of this thread could wind around one spindle. There was also a large customer service office and a research and development center with a small textile mill where new products were developed. In 1970, Hercules left the partnership, and the plant became known as Hoechst Fibers. And in the 1980s, using the same technology that it used for polyester staple, Hoechst began producing plastic for soft drink bottles. All together, more than $500 million was invested in the sprawling complex.

The biggest change came with the merger with Celanese Corporation in 1987. Although it was about 30 percent of the size of Celanese—which had polyester plants in Greenville, Florence, and Salisbury and Shelby, North Carolina—Hoechst purchased its competitor for $2.7 billion. The Federal Trade Commission ruled, however, that the new company, because of its size, violated anti-trust laws. Foerster and a group of three other Spartanburg men considered buying the Spartanburg plant and creating a separate company, but in the end, the plant in Florence was spun off instead.

Unfortunately for Spartanburg, Celanese's larger Charlotte offices became the fiber company's new headquarters. There was a steady erosion of white-collar jobs as the new company consolidated operations. Then, changes in technology made the manufacturing side less labor intensive, further reducing employment. Finally, textile customers began going out of business because of competition from foreign imports, cutting demand for Hoechst-Celanese products.

In the late 1990s, Hoechst began selling off the plant in pieces, and several operations were shut down or relocated. The polyester staple and resin business were all that remained in Spartanburg, owned now by a Mexican-American partnership, KoSa. More than 2,000 jobs had disappeared in fewer than 10 years.

The Draper Story
Rise and Fall of a Loom-making Giant

by Glenn Bridges

With the textile industry well established in Spartanburg by the 1920s, the nation's largest loom maker soon followed.

Draper Corporation, established in the early 1800s in New England by the Drapour family, realized the urgency of having spare parts readily available for Southern mills if it wanted to meet the industry's solid shift to that part of America. Highway travel between the North and South meant countless hours of lost time if the high-speed, hard-knocking machinery it built suddenly splintered a part.

Thus, in 1929 Draper opened a warehouse in Arcadia that not only served textile mills in the Upstate but also complimented a loom manufacturing shop it had opened in 1910 in Atlanta. Then in 1936, it began manufacturing loom parts at a new plant on South Pine Street in Spartanburg. From the outset, this sprawling factory had a large foundry operation—one of the largest in the Southeast—and from those castings, Draper manufactured more than 2,000 different parts used on looms.

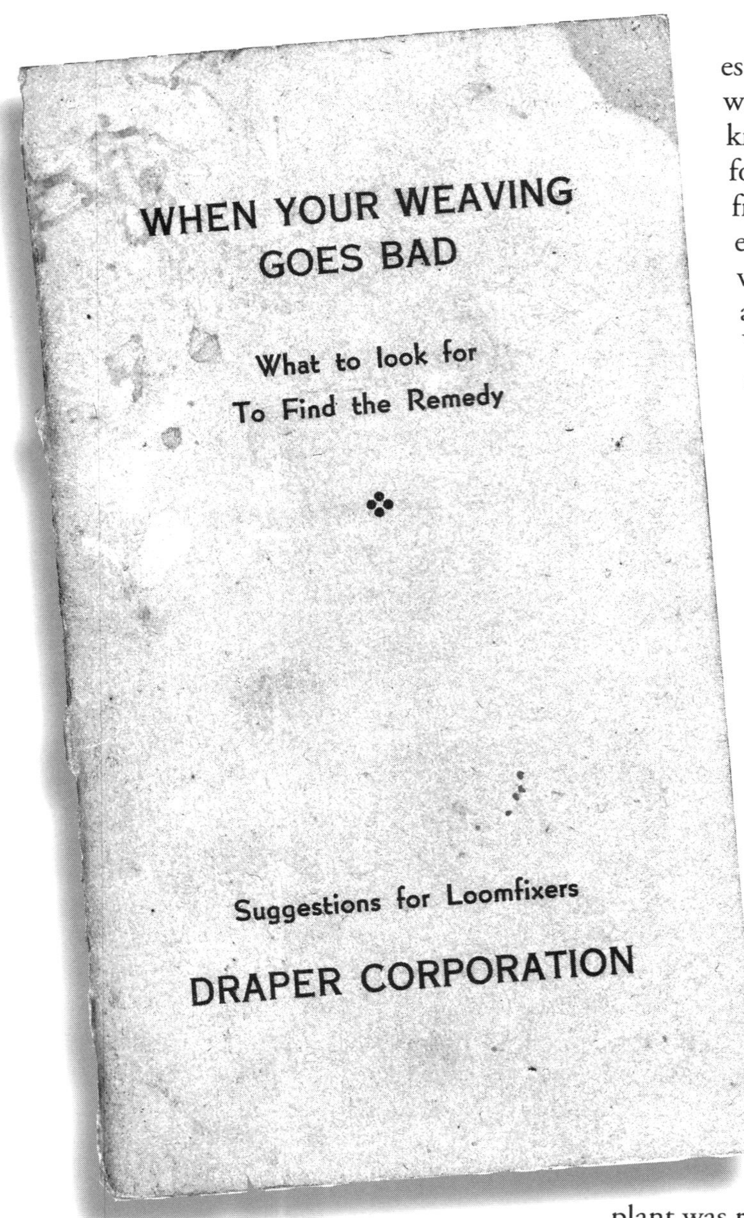

This 16-page manual was issued to a Drayton Mills employee in 1947. (Courtesy Spartanburg Regional Museum)

By the time Draper came to Spartanburg, it had already established a solid reputation as a leader in loom designs and was the owner of numerous patents. One of Draper's best-known inventions was the "temple." In the early years following the Industrial Revolution, most mills converted from a water-generated power source to automation through electricity. This greatly enhanced production, but the work was still tedious and slow—partly due to frequent stops to adjust the loom and re-stretch the cloth back into position. With the introduction of the temple, this process became automatic and meant the weaver could operate ten or more looms simultaneously.

Between 1886 and 1996, Draper Corp. built more than one million looms. Operating from plants in the South and Northeast, it controlled a substantial portion of the U. S. market and was hugely profitable. By 1960 the demand for Draper looms was so great that about 1,200 of them were assembled at the Spartanburg plant. Employment in Spartanburg ballooned to 1,200 people, and Draper was considered one of the premium places to work, offering some of the best wages and attracting the brightest talent in the area. The Spartanburg operation was not only providing a secure lifestyle for hundreds of Upstate families, it was also a leading contributor to local charities. At that point, just about every fabric worn or used in the United States was made by a Draper loom, including shirts, towels, draperies, blankets and bed sheets, according to Ron Brown, a Spartanburg resident who began working for Draper immediately after graduating from high school in 1953.

In 1966, Draper peaked at about 5,000 employees nationwide and was well known as a performer on the New York Stock Exchange. Employees wanted a bigger piece of the pie, and in September the Spartanburg plant was rocked by a two-week strike that put 1,000 workers on the picket line. The company's profitability also attracted the attention of Rockwell Standard, a manufacturer of carburetors and axles, later known as Rockwell Automation. The huge company expressed an interest in buying Draper, but Draper officials were not interested.

Refusing to take no for an answer, Rockwell made a lucrative pitch to individual stockholders until it secured enough interest for a takeover. The company felt it had the scientific know-how to re-invent the weaving industry, but it never succeeded. In fact, just the opposite happened. In the 1960s and '70s, the Swiss and Italians aggressively captured the market. Then followed the Japanese, making faster, more efficient and longer-lasting looms, which obliterated demand for Draper-made products.

In 1971 the Arcadia warehouse and offices consolidated with the South Pine Street facility, and a 90,000-square-foot expansion was built. The Arcadia facility was sold to Barnet Southern, a fiber reclamation company that eventually moved its U. S. headquarters to the city.

Draper loom, 1950s (Courtesy, Ron Brown)

By 1983, Rockwell had lost interest in textiles and was no longer keeping pace with the latest technology, so Brown and nine other investors secured the funds to repurchase the company. Unfortunately, it was difficult to regain the luster Draper enjoyed prior to Rockwell's takeover; in 1996 the company was acquired by Texmaco, a large Indonesian industrial firm with 26,000 employees, 15 manufacturing divisions, and almost $1 billion in annual sales.

The new arrangement as Draper-Texmaco was a perfect fit for the company's products—air-jet, water-jet and rapier looms. Under the DT umbrella, Polysindo now produces the fiber and yarn, Jaya Mills the cloth, and Perkasa Engineering, the machinery, meaning Draper-Texmaco has evolved into a vertical integration that serves the industry from start to finish. From the blue building on South Pine Street, the company makes and sells parts for textile carding and spinning operations, as well as a growing line of non-textile parts. Employees there also continue to make parts for the last Draper looms remaining in operation around the world—perhaps 2,000 of them.

Fred Dent
From Arcadia to Moscow

by Tanya Bordeaux Hamm

One Spartanburg textile leader made it to Washington, D. C., and around the globe in a post that merged his experience in the industry with his interest in government and politics, as he negotiated trade agreements overseas and led commerce activities in the United States.

Frederick Baily Dent, longtime president and chairman of Mayfair Mills in Arcadia, served in the presidential cabinets of both Richard Nixon and Gerald Ford, first as U. S. Secretary of Commerce, then as a special trade ambassador. As Secretary of Commerce, he was credited with organizing the Bureau of East-West Trade, a service agency that informs American businessmen of trade opportunities in Communist countries. He encouraged a greater national commitment to U. S. export markets, reducing the number of products that needed a special government export license from 550 to 73.

Dent's service in Washington came after decades as a textile businessman, beginning with the sales firm Joshua L. Baily & Company of New York, founded by his great-grandfather. He moved to Spartanburg in 1947 to learn the manufacturing end of the business and ended up staying at Mayfair Mills in Arcadia, becoming president in 1958 when the mill's president died. Under his leadership, the company expanded from one mill to six plants—five in the Upstate and one in Georgia—producing fabric for the apparel and household industries. "I came south to learn manufacturing at Mayfair, thinking we would be here a year or so, but we have been here ever since," he said.

In the 1960s, with the backing of U. S. Senator Strom Thurmond, Dent traveled overseas, negotiating trade agreements limiting imports into the United States, then was asked to be a member of the Commission on an All-Volunteer Army. Through his friend and fellow Upstate textile leader Roger Milliken, he helped to recruit business support during President Nixon's second campaign for the presidency.

These experiences gained Dent the recognition of the Nixon administration, and in 1973, President Nixon named him his third Secretary of Commerce. "Out of the blue, I received a call about whether I would be

Fred Dent with President Gerald Ford (Courtesy, Fred Dent)

interested," Dent said.

"I'll never forget our first cabinet meeting was one night late in January, when the president announced they had received a peace accord with Vietnam," he said. "It seemed like a magnificent start to a new administration. I knew President Nixon quite well, and he was a very, very able man."

After the Watergate affair and Nixon's resignation from the presidency, Dent continued to serve as Secretary of Commerce under President Gerald Ford. Then in 1975, he was named a special representative for trade negotiations with the rank of ambassador, a new cabinet post. Through his work in Washington, Dent is credited with introducing an energy conservation program for industry, which organized the National Energy Conservation Council, and with establishing private sector trade and economic councils in the Soviet Union and China.

"The first time we were going to Moscow, my assistant came in and said we needed a gift to exchange with (Leonid) Brezhnev. I had always wanted a clock that rings on the quarter hour, so that's what we gave him," Dent said.

Oddly enough, the gift that Brezhnev gave Dent was also a clock. Then, during a later meeting, someone noticed the ship's clock that Dent had given the Russian leader on his desk and commented, "You must think of America often." He replied, "Yes, but sometimes, it stops." Up until Brezhnev's death, Dent would often recognize the clock in the background of photographs of Brezhnev in the media.

Once when Dent traveled to the Soviet Union for one of the summit meetings, the Soviet group took the American ambassadors out fishing, and the Soviets jokingly gave Dent's group plastic worms as a symbol of technology (or lack thereof). "We had a very good relationship," Dent recalls.

While in Washington, Dent said his posts required around-the-clock dedication, but also brought many "perks," such as dinners at the embassies, visits to Camp David when the President wasn't there, and evenings in the President's box at the Kennedy Center when the President wasn't using it.

"Working with the various foreign governments on trade matters was interesting and challenging, and Washington was very attractive and an exciting place to be, but it was nice to get beyond the Potomac afterward," he said.

Born in Cape May, New Jersey, in 1922, Dent was raised in Greenwich, Connecticut., and earned a bachelor's degree in political institutions from Yale University. His Yale degree was mailed to him in the Pacific, in 1943, where he was serving in the U. S. Navy during World War II. On a 10-day military leave in 1944, he married Mildred Carrington Harrison, known as "Millie," and they went on to have five children.

Through the years, Dent became active in the Spartanburg business community as well as in the textile industry outside the Upstate. He served as president of the South Carolina Textile Manufacturers Association in 1960, as president of the American Textile Manufacturers Institute (ATMI) in 1968, and earned membership in the prestigious Business Council, made up of leaders in American industry. During his tenure as chairman of the Spartanburg County Planning and Development Commission (1958-1972), the local textile industry received $600 million in government funds.

After his service in Washington, in 1977, Dent returned to the private sector and Mayfair Mills, where his son, Rick, had joined the company. In 1988, Dent was named chairman of the board, with Rick assuming responsibilities as president and CEO. However, foreign imports eventually took their toll on Mayfair and, in 2001, the company filed Chapter 11 bankruptcy and closed its plants. Dent and his family have remained active in the Spartanburg community through the Arts Partnership of Greater Spartanburg, Spartanburg Day School, Spartanburg Technical College, Spartanburg Regional Medical Center, and the Brevard (North Carolina) Music Center.

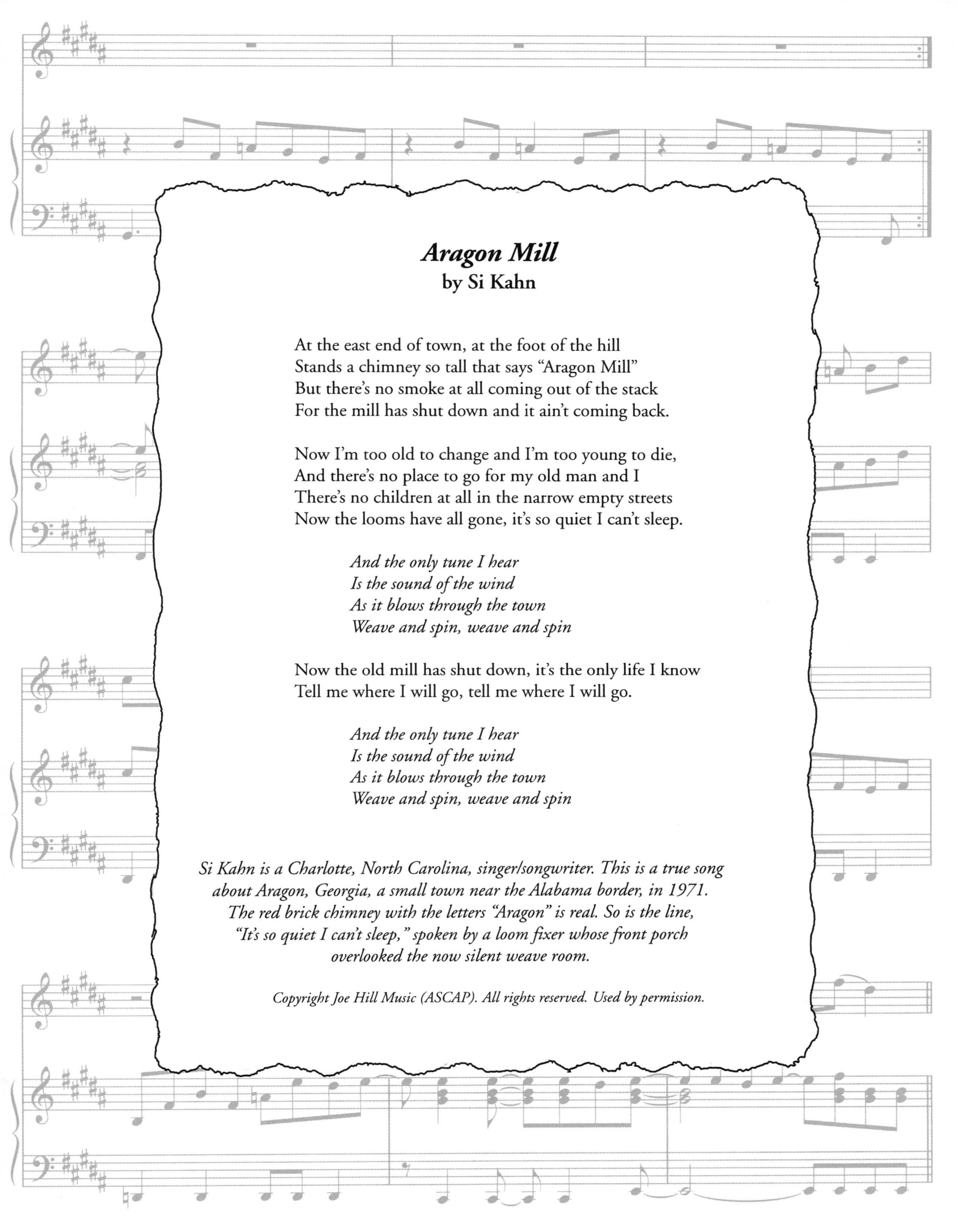

Aragon Mill
by Si Kahn

At the east end of town, at the foot of the hill
Stands a chimney so tall that says "Aragon Mill"
But there's no smoke at all coming out of the stack
For the mill has shut down and it ain't coming back.

Now I'm too old to change and I'm too young to die,
And there's no place to go for my old man and I
There's no children at all in the narrow empty streets
Now the looms have all gone, it's so quiet I can't sleep.

And the only tune I hear
Is the sound of the wind
As it blows through the town
Weave and spin, weave and spin

Now the old mill has shut down, it's the only life I know
Tell me where I will go, tell me where I will go.

And the only tune I hear
Is the sound of the wind
As it blows through the town
Weave and spin, weave and spin

Si Kahn is a Charlotte, North Carolina, singer/songwriter. This is a true song about Aragon, Georgia, a small town near the Alabama border, in 1971. The red brick chimney with the letters "Aragon" is real. So is the line, "It's so quiet I can't sleep," spoken by a loom fixer whose front porch overlooked the now silent weave room.

Junked loom parts on Union Street in Spartanburg
(Courtesy, Mark Olencki)

6 Textile Town in Transition 1975 to 2002

by Betsy Wakefield Teter

On a crisp spring morning in 1975, bulldozers crested a knoll above the huge Westgate Mall construction project on the edge of Spartanburg. They had arrived to dig the foundation for a new corporate headquarters for venerable Spartan Mills, moving now to a low-slung, modern building in a suburban office park developing on the new western frontier of the town that textiles built. This was a heady time for Walter Montgomery Sr. His company was the largest industrial employer in Spartanburg County and had 4,000 employees in the Carolinas and Georgia. Unlike Clifton Manufacturing, Spartan Mills was a survivor and was poised for the uptick in the U. S. consumer economy following the end of the 1974-75 recession. By the end of the year, his sales people and office staff would clean out their desks, drive through the tree-lined boulevard heading to the hilltop, and move into the new building which mirrored, not coincidentally, parts of the Milliken Research compound across town. This move was a tangible symbol that the rapidly modernizing Spartan Mills had arrived.

In fact, it seemed that the whole city had arrived. Community Cash, which had built its fortunes on the grocery money of now-mobile textile workers, had four new stores underway. Fiske-Carter Construction, the builder of mills and hundreds of mill houses in the early part of the century, was finishing up work on the new pedestrian Main Street Mall. A short item on the front page of the Spartanburg *Herald-Journal* announced that French tire maker Michelin had purchased 149 acres west of the city. Craft Drugs, another local chain propelled by rising textile wages, was opening new stores. Plans were being readied for a mega-development downtown called Spartan Square—a project that was expected to bring two skyscrapers, an ice skating rink, a civic center, and a two-level shopping mall. To top it off, a new arts center had opened its doors, Van Cliburn was entertaining hundreds at Converse's Twichell Auditorium, and a rising Spartanburg rock star, Toy Caldwell of the Marshall Tucker Band, appeared

Milliken associates at one of the first company "sharing rallies" (Courtesy, Milliken and Company)

Alvin Wood, president of the S.C. state board of the Carolina Brown Lung Association, makes an appeal on the behalf of Upstate workers with brown lung. (Courtesy, Betty Livingston)

on the cover of *Guitar Player* magazine.

As Walter Montgomery surveyed his new domain, there certainly was no sign that this would all be gone barely a quarter-century later. All of it: the employees, the factories, the corporate headquarters, the millions of dollars in spindles and looms. A sevenfold increase in international textile production between 1945 and 1975 meant that virtually all 200 of the world's nations were producing textile goods by the mid-1970s. As Spartan Mills opened its doors atop the hill at Westgate, the market for American-produced textiles stopped growing, never to resume. The piece of the pie commanded by the titans of Textile Town would get smaller and smaller and smaller.

The news on the short-term, however, was bright. The national recession ended by the second quarter of 1975, and that provided a boost for Spartanburg County textile workers. Their industry

In 1977 the Ramey Plant of Inman Mills, located in Enoree, produced more than ten times as much cloth as the entire company did in 1902. (Courtesy, the Spartanburg Herald-Journal)

Textile employment in Spartanburg County held steady at 19,500 jobs in the 1970s. But more than 5,000 jobs disappeared in the '80s; another 6,000 in the '90s.

had been operating for months at its lowest point since the Great Depression. So many plants cut back hours that the average workweek declined to 35 hours. In some pockets of the Piedmont, unemployment in the textile industry peaked at 30 percent. Some companies didn't survive. Burlington shut down the Valley Falls plant (though Milliken would later reopen it). Collins & Aikman closed a sewing factory in Cowpens. Most, though, rolled with the punches. Deering Milliken emerged with a new name, Milliken & Co., heralding the impact one man—Roger Milliken—was to have on the industry in the latter twentieth century. In 1977 U. S. textile makers experienced an 18 percent sales increase, and the good times spilled over to workers as well; that year, textile workers in Spartanburg County averaged $4 an hour, or $165 a week. While those wages were still lower than those paid by other emerging Southern industries, the rapid inflation in textile wages had begun. Fifteen years later, they would top $10 an hour.

Yet the voices of Southern textile workers could be heard in a rising din of complaints about working conditions, which had changed little since World War II. Studies in the 1960s showed that up to 40 percent of employees in textile weave rooms experienced "severe deafness" after ten years' exposure to the roar of looms. The problem had been exacerbated by the construction of new plants, like Saybrook in Inman and John H. Montgomery in Chesnee, with cement floors, low ceilings, and no windows. Federal legislation introduced during the Nixon administration compelled mills to reduce noise up to 20 percent. As a result of the federal mandate, supervisors got serious about requiring employees to wear earplugs and muffs by the mid-1970s. Even those who resisted were forced to comply; studies showed the soft plastic plugs cut noise levels in half. New European-made looms, gradually making their way into Upstate plants, helped, too. Not only were they more efficient, they were quieter.

Mill owners, including those in Spartanburg County, fought federal legislation on cotton dust standards, claiming the expense to upgrade their plants would result in only marginal improvements in flying lint. In a fierce lobbying battle with labor unions, they advocated the use of protective masks and the transfer of workers exhibiting signs of "brown lung" to other departments. Unions knew, though, that meant many of the highest paid workers—loom fixers and weavers—would suffer a decline in income. Backed by labor union members at J. P. Stevens, Burlington, and other large companies, cotton dust standards were imposed in 1980 and upheld by the Supreme Court in 1981. Several plants in Spartanburg County, however, were ahead of the game; both Beaumont and Whitney received new air filtering equipment in 1976.

By far the biggest change in the factories was the advent of automation and superior weaving, spinning, and carding machines. The Swiss introduced a loom that wove with a burst of air instead of wooden shuttles running back and forth

Where Mills Got Their Names

Arkwright Mills: Named for Sir Richard Arkwright of England who invented the spinning jenny in 1735.

Clifton Manufacturing: Named for the imposing cliff that fronts the factory.

Drayton Mills: Named for William Henry Drayton who founded the Spartan Regiment in the Revolutionary War.

Fingerville Mills: Named after its founder, Joseph Finger.

Glendale Mill: Named by Helen Converse after a bucolic town in Southern California. Original name was Bivingsville.

Mary Louise Mill: Named after one of the owners' daughters.

Mayfair Mills: Named after the elegant Mayfair Hotel in London.

Pacolet Mills: Named after the Pacolet River. The source of the river's name is uncertain. Some believe it is a Native American word for "running horse." Others believe it is a name left by a European explorer familiar with a mythic horse from a French fable. In that fable, which first appeared in English in 1550, "the horse of Pacolet" is a magic horse of wood that helps rescue the characters Valentine and Orson from the castle of Clerimond.

Riverdale Mills: Named because it sat next to the Enoree River and spanned a dale.

Rosemont Plant (Jonesville): Named after Walter Montgomery's wife, Rose Montgomery.

Saxon Mills: Named by John Law for his English heritage.

Saybrook Plant: Named in honor of the Connecticut area settled in 1635 by an ancestor of the Chapman family.

Tucapau Mills: Named after the Native American word for "strong cloth." Later known as Startex.

Whitney Mill: Named for Eli Whitney, who invented the cotton gin in 1793. The company kept a cotton gin outside the mill so that farmers could sell their cotton directly to the mill.

across the cloth. Milliken & Co., flush with cash, was quick to buy the latest and best of the European machinery—wide looms, open-end spinning machines, and even robotics—and fine-tuned them in its own research facility. Other local companies soon followed. "I was scared to death when I first looked at it," said Inman Mills employee Jessie Mae Gregory of her new air jet machine in 1985. For Ruby Rogers, age 63: "They are a weaver's dream. I'm just sorry this didn't come along 20 years ago."

Jobs, as a whole, became less manual. There was more walking between machines, less standing. The days of mill workers being called "hands" were over. Increasing automation brought computers to almost every machine on the floor. In the late 1980s, computer-aided design helped Milliken and Springs Industries' Lyman plant create exclusive print cloth and rug patterns and get them in production within weeks, rather than months. By the 1990s there were predictions of "lights-out factories" operating without workers. (TNS, a Japanese manufacturer that arrived in the Upstate in the 1980s, came closest; at its weaving plant north of Spartanburg it actually ran looms from Friday evenings to Monday mornings with no employees in the factory at all.) But with automation came job loss—often massive job loss. Between 1972 and 1985, automation was the primary factor in the loss of 700,000 U.S. textile jobs. *The Christian Science Monitor* attributed as much as two-thirds of the job loss to "technological backwardness," one-third to foreign imports.

Increased efficiency had other downsides. Because the new weaving and spinning machines could run continuously, many mills, including Reeves Brothers' Chesnee plant, moved to two 12-hour shifts, four days a week, rather than the traditional eight-hour shift, five days a week. In a flashback to the 1920s, some workers complained that they didn't get home until 7 p.m. Textile mill work, once again, interfered with family activities.

Walter Montgomery Sr. in his later years at Beaumont (Courtesy, the Spartanburg Herald-Journal)

Despite the improvements in machinery, the specter of foreign imports loomed over every plant, every day. In 1985, the typical Spartanburg textile worker was making $6.50 an hour. A Chinese worker was making 20 cents an hour. Imports of textile and apparel imports tripled from 1980 to 1990, and the market share of imports, less than 30 percent at the beginning of the decade, climbed toward 60 percent by 1990. Milliken & Co., as usual ahead of its competition, began realigning its products and shuttering factories across the South. The historic factories in Pacolet were among the first to go; by 1983, both weaving

and spinning had ended, and demolition of these landmarks soon followed. Victor Mill closed. The original Arkwright Mill closed. Butte Knit closed. Layoffs through the 1980s were constant. "In 10 years, if something isn't done to control it," Walter Montgomery Sr. said prophetically to the 1990 annual meeting of the American Textile Manufacturers Institute, "this business won't be here."

Meanwhile, a study conducted by USC College of Business Administration in 1985 found that laid-off textile workers were generally finding new jobs, though at less pay. Wages at their next jobs averaged 73 percent of their prior earnings. After layoff, white males waited an average of 3.7 months for new jobs, while women waited 5.7 months and blacks waited 6.2 months. But those who were left in the industry became increasingly loyal. In 1983, workers at Spartan Mills voted out the union that had represented them in some form since the early 1930s. Two years later, Startex, the last local left in Spartanburg County textiles, followed suit. Some apparel plants, including Spartan Undies, continued their affiliation with the International Ladies Garment Workers' Union—in fact, 75 members demonstrated in Morgan Square to protest imports—but as a rule, the romance with organized labor was over.

The new sweetheart was management. Company owners, like Marshall Chapman at Inman and Mac Cates Jr. at Arkwright, knew they needed the voices of workers if they had any hope of convincing Congress and President Ronald Reagan to support controls on imported textiles. As the U. S. textile industry unveiled its $100 million Crafted with Pride advertising campaign in 1984, ordinary workers took center stage along with Hollywood heavyweights Bob Hope, Carol Channing, Sammy Davis, and others. Thousands of Spartanburg County textile workers pledged money from their own paychecks to ensure that the flashy "Made in the U. S. A." television commercials appeared on the stations they watched in the Carolinas. At the suggestion of management, many donated one hour's pay a week, an amount that was matched dollar for dollar by the companies. Workers at Arkwright raised $1,370. At Mayfair, where 87 percent of all workers contrib-

Milliken & Co. Domestic Plants, 2002

Abbeville Plant...Abbeville, S.C.
Alma Plant...Nichols, Ga.
Avalon Plant...Toccoa, Ga.
Barnwell Plant...Barnwell, S.C
Cedar Hill Plant...Jonesville, S.C.
Cotton Blossom Plant...Spartanburg, S.C.
Cushman Plant...Williamston, S.C.
Cypress Plant...Blacksburg, S.C.
DeFore Plant...Clemson, S.C.
Dewey Plant...Inman, S.C.
Duncan Stewart Plant...LaGrange, Ga.
Elm City Plant...LaGrange, Ga.
Enterprise Plant...Marietta, Ga.
Excelsior Plant...Union, S.C.
Gayley Plant...Marietta, S.C.
Gerrish Milliken Plant...Pendleton, S.C
Gillespie Plant...Union, S.C.
Gilliland Plant...Laurens, S.C.
Golden Valley Plant...Bostic, N.C.
Hatch Plant...Columbus, N.C.
Hillcrest-Sommer Plant...Simpsonville, S.C.
Hillside Plant...LaGrange, Ga.
Honea Path Plant...Honea Path, S.C.
Humphrey Plant...Toccoa, Ga.
Johnston Plant...Johnston, S.C.
Judson Plant...Greenville, S.C.
Kex Plant...LaGrange, Ga.
Kingsley Plant...Thomaston, Ga.
Kingstree Plant...Kingstree, S.C
Live Oak Plant...LaGrange, Ga.
McCormick Plant...McCormick, S.C.
Magnolia Finishing Plant...Blacksburg, S.C.
Midway Plant...Union, S.C.
Milliken Sales...New York, N.Y.
Milliken Research...Spartanburg, S.C.
Milliken Design...LaGrange, Ga.
Milliken Packaging...Jonesville, S.C.
Monarch Plant...Union, S.C.
New Holland Plant...Gainesville, Ga.
Newton Plant...Hartwell, Ga.
Peerless Plant...Belton, S.C.
Pendleton Plant...Pendleton, S.C
Pine Mountain Plant...Pine Mountain, Ga.
Saluda Plant...Saluda, S.C.
Sharon Plant...Abbeville, S.C.
Sibley Plant...Lavonia, Ga.
Stearns Plant...Stearns, Ky.
Sycamore Plant...Clinton, S.C.
Tetra Pak...White Stone, S.C.
Unity Plant...LaGrange, Ga.
Valway Plant...LaGrange, Ga.
Winfield Plant...Winfield, Tn.

Milliken also serves foreign markets from plants and labs in Australia, Belgium, Brazil, Denmark, France, Great Britain, Germany, Japan, Portugal, and Singapore.

uted, some gave up an entire day's pay toward the effort. "If we're not going to stand up and fight for our jobs, who will?" asked Mayfair's Janet Adair.

Next followed an enormous letter-writing campaign. Plant managers persuaded thousands to take up their pens and write President Reagan, urging him to limit imports. Just as happened in the 1930s, the White House mailroom filled with letters from textile workers; little did they know, however, that Reagan would let them down, just as Roosevelt had.

In an effort to cover all bases, the South Carolina Textile Manufacturers rolled out the Gold Seal Merchant program to recognize retailers who stocked at least 75 percent American-made merchandise. In turn, they promised retailers, like Spartanburg's Carolina Cash, that they would urge the army of South Carolina textile workers, still 100,000 strong, to patronize those stores. Manufacturers purchased expensive ads in local newspapers and established Textile Week every October to promote their industry's importance to the communities across South Carolina. In 1986 they sponsored a parade and carnival in Morgan Square where everything was "made in the U.S.A.," from the cotton candy to the hot dogs to the cars. Almost overnight, Spartanburg—a community that ironically had benefited from globalization perhaps more than any other in the United States —was afire with protectionist activity.

Meanwhile, in Washington, D. C., a full court press was on, and both Democrats and Republicans from the Carolinas, including Liz Patterson of Spartanburg, fell in behind the textile industry. Although Reagan was elected in 1980 with the support of textile leaders, he disappointed them greatly with his veto of a bill that would have imposed higher tariffs and stronger regulation of textile imports. In August 1986, the owners of every textile plant in Spartanburg County joined 100 of their counterparts in the gallery of the U. S. House of Representatives for a critical vote to override the president's veto. Guardedly optimistic, they had no way of knowing that Reagan's classic arm-twisting had swayed the votes of nine Republicans in the 24 hours before the vote, turning what looked like a razor-thin victory for textiles into an eight-vote defeat.

The idea began to dawn on some textile leaders that salvation would come not from

A spinner at Mayfair Mills works the thread so it stays on the line. (Courtesy, the Spartanburg Herald-Journal)

Washington, D. C. Perhaps it would come from a decisive improvement in quality.

ON A FOGGY NIGHT IN SPARTANBURG in March 1984, Roger Milliken's $2 million Sikorsky helicopter came in low over the Research Park north of Spartanburg. Inside, Milliken's guests included Walter Montgomery Jr., Hamrick Mills President Jim Hamrick, concrete contractor Dick Pennell, and Milliken's young attorney George Dean Johnson Jr. Attempting to land in a low cloud ceiling, the helicopter pilot instead clipped the trees and came crashing down. Drenched in fuel, the occupants escaped with cuts and bruises, though the Sikorsky was a total loss. ("We've reordered," Milliken told a newspaper in the morning.) Some observers later pointed to this harrowing event as a metaphorical turning point for Milliken, insisting that this life-threatening episode was the impetus for new focus on quality manufacturing and employee involvement.

In fact, though, the seeds had already been planted. In 1981, Milliken, then in his late 60s, had made a fact-finding trip to Japan. There, he found "Japanese competitors turning out better product on ancient machines." Arriving back in the United States, he began asking why. Over the next five years, he intensively consulted with the world's gurus of quality production: Philip Crosby, Tom Peters, and Charles Deming, the American philosopher who shaped the Japanese post-war manufacturing system. He also enlisted the Japanese themselves—quality consultants with names like Ishikawa, Imai, Suzaki, and Taguchi. And there, in those meetings, was born a monumental transformation of the southern textile industry. Known as "the Deming Way," this management initiative gave more responsibility to the people running the machines. Employees at Milliken became "associates," at Inman, "members." They worked in self-directed teams and had the authority to fix or improve any process that wasn't working. Deming, like Milliken a Yale graduate, believed fundamentally that the average employee was competent and, given more authority, could improve the quality of production. He pointed the finger at management for all of the problems of American industry. Deming had resurrected the

Japanese economy by statistically analyzing the manufacturing process—with charts and graphs and numbers galore. By 1986, they filled the walls of Milliken plants. A new kind of "scientific management" was back, a remnant, perhaps, of the old stretchout stopwatch, except this time, the stopwatch was in the hands of the men and women using the machines. Management at Milliken (and by the end of the 1980s, Spartan, Mayfair, Inman and others) stood back and watched the numbers improve. The textile industry was on its way to becoming the most efficient industry in the United States.

Hand in hand with the quality initiative was an inventory reduction program called Quick Response. Throughout their history mills had built warehouses to store surplus goods "and kept making them until downturns in the economy forced them to curtail production." It was highly inefficient, wasteful, and sometimes financially debilitating. Taking a clue again from quality consultants, textile manufacturers realized they might have an edge over their foreign competitors if they could deliver exactly the goods major retailers like Wal-Mart and Kmart wanted, and deliver them *fast.* Domestic companies vowed to cut delivery times by 90 percent. If Sears needed baby blue sheets, for instance, computers would alert U. S. manufacturers, and they could beat their Asian competitors to market by a month or more.

Küsters Corp. employee Coy Wright works on a compensator, which maintains tension on cloth. (Courtesy, the Spartanburg Herald-Journal)

Such systematic change caught the eye of Wall Street investors, who quickly realized that they could buy publicly-owned Southern textile companies at depressed stock prices, chop them up into trimmed-down, efficient niche manufacturers and resell them at higher values. Many of the major Southern players had ballooned in size in the 1970s and '80s through expansion and were now unwieldy, bureaucratic companies. North Carolina's Burlington Industries, for instance, had 75,000 employees—almost twice as many workers as the city of Spartanburg had residents. By the end of the decade, Wall Street investment firm Morgan Stanley became the owner. Greenville's J. P. Stevens, a 175-year-old textile behemoth, disappeared into a private buyout that cut it up into three new companies. Georgia's West-Point Pepperell suffered the same fate. Fieldcrest bought Cannon. Springs bought Lowenstein. Between 1978 and 1988 mergers reduced the number of publicly-owned firms from 75 to 31.

Most Spartanburg companies, because they were family owned, sat on the sidelines and watched, but the long-term impact became clear. The colleagues they counted on to help fight imports in Washington, one by one, received pink slips. In their places were highly leveraged corporate raiders, like William Farley, with no prior connection to the industry. The brotherhood scattered. Meanwhile, buried under debt, the publicly-traded companies began cutting costs, which preempted research, worker training, and machinery purchases. An industry that might have held its own against low-cost imports was badly wounded.

And not all Spartanburg companies escaped the trend. Reeves Brothers, which had its start in the 1920s as a Milliken-style selling agent, changed from a publicly-owned company to a private one in a management-led buyout in 1982. Those managers then sold the company off to Schick (the company that formerly made razors) in 1987 in a junk bond deal worth $250 million to themselves. At the time, Reeves had 7,000 employees across the South, including hundreds at Chesnee and Woodruff who made apparel fabric. In order to make the deal work, lenders terminated the company's pension plan and absorbed the $20 million in assets that was invested there. Retired workers continued to receive their payment checks, but current workers got a partial lump sum and no more pensions. The American Textile Workers

Union led small protests in New York and Los Angeles at the investment houses that concocted the deal—to no avail.

Spartan Mills also had an ownership change up its sleeve. In the mid-'80s Spartan offered its employees, who now numbered 5,000, ownership in the company in exchange for substantial corporate federal income tax breaks. Called ESOP (employee stock ownership plan), the program annually gave vested employees shares of company stock with the understanding that they could cash out at retirement or termination. By making employees owners of the company, Spartan intended to increase their dedication and reduce turnover. It was a great addition to the company benefits package—as long as Spartan stayed solvent. In the case of insolvency, an outcome no one considered, everyone's money went down the drain.

Meanwhile, Milliken & Co. had its own bitter ownership fight in 1989. In a nasty battle with owners of Greenville's Delta Woodside Industries over a fraction of one share of Milliken stock, weaknesses were exposed in the giant Spartanburg company, which had become the largest privately owned textile company in the United States with $3 billion in sales. A year-long court battle revealed that Roger Milliken was not the majority owner of his company—a host of relatives scattered across the globe owned substantial pieces. A group of relatives in Pennsylvania wanted money out of the company, and another clan in Italy was moving toward the same conclusion. By selling a sliver of stock to outsiders, they hoped to force Milliken's hand. Through a series of family meetings, Milliken was able to quell the feud, at least temporarily, and the issue went dormant. But the episode was another ugly tear in the fabric of the southern textile fraternity. "The textile industry has, to a large extent, been something of a social club where the people were very gentle with each other and everyone played within the rules and didn't upset other people," erstwhile corporate raider Bettis Rainsford of Delta Woodside crowed during the Milliken stock episode. "We're not interested in being a part of that social club."

Outside the textile industry, the economy in Spartanburg was undergoing a dramatic shift. The successful recruitment of European textile machinery companies in the 1960s opened the world's eyes to Spartanburg's viability as a business location. "We don't just sell magnolias and moonlight," Chamber of Commerce executive Dick Tukey said in 1983. "We sell economic justifica-

Machine operator Max Barton works a Murata spinning frame at Spartan Mills. (Courtesy, the Spartanburg Herald-Journal)

tion." The steady stream of employees out of textiles became a selling point for business recruiters. These were the people whose labor had made multimillionaires out of ownership families; they had a strong work ethic, still rooted in the farm economy, and they were conditioned to wages much lower than in the Midwest and Northeast. Labor unions now represented barely 2 percent of local workers; a two-month strike at Kohler's 800-employee bathroom fixture plant in 1979 was the last major union activity in the county. And South Carolina, struggling with high rural unemployment rates, was willing to open the candy store for incentives: tax breaks, sewer lines, and sometimes, free land.

The textile industry, though, continued to be a draw for other kinds of affiliated businesses. Leigh Fibers and Wm. Barnet & Sons uprooted from New England in the 1960s and early 1970s to come to Spartanburg County to reprocess textile waste—such as rags and threads—into fiber and filament that could be recycled into other kinds of textile products. Leigh built a giant factory in Wellford that employed hundreds; Barnet bought the old Draper headquarters in Saxon and began to spread out around the globe. Meanwhile, a young Jimmy Gibbs began selling textile machinery parts

Left, Jimmy Gibbs (Courtesy, Spartanburg Regional Medical Center)

Bottom, This Pacolet mill was torn down soon after the photograph was taken. Residents placed a white horse on a former bridge support to commemorate the mill. (Courtesy, the Spartanburg Herald-Journal)

> **In 1964 less than five percent of mill workers in South Carolina were black. By 1976 nearly one in three textile workers in the state were black. African Americans made employment gains in the southern textile industry at a much greater pace than in most American industries.**
>
> *—Author Timothy Minchin*

in the mid-'70s. By the end of the century, his homegrown enterprise would become Gibbs International, one of the world's largest dealers of used textiles machinery and, indeed, whole mills. The success of his venture turned Gibbs into a prominent local real estate developer and substantial philanthropist.

The "march of industry" to Spartanburg County began in earnest in 1980 with a Borden potato chip plant, an R. R. Donnelly printing plant, and the $41 million conversion of Monsanto's yarn operation in Switzer to the world's largest electronic grade silicon wafer factory. An American-Japanese partnership, Alcoa Fujikura, broke ground in the late 1980s near Duncan in a formerly rural area that would eventually draw dozens of factories and thousands of jobs. Between 1982 and 1992, Spartanburg County attracted 16,000 new manufacturing jobs, drawing the attention of international journalists and business gurus like Harvard's Dr. Rosabeth Moss Kanter. By 1993, county unemployment was 4.7 percent, compared with national levels of 7.2 percent, and the industrial growth had spurred unprecedented retail and residential development. Communities such as Arcadia and Fairmont, and former mill communities like Fairforest's Clevedale, were surrounded by shopping centers, pavement, and $300,000 homes.

Yet there were plenty of mill villages untouched by the fingers of progress. In the Apalache community, for instance, none of the 143 mill houses was occupied by a black family, according to the 1990 U.S. Census. In remote

By the 1990s family life in the mills had extended to African-American families. Six members of the Rice family worked for the John H. Montgomery Plant in Chesnee in 1990. They are, from left: Jean Miller, Ivory Rice Jr., Sandra Bobo, Jack Manning, Bernice Rice, and Ivory Rice Sr. (Courtesy, the Spartanburg Herald-Journal)

Enoree, the per capita income in 1989 was $8,985, substantially below the county average of $12,218, and almost 40 percent of the adults had less than a ninth-grade education.

Pacolet Mills proved to be one of the most sustainable mill communities. The houses, which had been solidly constructed and well maintained during Victor Montgomery's tenure, retained their values in the 1990s. Younger families often spent $60,000 to purchase them, then paid to renovate them. A small handful of new homes were even built on the outskirts. Although there was no longer a mill nurse, two Spartanburg hospitals built satellite offices there. The community, which numbered about 3,000 at century's end, was held together with schools residents considered "good," more than 20 churches, and a sense of security. Unlike such villages as Startex and Glendale, Pacolet had a local government that paid for police, garbage pickup, and streetlights. Municipal taxes on a $50,000 house were about $40 a year, a bargain by most measures. Still, some Pacolet Mills residents longed for the paternalism of the old days. Not far removed from the days when company representatives would come change a light bulb in their homes, they made persistent demands of their under-funded local government. And many mourned the disappearance of local landmarks. In an effort to trim its property taxes and cut upkeep expenses, Milliken demolished not only the mills, but the hotel and the grand residences that lined Victor Park. The elaborate amphitheatre remained, like a Roman ruin, at the top of the hill, and community members did their best to keep it up. Pacolet endured, though much of its history existed only in photographs.

In contrast stood Clifton. Abandoned by the corporate fathers for 30 years, some Clifton streets looked more like those in a third-world country than they did in the rest of Spartanburg. Those who could, built new houses on the outskirts—a total of 285 new homes between 1960 and 1980—and retained a fierce community pride and connection. Left behind on the mill hill was a deteriorating housing stock that was split between an aging population and an increasing number of transients. A survey in 2000 revealed that one in five Clifton residents had been the victim of a crime during the preceding 12 months. As late as 2002, there were still people living in Clifton so poor they had no electricity and running water.

Meanwhile, the textile companies still hanging tough in Spartanburg County were spending millions on modernization. Between 1984 and 1996, Inman Mills invested $120 million in new machinery. (It also avoided an ownership crisis of its own in 1986 by buying up stock from far-flung Chapman relatives and consolidating it in the hands of five owners, rather than 160.) The textile industry's *Daily News Record* called Inman, "One of the more successful gray goods operations in the United States and one of the best kept secrets." By the 1990s, each of Inman's five plants offered interactive video training for employees, from electronics to wastewater to welding. Like many other Southern textile companies, Inman doubled its safety efforts; when the Ramey plant in Enoree hit 2 million hours without a lost-time accident, company president Marshall Chapman and his wife Liz danced on the roof of the plant while an oompah band played *Nothing Could be Finer*.

Over at Milliken, spending on worker training and education tripled between 1983 and 1994. Each of the 14,000 employees got a Dale Carnegie course in speaking skills. Milliken built its own "university" at the research campus and required all associates to take a minimum 40 hours of classroom training a year. The company's employee involvement initiative was so successful that the Internal Revenue Service borrowed Milliken's "employee suggestion program" as a possible model for "improving its own flawed effort." As a reward for a winning suggestion, Milliken associates were sometimes given free dinners and other kinds of bonuses. Those suggestions ranged from how to win a new customer to how to improve traffic flow out of the parking lot. In 1995, *Management Design* magazine reported that Milliken's internal Employee Satisfaction Morale Index indicated "definite improvements have been achieved but there is still some way to go." As they had for 40 years, employees complained about grueling 60-hour weeks and sustaining the work pace Roger Milliken demanded of them. Some characterized the micro-management style of the company chairman, then in his eighties, as excessive and idiosyncratic.

There was no arguing with the success of the company though. Unlike its domestic competitors, Milliken was debt free and often could turn on a dime. Between 1985 and 1990 it shuttered eight plants in the Carolinas and Georgia as it moved away from the basic apparel fabrics market now dominated by imports. The new thrust was packaging, chemicals, and international operations. In 1990, Milliken rocked the textile world by starting construction on a carpet plant in Japan; by 2002 there would be Milliken operations in 16 foreign countries—serving, Milliken folks were quick to add, foreign markets only. Roger Milliken refused to ship foreign-made goods into his own country.

Six out of every seven apparel jobs in Spartanburg County disappeared between 1971 and 2002.

Despite constant reinvestment at Milliken and other local textile plants, the worst was yet to come in the battle against imports. Adding insult to injury, key textile partners dropped out of the "Buy American" campaign, originally conceived by James Chapman and funded heavily by all Spartanburg County mills. Greenwood Mills and Springs Industries were among the first deserters, saying they believed the future lay in a global economy. Many of these rapidly globalizing textile firms also supported a controversial new government program that allowed them to ship cloth to the Caribbean to be sewn in low-wage factories, then sent back to the United States as "American made" sheets, towels, and apparel. In the mid-1980s, Spartanburg still had several viable sewing factories: Raycord, S&S Manufacturing, and Woodruff's Enro Shirts employed hundreds each. But competing against the Caribbean, where factories in the Dominican Republic paid 95 cents an hour—and in Haiti 65 cents—was an impossibility. Sewing jobs, though not among the highest paying in Spartanburg County, completely melted away.

Machinery is moved from a top floor at Drayton Mills in preparation for the closing. (Courtesy, the Spartanburg Herald-Journal)

Then came NAFTA. In the early 1990s, countries worldwide decided to phase out "the Multi-Fiber Arrangement," a leftover of the presidential administration of Gerald Ford. Exporting nations had considered the system of tariffs and quotas "too harsh," importing nations, "too lenient." In its place, President Reagan promoted the North American Free Trade Agreement. But the textile brotherhood had changed so dramatically that a majority of textile owners this time supported the removal of trade barriers. They wanted to be able to export and import throughout North America without financial penalties. Over the strident opposition of Roger Milliken, Walter Montgomery Sr., and Mac Cates, who worried that their competitors could pay 20 employees for the price of their one, NAFTA became law and took effect January 1, 1994.

The long, sad decline of local textiles began in earnest with the shutdown of Drayton Mills in December of that year. "I've spent the better part of my life here. It's like leaving home," said Iris Dill, a 30-year employee of the mill. Less than a year later, the town of Woodruff lost all three of its factories in a span of just a few months; in fact, Mt. Vernon and Reeves Brothers cruelly announced shutdowns in the same week.

There was a frantic production shift at several companies—away from apparel fabric and, instead, to industrial fabrics and material for home furnishings. Spartan Mills, for instance, began making road substrates and beach re-nourishing fabric at its Powell plant. The transformation was widespread: in 1990, U.S. producers sold 75 percent of their product to apparel makers, such as Haggar and Lee; by 1996, they sold an equal amount to home furnishings. There was a new kind of spinning and weaving under way, and it was much more like the spinning and weaving of a punch-drunk boxer: backpedaling, reeling, and in some cases, crashing to the canvas. Reeves Brothers exited the apparel fabric business altogether, closing not only Woodruff and Chesnee, but plants throughout the South. Employment dropped quickly from 7,000 down to 900, leveling out only when Reeves had settled into the more profitable business of manufacturing highly specialized fabric for aerospace, marine, military, and other uses. "We knew this was coming, but it's still hard," said a 31-year employee of Reeves' Chesnee plant when the 400-employee apparel fabric operation abruptly shut down in 1997.

It seemed textile workers were always waiting for the other shoe to drop. At Spartan Mills, they were transferred from one plant to another as loom after loom ceased. Jackson Mills saw the handwriting on the wall and closed its doors in 1997. Still, it was hard for some workers to realize the seriousness of the problems in the U. S. industry. "I'm signing up to learn computers tomorrow," said a 26-year Startex Mills employee when a companion plant, Startex Finishing, announced its closing in 1997. Unfortunately, 15 months later, her plant closed, too. "Textiles are dying out," lamented Tommy Richardson, a 28-year veteran of Beaumont Mills.

Some companies managed to hold their own, though not particularly profitably. Despite the bleak outlook, Inman and Mayfair reduced employment by just 20 percent between 1990 and 1997. And Milliken blazed ahead, picking up a tennis ball fabric plant in Great Britain, an air bag manufacturer in Kentucky, and a German business that made electroconductive powder. In the latter 1990s, Milliken actually added 2,000 employees, as its product line grew to include 48,000 different items—chemical, plastic, and textile. Buoyed by a host of international quality awards, the company also began setting the national standard for environmental stewardship and workplace safety. Roger Milliken bragged that he had a total of 300 employees on the payroll whose job was connected to protecting the environment.

But over at Arkwright, one family was ready to give up the ghost. After being in the textile business for more than 130 years, the Cates family sold its last remaining plant, Cateswood, to Greenville's Mt. Vernon Mills, known for eking out one of the highest profit margins in the industry. In a bow to employees who had suffered through short workweeks and idle departments, company president Mac Cates humbly conceded in late 1998, "This should ensure that our employees have security over the long haul."

Upset that the American Textile Manufacturers Institute wasn't fighting imports sufficiently, Roger Milliken protested by withholding his dues to the trade organization he helped shape. In October 2000, ATMI voted him out—and said good-bye to the hundreds of thousands of dollars Milliken paid in dues (which were based on a sliding scale according to the size of the company). Milliken's departure and the bankruptcy of such major firms as Burlington Industries and Martel Mills put ATMI on a shaky footing, and several top staff members were later laid off. In an even stranger twist, Milliken moved toward an active import-fighting partnership with his old nemesis, the textile labor union. Indeed, as another trade battle raged in Congress, some Milliken-allied textile leaders were heard to remark that if they had cast their lot with the union earlier—as had such industries as steel and automobiles—they might

Roger Milliken and John Hamrick, chairman of Gaffney's Hamrick Mills, at the funeral for Walter Montgomery Sr. (Courtesy, the Spartanburg Herald-Journal)

Spartanburg textile companies weren't the only ones in dire financial straits. Among the other giants filing bankruptcy in 2001-2002 were Burlington Industries, Guilford Mills, and Martel Mills.

> ## A Tribute to the People of Beaumont Mills
>
> We are Beaumont mill people.
> Now we are a vanishing breed.
> The work we used to do, is done
> By others, half a world away...
> There were no people like
> Beaumont people. You had to
> Have been there and lived it, to
> Know what it was like.
> Our people came from the land
> To go to work in the mill and
> Move into the village. They
> Were good, proud, happy people.
> They worked, laughed, loved
> And raised their children to be
> The best.
> The Beaumont people took pride in
> Their work, gave their best in
> World War II. We were often
> Called lint heads, but the ones
> Who called us names never
> Knew what it means, or how
> Great it was to be...
> Beaumont Mill people.
>
> *The poem was written by Beaumont resident Sarah R. Whitlock upon the closing of the mill in 1997.*

have saved more jobs.

Still the worst was yet to come. A steady, deliberate slide in the value of Asian currencies, which began in 1996, picked up speed. By 2001, the world's apparel manufacturers could buy Pakistani fabric 47 percent cheaper than five years earlier. They could buy goods made in Thailand at a 76 percent discount over 1996 prices. Meanwhile, prices of goods made in the United States were unchanged—or in some cases, climbed higher. Companies like Spartan Mills and Mayfair Mills watched their customers evaporate almost overnight. Even though profits for some companies had been at record levels as late as 1999, the bottom fell out. And those who had placed orders for expensive new machinery were in serious trouble.

Fighting for its life, Spartan Mills already had closed Whitney and Startex and shut down weaving at Chesnee in 1996. It put a plant in Augusta, Georgia, up for sale and tried to change its product line by purchasing a knit fabric plant in Jefferson, North Carolina. Walter Montgomery Sr., by then an icon to his local employees, died in October 1996 at the age of 95, and it seemed the wind went out of the company's sails. Three years later, Walter Montgomery Jr. brought in a 46-year-old executive from Springs Industries to run his family company. Saying he wanted to reposition Spartan as a global marketing company—not just a manufacturer, but a selling agent for multiple product lines—new President Barry Leonard immediately closed what

Walter Montgomery Sr. (Courtesy, the Spartanburg Herald-Journal)

was left of Beaumont and Chesnee and ended both spinning and weaving at the historic downtown plant. Gone from the Spartan product line were greige fabrics and yarn; in came shower curtains, ceramics, and shelving units. In January 2000, Leonard renamed the 110-year-old company "Spartan International" and eliminated the helmeted Greek warrior as the company's symbol.

International or not, Spartan Mills was in deep trouble. So were Mayfair and Inman Mills, two of the more stable companies in the region. In 2001, Inman ended production both at its original plant and at Riverdale, leaving open three newer plants. Mayfair shut down three of its six plants, including the original plant in Arcadia, and hoped for a miracle from its lenders, who had the power to reduce the terms on multimillion-dollar loans on new machinery. And demolition, which had already wiped out Whitney, Startex, Pacolet, and Arkwright, came to Beaumont. Each time that happened, the machinery was sold off to manufacturers elsewhere in the world (except for Milliken, which usually warehoused its souped-up machines for fear that competitors would steal its trade secrets). Crews pulled out the valuable heart pine beams and shipped them overseas as well. Because the beams came from longleaf pine trees, which take 150 to 400 years to mature, they commanded a high price; such trees were standing in Spartanburg in 1885, but no tree farmer could justify growing them since. The bricks, because they were old, also sold at a premium. Even the stainless steel, copper, and aluminum were sold.

On May 3, 2001, Spartan's lender, G. E. Capital Corporation, seized all assets of the company and promptly shut down every loom, every spindle, every carding machine. In a matter of hours, every factory went dark, and 112 years of history was over. The overnight collapse robbed the 1,200 Spartan employees of the pay they would have received under a customary 60-days' notice. Employees also were left without life or health insurance. Many of the older employees would never be able to economically replace those. While primary 401-K retirement accounts remained intact, those who had counted on cashing out tens of thousands of dollars of company stock through the much-heralded ESOP program saw those investments drop to zero. Their years of dedication in exchange for a piece of company ownership financially had gone for naught. Carrying his belongings out of the company's offices, Walter Montgomery Jr. told the local newspaper, "This is a very agonizing time for me, as I am sure it is for all Spartan associates." Some associates, in fact, were agonized enough to file lawsuits, saying the method of severance blocked them from getting what they deserved financially.

FOUR MONTHS LATER, THE END CAME for Mayfair, too. A bankruptcy shuttered the remaining mills, and sent 600 people—including fifth-generation company president Rick Dent—looking for new jobs. Among those newly jobless was Am Nguyen, a former South Vietnamese Army officer who was airlifted by helicopter from his home country when his government fell on April 30, 1975, and improbably wound up in a Spartanburg County textile mill. When the Hub City Writers Project announced plans to create the book *Textile Town*, Am Nguyen pulled a dust-covered photograph out of the attic in the house where he lived and brought it to the public library. He and his American wife waited patiently to donate the photograph to the history book project. But no one in the room could identify the mill in the badly yellowed photograph. Exposure to sun had turned the mill ghost-like, and the image, once sharp, was fading away. Nguyen was disappointed. Did no one know the history? Did no one know who tended the looms? Wasn't there some kind of story there?

Yes, Mr. Nguyen, there was a story there. And it was quite a story.

Herald-Journal

Mayfair Mills files for bankruptcy

TEXTILES HIT AGAIN: Company will close its final three mills; 780 workers will lose jobs

Herald-Journal

Spartan International closes

END OF ERA: 1,200 lose jobs as headquarters, 6 facilities shut down

'A very agonizing time' for mill

Hodges vows veto of House lottery bill

The Aug. 15, 2001, morning newspaper described "the passing of another legend." (Courtesy, the Spartanburg Herald-Journal)

A May 4, 2001, newspaper article called the closing "a sudden, brutal end." (Courtesy, the Spartanburg Herald-Journal)

Voices from the Village

Beatrice Norton

Beatrice Norton, born in 1919, moved from village to village—Tucapau, Fairmont, Glendale, Arcadia, and Saxon—as her parents "got tired of working or had a dispute with somebody in the mill." She started work in Arcadia at age 12, married at 15, and eventually settled on the mill village in Saxon to work in the mill and raise her family. After retiring with disability for "nerves" and breathing problems, she joined the national campaign to address brown lung disease during the late 1970s. Here she reflects on that work:

The doctor said I had brown lung, and I wanted to know what it was, and I asked him and he said, "It's cotton dust." I had it on my vocal chords down into my lungs. He said the older I get, the worse it would be. And I believe it now, 'cause I can't hardly talk.

There was a man who come down from New York, checking diseases and tryin' to keep up with the names. Jerry Wingate was his name. He wanted me to go from one place to another [talking about brown lung]. Went to Georgia, talked to the governor down there. Went to Florida, talked to people down in there. Went everywhere, from New York on down to Tennessee. I went to different places, and I talked to people about the brown lung, about what it do to you. That's one reason right now I can't talk like I should. My lungs is weak, and I can't do like other people. I get to workin' at doin' something, and I just get so weak I just have to give it up and sit down and rest a while.

Beatrice Norton at home in Saxon (Courtesy, Mark Olencki)

I did think for a while it was their fault [the mill owners], but I got to thinking about it, and I said, well, we didn't know, and they surely didn't know. I said, well I don't know whether they knew it was the cause of us getting lung trouble, but I said, they are going to find out. I am going to let them know!

I made a speech down there in Georgia. I went down there, and in the capitol, I was talking about brown lung and what it done to the people, how they got it, and what should be done. There was a chair up there and a man sittin' in it, had a fence around it, you know, and he had a little boy go up and down the steps takin' notes here and yonder. He wrote a note and had that boy bring it to me. I looked at it, read it, and laid it down. I just kept a-talking, kept a-talking. He wrote another note, and the boy come and laid it down. I read it. I just kept a-talking. He wrote another one and it said, "SORRY, LADY, BUT TIME IS UP!" And he underlined that three times.

I held that paper up in that auditorium. I says, "Well, ladies and gentlemen." I says, "This is it. That man up yonder on the throne who knows so much and has got so much, he's telling me to shut my mouth and get out of here."

That crowd raised up and says, "Give her more time! Give her more time! Give her more time!"

I says, "No, thank you. I don't want it. I think I've said enough. You know what brown lung is." Now I says, "It's up to you. If you want to go on back in the mills and keep your mouths shut and die with it. Brown lung'll kill you. Or you get onto that thing settin' up there in the throne and make him do something about it."

I knocked the place out from under him. Oh boy, when I started out that door, that man come down. He had something like an apron around him, or something. I don't know what it was. He come around with his hand out. He said, "I want to shake your..."

I didn't put my hand out. I said, "What in the name of...is this for?"

He said, "I didn't know that brown lung was killin' people."

I said, "Well, why didn't you go get off your throne and find out!" And I turned and walked away.

People on the mill village had been taught not to open their mouths and say nothing about their superiors that they were working under, be it the boss man, the second hand, the overseer, or the owner. We had to watch out for them, not let them catch a word against them 'cause we would be fired, run off the mill village, and it would be broadcast where we couldn't get a job.

—Interview by Betsy Wakefield Teter

Beatrice Norton, fourth from the left, traveled to Washington, D.C., to lobby for brown lung legislation. Jerry Wingate, left, organized the local campaign. (Courtesy, Betty Livingston)

VOICES FROM THE VILLAGE

Charles Sams

Charles Sams was born in 1935 and grew up in Glendale, one of seven children in a three-room house. After beginning millwork at age 16 at Glendale, he served in the Korean War, then came home to work 15 years at Converse. Here he talks about the last 18 years of his working life spent at Spartan Mills—the day in 1983 their textile union (Local 1881) was decertified, modernization of the plant; and the company's losing battle against imports.

I remember when we was over at Spartan Mills—they was closing up mills everywhere at that time—it's been about 10 years ago, a little more than that, they had a little union over there. They always did have a union over there. "Old man" Montgomery said if we didn't drop out of that union, he was going to close the gates up out there. So if we dropped our union, he'd keep it open as long as he could. So he kept it open about as long as he was there. You know, he's dead now. There was a lot of them didn't want to drop out, but I dropped out because I knowed he was going to shut the gate if we didn't. Enough of 'em knowed it so that they dropped out. He kept it open.

People knowed the union was goin'out, you know. At that time they was just closing up [everywhere]. They'd just close the mill. They'd just shut the doors. And I guess he woulda, too. He told us he was gonna run that mill if he could, you know, for us. He'd like to have a job for the people, but if they wanted to have a union and make it rough, he couldn't do that because he was having a hard time staying in competition. Which he was. He said he'd just close the thing up if we stayed in the union and give him a hard time. Some of 'em got real angry, the ones that was the leaders in the union. They thought we shouldn't have dropped out. But the majority of people knowed that he was telling the truth, and they dropped out because they wanted him to stay open. And it did stay open five or 10 more years. You know, places was closing pretty fast then.

Charles Sams
(Courtesy, Mark Olencki)

You know, things was getting in pretty bad shape, and they was beggin' 'em in Washington to do something about shippin' [imports] and not let other countries have all our jobs. Mr. Montgomery said we couldn't compete with them being able to ship cloth over here. They could sell it cheaper [than] what we can make it. And so he started havin' meetings, and we went to his meetings quite a bit. They would talk to us and tell us how things was comin' along, that they was trying to get them in Washington to quit this NAFTA. But Washington, they was on that road, and they wouldn't turn back, and it just made it rough on driving people out of work over here. But yeah, he would come down and talk to us, and he would keep us informed what we was goin' to have to do, and they'd cut back, and he was goin' to try this and he was goin' to try that. We knowed about as much about it as he did. He was a real nice owner. He paid us good, and he was a good somebody to work for. He made sure we had breaks. It was a good company to work for.

I never did like to belong to the union, but I believe the union did help poor people in its time. Like when Roosevelt told them to unionize, I believe it done a lot of good for the poor people. It got 'em a lot of money and they was doing good, but since we joined this worldwide thing, America ain't goin' up, America is goin' down...

They cut back a good bit and modernized. Ever' time they'd put in new equipment, it didn't take as many people to run. It was kind of computerized a little bit and all. They cut out some of the machinery, you know. They done away with some of them old machines and the old help, too. They got down where they was runnin' what we call a skeleton crew, less than 50 people I guess runnin' that thing [Spartan No. 1]...

He told us it was goin' to shut down about three months before he shut that one down [No. 1]. Things were getting so bad, he didn't modernize that one like he did the other one except for the looms down there in the bottom. I thought they would run the other one [No. 2] on because they had it modernized, but I see that they shut it down, too. I thought he had it modernized where he could stay in competition because he was making perfect cloth and a lot of it...

I think American's goin' to come down to the standards of other countries. I think we was up higher than other countries. It's a thing that's goin' to come. Ain't nothing nobody can do about it. Our country's going downhill, and I don't see no way to stop it. I don't worry too much about it. I believe in the Bible and I believe this is all the work of the Lord. It's just comin' on down to the Bible being fulfilled.

—Interview by Betsy Wakefield Teter and Don Bramblett

Behind Spartan Mills (Courtesy, Mark Olencki)

Mike Morris

Mike Morris came to work at Riverdale Mills in 1985 as a night supervisor. During the next 15 years, he rose through the management and became part of the Enoree "family" as the old mill maintained its production of yarn despite a heavy barrage of imports. Here, he describes the people of Riverdale Mills and the day—Monday, June 25, 2001—when he had to inform them that the mill was closing after 113 years of operation.

Mike Morris at Inman Mills' modern Mountain Shoals Plant, also in Enoree (Courtesy, Mike Corbin)

I found that they were a very proud group, "normal folks," as we could call it. The difference was that they just had been there all their lives. They had been in that one facility all their lives, and I looked around, and I said, "Man, everybody's here." They knew all the history, all the stories, which I found interesting…I found that most of the people had worked here for generations. We had husbands, wives, mothers, fathers, daughters, sons, you know, up two or three generations, still employed. I also found that they were somewhat territorial. They did not like people to break down their barriers. You worked in this area, you didn't mess in the other area, okay? And being territorial myself I kind of liked that…

Most of your people had been on the same shift for years and years and years, and they never left. Some people was going to stay on third shift their entire life. Some people wanted second their entire life. Several people retired with more than 50 years' service while I was at Riverdale…The length of service of Riverdale from my calculations would have been over 20 years. The average age would have been 48. All of us fell within the same age category so much that we were concerned what was going to happen when everybody reached retirement age, because we were all going to be there at one time. I think that's one thing you would find at Riverdale that was unique, that the people who came there stayed…

On this particular Monday, we had been fighting [to stay in business] and had made several reductions in order to keep afloat, and we had reduced the cost at Riverdale. People were fighting tooth and nail. I had never seen a group of people come together and do anything in my life like those people had done. They had cut the costs per pound significantly, which no one thought could be done. The unison in there—it wasn't me or you, it was us together. It was phenomenal. I called in on Monday morning to George, our vice president, as we do every Monday to check how the previous week had been. George said, "We'd like to get together about 10 o'clock this morning, could we?"

I said, "Yessir," and hung the phone, put it back on the receiver, and I just set there. And I didn't say anything to anyone.

Finally, I said, "I've got to go to Inman. Some of the managers are meeting."

They said, "On a Monday?"

And I said, "Yeah."

They said, "Is everything all right?"

I said, "Sure, everything's fine." So I got in the vehicle and I started up 26, and I called my wife. I said, "Honey, I'm going up to Inman. We're goin' to close Riverdale."

She said, "They already told you?"

I said, "No, but you never go on a Monday morning to Inman. I know what it's going to be. I'm positive." So I got to Inman and I ran into George, and we walked out in front of the corporate office, and we walked out in front of the old Inman plant, and we were discussing business and how it was doing. And we got about half way in front of the old Inman plant, and he turned to me and he said, "You know what this means, Mike."

And I said, "Yessir. We're going to close the plant."

He said, "It's what we're going to need to do. We've tried to figure out what we need to do, but this is the only option we've got now. We're going to need to close the plant."

So it came down to making the announcement. If there was any good news or bad news, I wanted to be the guy that did it, so we made a decision. Of course, everybody knew. One thing I was always proud of, I had never lied to anyone. So as soon as I came back on that Monday, everybody wanted to know if everything was okay, and I didn't lie to them. I told them, "Everything is going to be fine," and then the rest of the day I tried to find some place to hide so I didn't have to answer any more questions. And I got the management together that afternoon, and I told them. We did not tell supervision. I only told upper management that we were going to stop the plant off.

We caught the third shift coming in and the first shift going home. We actually herded them into the room. And I was up front and I talked to them a minute. And I said, "I've had some good things to tell you in the past, and some bad things I need to share with you. There's no way for me to say this, except effective on this date, we will officially close Riverdale," and that is exactly how I did it. If I did it any other way, they would have not appreciated it. They would have thought it was sugar-coated. No one said a word, and I walked down through the middle of the aisle where we had people on both sides, and I started talking to them individually. I started calling their names, saying you been here this many years, you been here this many years, and you know we did a good job. "You've done it. You have bought this much more time we would not have had. You bought it! You did it! So you've got some things to be proud of. And the only thing I'm gonna say is you've got 60 days, and we're gonna do this 60 days just like we've done it for the last hundred years. We're not gonna do it one bit different, and when we go out of here we're going to hold our heads high."

Toughest thing I've ever had to do in my life without crying, to be honest. Because looking around, there were well over a hundred people in there, I knew everybody's family, everybody's kids. I also knew the background—some of 'em may not ever go to work again.

—Interview by Mike Corbin and Betsy Wakefield Teter

Aerial view of Riverdale Mills (formerly Enoree Manufacturing) taken in the early 1960s. (Courtesy, Anthony Tucker)

Roger Milliken
Legacy Crafted with Pride
by Jim DuPlessis

Roger Milliken in 1971. At right is Andrew Teszler. (Courtesy, Wofford College)

Roger Milliken was the builder of a textile empire, a defender of an industry he saw beset by unfair trade practices, a Republican Party organizer, and, until late in his life, a virulent foe of labor unions. It's impossible to understand industry or politics in South Carolina in the latter half of the 20th century without reckoning with "Big Red," the chief executive of Spartanburg-based Milliken & Co. Milliken earned his nickname as much for his influence as his 6-foot-2 height and wavy red hair that long ago turned to gray.

In more than half a century at the company's helm, Milliken built his family's company into one with an estimated $4 billion in sales—even as many of his competitors were sold or closed. As a political activist, he helped turn South Carolina from a state dominated by Democrats to one dominated by Republicans. But, Milliken, who turned 87 in 2002, entered the new century with waning support among Republicans for his consuming passion to protect textiles and other manufacturers from what he considered unfair imports. Milliken, who left an indelible mark on labor history by closing the company's mill in Darlington in 1956 after workers voted to organize, found his longest-standing allies in his trade war were labor unions.

For all his emphasis in latter years on team building, Milliken & Co. was Roger Milliken's company, leading many to wonder what would happen when he retired or died. And, if he had his way, death would come first. "I'm going to keep on doing what I'm doing," he told a *Wall Street Journal* reporter in 1995. "I'm going to die in the saddle, fighting for American manufacturing supremacy."

For South Carolina, it's ironic that one of its greatest business leaders is a Yankee. Roger Milliken was born in New York City Oct. 24, 1915, went to the exclusive Groton preparatory school in Massachusetts, and graduated from Yale University in 1937. But Roger Milliken did what few of his Northern colleagues did: he moved his family to Spartanburg in 1954 (he said it was closer to New York by train than either Greenville or Clemson, plus it was "an ownership" town), set up the company's headquarters there, and rolled up his sleeves for a job that he considered unfinished nearly half a century later.

Gerrish Milliken, father of Roger (Courtesy, Walter and Betty Montgomery)

The legacy of the 137-year-old company founded by his grandfather, Seth Milliken, must have weighed heavily. Roger Milliken was old enough to have a memory of his grandfather; he was a young man during the bitter labor struggles of the 1930s, and the responsibility for running the business fell on him unexpectedly at an early age. Though *Forbes* magazine estimated he was worth $1 billion, in no account or personal experience of Milliken has there been a sense that he has been motivated by personal enrichment. His home in Spartanburg is modest. For all his secrecy, he's been one of the more accessible, visible, and publicly involved business leaders in the state. A reporter wanting to reach Milliken himself could usually find him and be obliged with at least a polite "no comment." Money itself did not seem to be the prime motivator of this man. Instead, he seemed to have an unquestioning faith in his company and his politics, coupled with a terrifically focused will to advance them.

"Having inherited an opportunity, I've always felt I had a responsibility to do the very best that I could with it," Milliken said in 1995. "I've devoted my life to doing that, and I've always been tremendously intrigued and challenged by trying to find a better way." Milliken has received a bounty of awards. In 1985, he received the Neville Holcombe Distinguished Citizenship Award from the Spartanburg Area Chamber of Commerce and was named to South Carolina Business Hall of Fame. In 1989, he received the Malcolm Baldrige National Quality Award. In 1998, he was named to the South Carolina Hall of Fame. In 1999, he received the Lifetime Achievement Award from the Northern Textile Association and was named Leader of the Century by *Textile World* magazine. In 2000, he was named to National Business Hall of Fame. Since the early 1990s he's emphasized environmental measures the company has taken—from reducing landfill waste by recycling to encouraging the planting of trees—and his company has won a bevy of awards for that.

One of his most visible achievements in the region has been the Greenville-Spartanburg International Airport. In 1957 Greenville businessman Charlie Daniel, founder of Daniel Construction, enlisted Milliken's support for a "jetport" that would be built halfway between Greenville and Spartanburg. Plans were unveiled in 1958 and it opened in 1962. Milliken's term as chairman of the Greenville-Spartanburg Airport Commission has stretched into its fifth decade, an accomplishment rivaling Strom Thurmond's years in the U. S. Senate.

Roger Milliken was named Textile Leader of the Century by Textile World magazine in 1999. (Courtesy, Milliken & Co.)

The company has a reputation for spending heavily on equipment and research. Milliken's commitment to research—though known to the public primarily by anecdote—has been an obvious exception in an industry that has come to depend on research by universities, fiber producers, and machinery manufacturers. The importance of research to Milliken is also evidenced by his choice for a successor. While the industry has tended to groom executives grounded in sales, Milliken picked Tom Malone, who joined Milliken research group in 1966 after earning a doctoral degree in chemical engineering from Georgia Tech.

The true picture of Milliken & Co. has been obscured by what some of his relatives have called his "fetish for secrecy" about the company. Yet the Milliken code of privacy has its exemptions. In the 1990s, Milliken was sued by two much smaller textile companies that claimed Milliken used a young man posing as a graduate student to spy on them. The court records included an investigation contract signed by a Milliken division chief that stated, "Milliken's identity will remain absolutely anonymous." Johnston Industries, based in Columbus, Georgia, estimated it lost $30 million in sales from the ruse. Milliken denied the allegations and both suits were later settled without disclosing the terms.

Milliken has become legendary for his hands-on management of the family textile business. When one ex-Milliken executive was asked what his title had been at Milliken, he replied: "Titles are a little bit informal in Milliken. There's Mr. Milliken and then everybody else." Milliken represents the third generation of control in a business that has been handed from father to son. But, early on, Roger Milliken decided that none of his five children or any other family member would inherit control. Instead, in 1983, he appointed Tom Malone president and chief operating officer. Milliken took the newly created positions of chairman and chief executive officer. He also reconfigured his board of directors so outside business leaders would always have a majority. His children's lack of involvement in the company was "a wise policy," he said in 1995. "Our company is too large, and whether true or not, any successes they had would be attributed to nepotism and not ability." Milliken has called Malone, aged 63 in 2002, "the future leader of our company."

After Roger Milliken received his bachelor's degree in French history from Yale, he went to work for Mercantile Stores, a large department store chain that the family partly owned. In 1941 he was appointed a director of Deering Milliken and went to work for the family textile business in its sales office in New York. He was only 32 when he suddenly was thrust to the top of the company. Roger was with his father, Gerrish Milliken, and his father's friends on a golf course at Southampton on Long Island in 1947 when Gerrish had a heart attack and died. Roger was named

Roger Milliken and author Tom Peters (Courtesy, Milliken and Company)

president later that year.

Milliken's early opposition to unions became enshrined when he closed a mill a day after its workers voted for the union—a case that reached the U.S. Supreme Court and was not settled for 24 years. Workers at Darlington Manufacturing Co., a textile mill controlled by Deering Milliken, began an organizing drive in March 1956 with the Textile Workers Union of America. "The company resisted vigorously in various ways, including threats to close the mill if the union won a representation election," according to the U. S. Supreme Court's 1965 ruling against Milliken. On September 6, 1956, the union won an election by a narrow margin. The next day Roger Milliken decided to close the mill, testifying later that the vote indicated there was little hope of lowering costs by introducing new machinery.

That part of the story went quickly. The National Labor Relations Board and the courts would then take a quarter century to mete out justice. The Supreme Court affirmed the NLRB's finding that the closings were illegal and "the closing was due to Roger Milliken's antiunion animus." The case continued on. Milliken settled with the NLRB in 1980, agreeing to pay $5 million in back pay to Darlington workers and survivors of the 144 who had died. Ironically, a history of the NLRB states that the ruling showed a company "cannot legally shut down one of its plants to discourage unionism at its other plants." In practical effect, the delay of justice proved the opposite. The awards to workers or their survivors ranged from $50 to $36,000—an amount that was "virtually insignificant" and a quarter of what they had sought.

Just as Roger Milliken will be remembered for Darlington, the flip side of his personality was shown after a fire destroyed the company's carpet mill in LaGrange, Georgia, on January 31, 1995. Milliken marshaled all the forces of his company, and longtime contractor, Fluor Daniel of Greenville, to rebuild the sprawling plant. The plant's 680 workers were moved to other plants—most in LaGrange, but some to the United Kingdom and Japan—so that they could keep their jobs. Six months later the rebuilt plant opened, a feat unheard of in the construction industry.

After moving to Spartanburg, Milliken became one of the pioneers of the modern Republican Party in the South. From 1956 through 1984 he was a South Carolina delegate to every Republican National Convention.

His prominence in the party's state convention in 1960 made some wonder if he was being groomed as a potential challenger to then Democratic Governor Burnet Maybank or U. S. Senator Olin D. Johnston, a Democrat from Spartanburg who had strong support from textile workers and their unions. Ironically, the Republicans' push to power in the South was then called "Operation Dixie," the same name that had been used by textile unions in a Southern organizing campaign that began in 1945 and fizzled out in the early 1950s. The Republican operation would prove more successful, and Roger Milliken would have an important role.

Milliken pushed U. S. Senator Barry Goldwater to run for president as early as 1960 and supported Strom Thurmond's decision to abandon the Democratic Party for the Republicans in 1964. Former South Carolina Governor Jim Edwards said Richard Nixon and Ronald Reagan would not have been elected without Milliken's help. Milliken's fervor was demonstrated in 1972 when he carried $363,000 in cash and checks from textile executives to President Nixon's re-election campaign on the last day before such contributions became illegal.

Despite his years of support for Republicans, by the late 1980s he found himself completely at odds with the party on his most important issue: trade. He and the American Textile Manufacturers Institute campaigned hard for three bills to limit textile and apparel imports from the early 1980s through 1990. The bills were generally backed by Democrats, but opposed by Republicans and vetoed by Republican presidents Ronald Reagan and George Bush.

Milliken's efforts to control imports became a public campaign in 1983 when he launched the "Crafted With Pride in the U. S. A." advertising campaign in the wake of a flood of new textile and apparel imports that industry officials blamed for a wave of plant closings in the South. The campaign was institutionalized as the Crafted With Pride in the U. S. A. Council, Inc., which included textile and apparel companies and labor unions. Among the unions was the Amalgamated Clothing and Textile Workers Union, the successor union to the one Milliken fought at Darlington.

In late 1985, the group launched a nationwide television campaign with ads showing a celebrity pointing to a "Made in the U. S. A." label and saying, "It matters to me!" Six years and $100 million later, some textile mills and apparel groups became free trade advocates and dropped out of the Crafted With Pride council. The group, now funded almost exclusively by Milliken, has recently abandoned its focus on textiles and apparel and moved to a broader concern for saving all U. S. manufacturing industries, running expensive ads prominently in *The New York Times* and *Wall Street Journal*.

By the late 1980s Republicans were overwhelmingly becoming free-traders, and the issue of trade protection was having less power at the polls. His very success in promoting Republicans hastened the process—even at home. Liz Patterson, a Spartanburg native, daughter of the late Olin Johnston, a Democrat representing Spartanburg, Greenville, and Union counties, and a consistent supporter of textile trade protection, was defeated in 1992 by Bob Inglis, a 33-year-old Greenville lawyer, political neophyte and a pious free trader supported by Milliken. Other long-time allies were also disappearing, such as North Carolina Republican Jesse Helms, who was retiring from the U. S. Senate as the new century began.

Milliken had long been considered one of the most powerful executives within the American Textile Manufacturers Institute. But even that trade group split with Milliken by supporting the North American Free Trade Agreement (NAFTA) in 1992.

In the early 1990s Milliken shifted his time and money to Reform Party candidates Ross Perot in 1992 and Pat Buchanan in 1996 and 2000. In 1994 he gave $2.2 million to foundations controlled by Buchanan's sister that were used to buy anti-free trade television ads and lay the groundwork for Buchanan's first presidential campaign, according to an article in *The New Republic* in January 2000.

Roger Milliken and President George Bush (Courtesy, Milliken and Company)

Business guru Tom Peters had championed Milliken's quality program in *A Passion for Excellence*, which Peters co-wrote. But in a 1993 column Peters chided Milliken for opposing NAFTA, "which I guess he sees as a threat to the billion or so bucks his family has pocketed courtesy of the substantially protected U. S. textile marketplace."

Where Peters saw an invisible and benevolent hand leading us to the green pastures of free trade, Roger Milliken saw a world of hard-won gains threatened by outside forces that he must control or eliminate. It's us against them. The "us" in Milliken's case was, from broadest to narrowest, the United States, the textile industry, and Milliken & Co.

The day before the atrocities of September 11, 2001, Milliken gave a 25-minute speech in Lowell, Massachusetts, the birthplace of the American textile industry, where he said manufacturing is an essential component of the economy and one that is in "serious trouble." He tied the decline in manufacturing to higher consumer debt and erosion in the standard of living. "Despite strong productivity growth, [workers] have not seen an increase in their real income over the past 20 years." He said some argue that the current period represents a transition from an industrial economy to a knowledge-based economy, much like the transition from a farm economy to an industrial one in generations past. The difference, Milliken said, is that we still have strong farm production and an agricultural surplus.

"During the current transition, the U. S. is losing both its manufacturing plants and the products manufactured in them, as well as the jobs they provide—thus putting at risk our leadership position as the strongest manufacturing economy in the world," he said.

Foreign companies should be encouraged to invest in the United States, he said. Also, "They should be reminded that since the American market is by far the most important in the world, entry is not a right, but a privilege. In other words, there should be a price and a reward for doing business in the United States...making meaningful, long-term contributions to America's continued security and prosperity, and preserving the global environment."

No doubt, it would be for those types of contributions that he would wish to be judged himself.

Brown Lung
The Battle to Breathe

by Robert E. Botsch

The medical term is "byssinosis," and it dates back to 1877 when British physician Adrien Proust described a lung disease afflicting textile workers. In 1971, the disease entered the consciousness of the American South when consumer activist Ralph Nader coined the term "brown lung" to describe it. He hoped to bring popular attention to the problem so that the same kind of action could be taken to help sick workers and clean up the mills as had taken place in helping coal mine workers with "black lung" disease.

Mill workers in Spartanburg (which in 1980 had more textile workers than any other county in the state) and in other mill towns across the South, had their own terms for their breathing problems: card room fever, cotton fever, cotton cold, and dust fever, among others. It afflicted roughly a fourth to a third of the workers in the dustiest parts of the mills. At first workers felt only some chest tightening and coughing, typically on Monday mornings after being away from the mill for a couple of days. After years of exposure the condition became chronic. The lungs were permanently damaged. Eventually some victims could not even walk, let alone work. Constant coughing made sleep difficult. For generations they died premature deaths. Modern medicine enabled them to live out their remaining years in wheelchairs, tied to oxygen bottles.

Doctors, often hired by the mills, typically diagnosed the disease as chronic bronchitis or some other lung disease. Sick workers retired and survived on Social Security disability payments from the federal government. Even if they were diagnosed with byssinosis, state laws made collecting workers' compensation virtually impossible.

The workers' compensation insurance system was designed to give faster relief to workers hurt through their employment than would be the case if workers had to sue in the courts. A government board—in South Carolina, the Workers' Compensation Commission (formerly named the Industrial Commission) —would decide that the worker had been disabled because of work, and then the company's insurer would begin compensation so that the person could replace lost wages and pay medical bills.

That's the way it worked in theory. In practice, companies and their insurance providers contested most claims and used all their political power to fight any change in law that would make compensation easier. They denied that the dust had any causal connection to lung disease, in much the same way that tobacco companies denied the link between smoking and cancer. They also fought any change in law that would force them to reduce dust in the mills.

In the middle 1970s some young organizers began to try to change the situation. Motivated by the idealism of the times and a belief that democracy should work for the workers as well as the wealthy, they set out to create member-run organizations to change the balance of power.

They received funding from a variety of charities, liberal church groups, and unions. Unions hoped that organizing retired workers for the purposes of winning compensation and cleaning up the mills would give them an inroad to younger active workers. Using Nader's term (many of the organizers had worked with Nader in the Black Lung Movement), they called the new organization the Brown Lung Association (BLA).

The BLA started in Columbia, where active unions provided a base from which to build. The first effort had been in Greenville, but too much opposition existed there. After organizing a small chapter in Columbia in the spring of 1975, attention turned to Greensboro, North Carolina, and then to Spartanburg. The model for organization was typical of that used in other communities. They contacted retired people and asked about breathing problems. They invited them to a "breathing clinic." Fifteen Spartanburg residents attended a clinic in Columbia on September 15, 1975, and eleven were diagnosed as having byssinosis. Ten of these filed for workers' compensation as part of a mass filing of about 70 claims in both North and South Carolina on November 12. Spartanburg's first public meeting took place a few days later and was attended by about 50 people in the small union hall in town. (Five days earlier they had been turned away from a local church. The church's board, on which mill managers reportedly sat, forced them to move.)

The Spartanburg chapter was active until the organization went into decline in the early 1980s. Members participated in a wide variety of lobbying activities. For example, they invited state legislators from the county to talk to them about the workers' compensation laws. When none came, they drove to Columbia and met the legislators in their offices. They won support for legal reform. Organizer Frank Blechman described their lobbying technique as a "no-demand" confrontation: "We take a group, visit a guy (such as members of the industrial commissions or legislators or Labor Department officials who are in charge of inspecting the mills) and just smile and say hello. We cough some, and wheeze a little. When the guy is braced to fend off demands, it scares 'em shitless. They know what we want, but we just wait until they make an offer, and then we laugh and keep on waiting, making them sweat until we get something acceptable."

A group of Brown Lung activists, including Beatrice Norton, left, testify at a hearing in the 1970s. (Courtesy, Betty Livingston)

By August 1976, Spartanburg had about 100 active BLA members who had worked in 15 different mills and a mailing list of 200. About 45 had been screened as having byssinosis, and 10 had filed for workers' compensation. The Spartanburg chapter was one of five in the state when the organization peaked in the late 1970s.

Money for the BLA came from membership and grass-roots fundraising (like selling cookbooks and raffle tickets), private foundation grants, and government grants. Membership generated little money because of the relative poverty of members. Foundations generated about half the operating budget. Government help through VISTA (Volunteers in Service to America) provided staff for about 20 positions. A four-year OSHA grant in 1978 of just over $320,000 was aimed at education of active workers about cotton dust and workplace safety.

The BLA always struggled, but the recession of the late 1970s and the election of Ronald Reagan to the White House created a double whammy. The recession dried up foundation money, and the new conservative president cut off VISTA positions and blocked the last year of the OSHA grant. The budget dropped from more than $200,000 to $24,000. National staffing dropped from 35 to 10 by the end of 1981. To add insult to injury, in 1982 OSHA charged that the group had misspent monies and demanded that they return $220,000. Although the suit was eventually settled, the organization was finished.

Within three years all significant activities had stopped. Sick and dying members were unable to carry on without outside help. The group had never been successful in recruiting active workers, most of whom were either afraid to join or felt they would not be affected by brown lung.

What did the BLA accomplish? They certainly did not break down the individualistic Southern working class culture that rejected collective action. They did not alter the power structure in any long-term way. For a brief period they captured sympathetic media attention, they helped provide pressure and support for significant reforms in state workers' compensation laws, and they provided human evidence for cotton dust standards that were eventually enforced during the Carter years, which accelerated the modernization of textile plants. Finally, they gave some dignity and a small measure of power to a group of dying textile workers.

A Dynasty in Inman
Four Generations of Chapmans at the Helm
by Lisa Caston Richie

The Inman Mills story is all about family—generations of Chapmans who have shepherded the company through good times and bad. "What makes the Inman story all the more remarkable is the extent to which it has been dominated by the members of one family," noted U. S. Sen. Strom Thurmond, speaking to his fellow senators on the occasion of the company's 75th birthday in October 1977.

James A. Chapman founded Inman Mills in 1901, and the leadership of the company has remained in the family for a century. Orphaned at the age of three, he was raised in Spartanburg by aunts and uncles who played a major role in his education. These relatives provided the assistance he needed to attend Wofford College and Harvard Law School, where he graduated *magna cum laude* in 1886. Following graduation, he practiced law in New York City and Middlesboro, Kentucky. At the turn of the century, several textile mills were being built in the Spartanburg area, and Chapman decided to leave his practice in Middlesboro and return home to start a new textile company. In 1901 he raised $150,000 in cash, no small feat at the time, and bought over 600 acres of land near Inman where he would later build the company's first plant. His uncle, Alfred Foster, took 20 percent of the common stock in the new venture, Inman Mills. Among the first directors was Albert Twichell, president of Clifton and Glendale mills.

INMAN MILLS.

The mill began operating in 1902 with 15,000 spindles and 400 looms. Six feet tall, Chapman often wandered among his employees in a Panama hat. He was described as a quiet, modest man who was a positive force for progress. His first year at the helm, he made $1,800. By 1909, his mill more than doubled in size, increasing the production capacity to 33,000 spindles and 840 looms. Growth continued, and in 1928 a four-floor addition was made to the original plant. By 1952, the spindles were upped to 56,000 and the mill was completely air-conditioned; in fact, Inman Mills was the first mill in the country to have its carding, spinning, spooling, and warping facilities completely air-conditioned and refrigerated.

The new mill struggled for the first 15 years, but gained momentum as the years passed. Chapman was joined by his son, James, in 1916, and the two-man team headed the company for 20 years.

A graduate of Wofford College, the second James had experience in banking before joining his father. He became mill superintendent at the unusually young age of 24—at a time when the mill was still running one shift. "I took to raising chickens to keep myself occupied," he once said; in fact, he won numerous medals and ribbons at county fairs for his prize chickens. He went on to be elected president of the Southern Textile Association, the South Carolina Cotton Manufacturers Association, and the American Cotton Manufacturers Institute. No other textile executive had served as president of all three associations.

In addition to these posts, James (known to his employees as "Mr. Jim") was also an original trustee of the Spartanburg County Foundation and the founder of the Inman-Riverdale Foundation in 1946. This organization has awarded hundreds of college scholarships to the children of its employees and substantial funds to churches, schools, and Spartanburg County charities.

While leading the mill with his father, the two were asked to manage Riverdale Mills in Enoree, 25 miles south of Spartanburg. In 1928, they agreed without ever seeing the plant. Years later, "Mr. Jim" would say that if he had seen the run-down condition of the mill, he never would have agreed to it. Riverdale would later merge with Inman Mills and become the company's second plant.

Robert Hett Chapman, the youngest son of founder James A. Chapman, joined Inman as vice-president and assistant treasurer in 1937. He attended Wofford College for two years, but graduated from Cornell University in 1917 with a degree in mechanical engineering. He had much experience before joining the team at Inman, including serving as head of the mechanical engineering department at J. E. Sirrine in Greenville. Throughout this time, he designed and supervised the construction of steam plants all over the Southeast. He also served as chairman of the board of the J. E. Sirrine Textile Foundation, which worked with Clemson College, and as chairman of the Inman-Riverdale Foundation.

James Chapman Jr., 1970 (Courtesy, the Spartanburg Herald-Journal)

James A. Chapman Jr., son of the second James A. Chapman and the third generation of his family to lead the company, was born in 1921, less than 300 yards from the plant his grandfather built. He attended the village school until fourth grade. "All the other boys wore overalls to school," he recalled in a newspaper interview in 1983. "I remember I felt set aside because mother wouldn't let me wear them." The Chapmans moved to Converse Heights when he was 10. Later he raised his young family in Enoree as he served as plant manager at Riverdale in the 1940s and 1950s. A graduate of Davidson College, he served as president of the Southern Textile Association and the South Carolina Chamber of Commerce.

His younger brother, William Marshall Chapman, served as vice president and assistant secretary at Inman. An honor graduate of Clemson College, Marshall, as president of the Palmetto Council Boy Scouts of America, led the organization to win the coveted Campbell Trophy for the most outstanding progress in their region. Another brother, Joseph W. Chapman, served as assistant plant manager at Riverdale. He also graduated from Clemson and worked for Springs Cotton Mills before coming to Inman. Meanwhile, Robert's son, Robert Hett Chapman Jr., served as the cotton buyer at Inman. He also had much experience before joining the family business, working in the Cannon Mills cotton department in Kannapolis, North Carolina, and in the brokerage business with E. P. Murry in Spartanburg.

All of the Chapmans have been known for their determination and their pride in the community. When met with the challenge to modernize, change, and rearrange equipment in the Inman plant, they were often unwilling to reduce jobs. In 1959, to prevent this from happening, and to create more jobs and products, the modern Saybrook plant was erected. The new facility cost $5 million and was the first new mill built in Spartanburg in 24 years. Later expansions included the Ramey Plant, built in 1966, and Mountain Shoals in 1990, both in Enoree, just upstream from Riverdale. At its height, Inman employed 1,700 people.

The second James Chapman died in 1964. The third James Chapman served as president from 1964 to 1983. Upon his death, brother Marshall was president until 1991. Robert H. Chapman III then assumed control as president and treasurer of Inman Mills, and cousin Norman H. Chapman (son of Joseph) became executive vice president. They have a history of renowned leadership to follow.

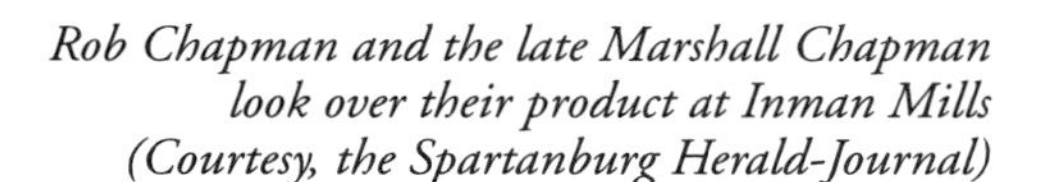

Rob Chapman and the late Marshall Chapman look over their product at Inman Mills (Courtesy, the Spartanburg Herald-Journal)

Woodruff's Ambassador
A Native Son's Tragic End
by JoAnn Mitchell Brasington

What started in 1985 with handshakes over blueprint drawings of a new textile plant in Woodruff ended six years later with a gunshot in San Fernando, California. A native son's budding baby products empire collapsed when he was killed by his partner.

Woodruff native Aaron Thomas bought his way into the textile business in 1978 with the purchase of Tailored Baby Inc. He soon had the cottage industry making some of the first covers for infant car seats. A salesman at heart, Thomas aggressively marketed the new product and added to the company's product line. He, his wife, and partners Tom Garrick and Walter and Victor Soto, participated in the expansion. Five years after obtaining Tailored Baby Inc., Thomas and his partners acquired Relative Industries Inc., Best Baby Inc., Best Industries Inc., Rainbows and Lollipops Inc., Amigo Corp. Inc., and Zittenfield, Inc., all makers of baby accessories and the sole suppliers of baby accessories for Sears.

Aaron Thomas, president of Ambassador Corp. (Courtesy, the Spartanburg Herald-Journal)

Well established, Thomas found himself in a position to help his hometown. "We're here to build a great company," he said at the groundbreaking for Ambassador Corp. Inc. in 1984. "It's going to be nice to read 'Made in the U. S. A.' in Woodruff, South Carolina."

When the plant opened in 1985, it, like Thomas' other plants, produced infant accessories like comforter sets, diaper bags, crib sheets, and diaper stackers. The 50,000-square-foot facility employed 225 people in a town where two failing major textile mills once employed thousands.

"It was a gold mine—a good running plant," said Betty Kelly, former manager of the Ambassador plant. "Aaron was a go-getter. He kept work coming in...When he was killed, the whole thing fell apart."

The company's quick growth and outward success masked internal problems. First, a suspicious fire burned part of the California plant. Then, after an argument, Thomas was shot and killed by his partner Victor Soto in 1991 at their plant in California. The company collapsed.

"I saw it (Ambassador) built from the ground up," said Kelly. "He wanted to help the people of Woodruff...After his death his wife just couldn't keep it going, and I don't blame her."

Kelly remained at the plant until 1994, finishing sub-contracted work for Gerber, Riegel, and other of Ambassador's rivals. She now runs WK&B Manufacturing in Woodruff. The company has five sewing machines, employs 60 people and still makes infant accessories for Riegel Manufacturing. Victor Soto, claiming self-defense, was never charged with the death of Aaron Thomas. However, he was sentenced to six years in a California prison for arson in connection with the earlier fire at the plant.

Uprising of '34
A Documentary Blackout
by Ray Merlock

When South Carolina Educational Television elected not to broadcast a documentary about the textile strikes of 1934, its decision was every bit as controversial as the program itself. Called "Uprising of '34," the award-winning, 90-minute documentary ran on public television stations nationwide in June 1995, but not in South Carolina.

Because the documentary centered on actual events that occurred in the 1930s in South Carolina (as well as North Carolina, Tennessee, Alabama, and Georgia) and because the program contained a number of interviews with Upstate South Carolina residents, this seemed a strange decision. Indeed, the film contained interview footage of Walter Montgomery Sr., president of Spartan Mills in Spartanburg, which had been provided by WRET in Spartanburg and the SCETV Commission. The Spartanburg County Public Library also received an acknowledgment in the credits.

The documentary's producer was disappointed that his work was not broadcast in South Carolina. "Textile history has been kept primarily from the standpoint of the owners," George C. Stoney told a reporter for the Spartanburg *Herald-Journal*. "It is their mansions that are preserved, their legacies to the universities that are honored. We're not saying history is wrong; we're just saying it's incomplete." Stoney expressed disappointment the film was not being shown in "the heart of the textile belt" and "whether there's a conspiracy, who knows?"

Faced with immediate criticism from labor unions, SCETV officials were quick to deny that a conspiracy existed to keep local textile workers in the dark about past labor unrest. Instead, they blamed the pre-emption on "a scheduling problem." Rather than carry "The Uprising of '34," part of PBS's "POV" series, SCETV aired an episode of "Frontline" that focused on using psychiatric evidence in court. A representative of the Spartanburg-based ETV Endowment assured concerned parties that only about $30,000 of the endowment's $10.4 million budget "comes from textile firms ...the endowment does not make programming decisions." Yet some people pointed fingers at SCETV President Henry Cauthen, whose late father was a former executive director of the South Carolina Textile Manufacturers Association. SCETV, however, maintained that Cauthen "had nothing to do with the decision not to air the film." Still, SCETV added three months later, it had no plans to ever air the program.

In April 1996, a group of 29 donors contributed $2,500 to purchase 90 minutes of airtime on WYFF, the Upstate NBC affiliate. Around $900 of the funds came from donations supplied by South Carolina union chapters. WYFF ran a disclaimer before the late-night broadcast, identifying it as "a paid broadcast program." The decision of the station to accept the money and air the film did not, according to WYFF's program manager, necessarily constitute "an endorsement by WYFF one way or another." Since that time, the documentary has been shown in churches and small gatherings throughout the Upstate.

An obvious question would be, "Is the documentary biased?" Actually "Uprising" is an unusual documentary film in that it has no voice-over narrator. The film consists of remembrances, testimony, and interviews with individuals who worked in the textile mills in the 1930s and with their descendants. Newsreel and documentary footage is edited into and around the personal commentaries spoken by people in their homes, on their porches, or in their offices. Some of those interviewed were directly involved in the 1934 Southern textile strike.

Although the majority of the voices denounce the working conditions in the mills and the living conditions in 1930s mill towns, there is some effort to present varied, diverse views. Mill owners and the sons of mill owners are also heard from, as are clergy, labor officials, and government officers. One section of the film deals with the "stretchout," a period in the 1920s and '30s when employees' workloads were increased dramatically. Walter Montgomery Sr. states, "We don't ever want to employ someone and have them work too much—but what's too much? You've got to be efficient in your set-up." While that may be true, according to one voice in the documentary, "the more that happened, the more that invited the labor unions in."

About allegations that union members were "blackballed" after the General Strike, Montgomery states, "When you employ somebody, you look 'em up. We paid much more attention to that than we ever did. If we knew the person wanted to organize this, I certainly wouldn't be in favor of bringing them into our company. Why bring him in? Why look for trouble?"

There is also an attempt to trace current Southern attitudes about mill communities, labor unions, and labor strikes to the events of 1934 and their aftermath. While the film might not necessarily be biased, it is obviously a documentary and—as can be said of nearly all documentaries—one with a point of view.

Spreading the Wealth
A Century of Textile Philanthropy

by Betsy Wakefield Teter

Profits made with the motion of spindles and looms not only created several family fortunes in Spartanburg County, they fueled more than a century of philanthropy. The mills and the families who ran them built colleges, enriched hospitals, fed the hungry, healed the sick, and spurred public art. They also constructed churches, built ball fields and gymnasiums, sponsored scout activities, created parks, and underwrote college tuition for hundreds of young people. Indeed, the entire social service framework of the community is built on the generosity of generations of textile people.

And while many of the textile companies did not survive the competitive pressure of the late 1990s, the endowments they created did. The money they socked away for continuing charity totaled more than $25 million in 2002.

The roots of this philanthropy can be traced back to a major donation by Edgar Converse in 1889 that made possible the founding of a women's college in Spartanburg. Converse College, named after him, has continued to be a focus of charitable giving by such textile families as the Montgomerys and the Millikens throughout the past century. One of the Milliken family foundations, for instance, underwrites $250,000 in college scholarships each year to Converse students. The Spartan Fund, associated with Spartan Mills, consistently gives about $200,000 a year. During difficult financial times for the college, the Montgomerys have given much larger sums to ensure the college's viability.

The Montgomery family also was integral to two charity projects in the early 1900s that served the health and educational needs of the fledgling Spartanburg community. The first Walter S. Montgomery (1866-1929) provided funding and a location for the Pellagra Hospital in 1913 so that the important work of finding a cure for the disease that afflicted the poor could be done in Spartanburg. Together with John Law, Montgomery also capitalized in 1911 the Textile Industrial Institute, an early work-study program for textile workers that evolved into Spartanburg Methodist College. "The pebble he dropped in the sea of this time sends wavelets of good influence on and on and on," wrote school founder David Camak of that donation. Later, local mills "assessed themselves" a certain amount per spindle, per year, to keep the school afloat.

Top, St. Luke's Free Medical Clinic
Bottom, Milliken Fine Arts Center, Converse College
(Courtesy, Mark Olencki)

Much of the "welfare work" of the textile companies during the 1920s also can be considered philanthropy. In an effort to recruit workers and stabilize their labor forces, many companies paid for band instruments, baseball uniforms, women's clubs, and scouting programs. Pacolet Mills, under the leadership of Victor Montgomery, was one of the leaders in the South in this effort, providing day nurseries, an elaborate amphitheatre for music and drama, a greenhouse and tree nursery for yard beautification, and sponsorship of a variety of community clubs.

Textile philanthropy was kicked into a much higher gear with extraordinary profits made during World War II. The National Defense Appropriation Act of 1942 stated that extra profits made under war contracts had to be returned to the federal government; Inman Mills, for instance, returned $235,788 between 1942 and 1945. After that, mills decided to spend that money at home for charitable purposes rather than send it to the federal government through rebates and high corporate income taxes.

There was a flurry of foundations created in the mid-1940s, including Inman-Riverdale Foundation, the Milliken Foundation, the Arcadia Foundation (Mayfair Mills), and the Arkwright Foundation (funded by the Montgomery-owned mills and the Cates family mills). Excess company profits began flowing into these new charity vehicles. The minutes of Beaumont Manufacturing on November 24, 1947, for instance, show a $100,000 gift to the Arkwright Foundation, as well as $15,000 to Spartanburg County Foundation, $10,000 to Beaumont Methodist Church, and $10,000 to Roseland Presbyterian Church.

The Spartanburg County Foundation was created in 1943 from an idea hatched during a hunting trip to Bamberg County. While shooting birds with Jim Hanes, president of Hanes Underwear, Walter Montgomery Sr. and Mac Cates learned that Winston-Salem, North Carolina, had created the first community foundation in the Southeast. When they returned to Spartanburg, they used Hanes's advice, and created their own. The first major move of the young Spartanburg County Foundation was to spend its assets buying part of the Camp Croft property that was being unloaded by the U. S. Government. While much of the land became a state park, a significant amount became the property of the Spartanburg County Foundation, whose trustees knew that its rail and gas lines made it extremely attractive to industry. As that land was sold off to industry, the coffers of the county foundation swelled.

Top, Milliken Science Center, Wofford College
Bottom, New home of Mobile Meals of Spartanburg (Courtesy, Mark Olencki)

In its early days the County Foundation funded community studies, scholarship programs for graduates, Bibles at high school graduations, medals for valedictorians, and even uniforms for baseball teams. There was also an emphasis on buying equipment for Spartanburg General Hospital. "Basically, this philanthropy mirrored what the mill companies used to do for their people," said Jim Barrett, who led the foundation from 1985 to 1998.

Through most of the 20th century textile companies were under tremendous pressure from labor unions and others to prove that they cared about their workers. Donations to charities were one way to diffuse the constant sniping about wage rates. The sell-off of the mill villages provided another outlet to give back to the communities. Many mill companies, including Inman and Mayfair, donated significant tracts of land for new schools in the 1950s. Such giving continued throughout the 1960s and '70s as local textile leaders held regular luncheons at the Piedmont Club to discuss, among other topics, targets for charitable giving.

Their charity only accelerated in the latter part of the century. In fact, Barrett believes that in terms of "relative ability to give," charitable activity in Spartanburg County outstripped similar Southeastern communities four to one. One reason: Textile companies in Spartanburg County were family-owned; publicly-owned companies such as Guilford Mills, Burlington Industries, and J. P. Stevens were controlled by stockholders who wanted to see strong profits to boost share prices. Later, leveraged buyouts and mergers hamstrung the charity of those companies even more.

By the turn of the 21st century, textile foundations were directly giving more than

$3 million a year to local charities. For instance, in 1999, the Spartan Fund of the Arkwright Foundation gave $105,000 to St. Luke's Free Clinic, $114,000 to Spartanburg Methodist College; $40,500 to the Arts Partnership, and $10,500 to the Salvation Army, among the 107 charities it supported. The smaller Arkwright Fund made major gifts to Spartanburg Technical College ($23,400) and USC-Spartanburg ($10,000), among many others. All together, the Arkwright Foundation held nearly $6 million.

In 2000, the Inman-Riverdale Foundation gave a total of $803,000 to nearly 200 charities. Some of the largest donations went to Spartanburg's First Presbyterian Church ($125,000) and the School for the Deaf and Blind ($50,000). There were also significant donations and pledges to Mobile Meals ($45,000), Spartanburg Regional Medical Center ($30,000), and the Arts Partnership ($32,000). Assets in the foundation topped $11 million, providing a base for giving long into the future.

Mayfair Mills' Arcadia Foundation held $2.2 million in 2000 and made more than $100,000 in gifts. Among them: $8,300 to the Boys and Girls Club, $20,000 to Wofford College, and $5,000 to the Spartanburg Day School. The Alfred Moore Foundation, created in 1953 from the profits of Jackson Mills, held assets of $3.3 million and made $190,000 in gifts, including multiple college scholarships. Other textile-related foundations, including the Barnet Foundation and the Zimmerli Foundation, also became active in the late 1990s, funding, among other things, parks, fountains, and library construction. Meanwhile, the Jimmy and Marsha Gibbs Foundation, created primarily through the sale of used textile machinery and mills, gained increasing visibility with substantial donations to Spartanburg Regional Medical Center and Wofford College. In 2000, it gave away more than $750,000.

Finally, the two Milliken foundations—the Milliken Foundation of New York and the Romill Foundation of Wilmington, Delaware—continued to provide vital support, with approximately one-third of their $3 million in total donations during 2000 going to Spartanburg County charities. Among the largest were: $305,800 to Wofford College, $250,000 for Converse scholarships, $132,000 to the Arts Partnership, $50,000 to the Charles Lea Center, and $75,000 to St. Luke's Free Medical Clinic. Well into his 80s, Roger Milliken also established the Noble Trees Foundation, which helped to spur a community-wide emphasis on green space, landscaping, and tree cultivation.

Mill Villages
Looking to the Future

by Thomas Webster

Spartanburg's mill villages are one of the region's most tangible reminders of a golden age that has come and gone. It would be easy to make the mistake of relegating them to the status of relics.

Yet these villages exist not as artifacts but as places that are still alive and evolving. Still harboring within them many of the people who once played a vital role in the area's textile history, they also are serving as homes for those who will populate the future. Today's mill villages are an interesting mixture of past and future. In some regards, they are uncannily similar to previous eras, and in many ways, the marks of change are evident.

The most recent U. S. Census figures indicate that the racial composition of most of the Spartanburg County's mill villages has changed very little. In Drayton, for instance, about 93 percent of the mill village population is Caucasian, roughly 7 percent is African-American, and other minorities represent a fraction of a percent. In Fingerville, minorities account for only 3 percent of the current mill village population. Inner-city mill villages, however, have been much quicker to integrate. The old Spartan Mills village is now predominantly black, as is Arkwright.

That said, however, the "other" minorities category, representing a mix of origins not present earlier in the century, is both growing and representative of change, albeit small. Areas previously endowed with only Caucasians and African Americans are beginning to see Asians and Latinos as well.

Census figures also show that the population of villages, in most cases, is getting predominantly older. Mary Radford, a longtime Fingerville resident, recalls a time when the Fingerville mill village was alive with children. Today, she says, the village is nearly devoid of children. She observes that it is now populated instead by retired people, many of them the same lively children of her distant memories. Indeed, a survey of 11 local mill villages conducted by the Spartanburg County Planning Commission in the late 1990s, showed that most mill village houses were occupied by two-person families. The majority of these residents were married, and many were nearing retirement age.

Another trend indicated by the report, *Spartanburg County Mill Villages*, is the level of education in the mill village populations. The vast majority of the people living in the villages had only a high school degree or less. In Enoree, for instance, of the 523 people living in the village, the largest percentage of residents—160 people—had less than a tenth-grade education. This has a direct impact on the nature of their employment, relegating many of the residents to an employment status and salary lower than that of a managerial position.

Information such as this indicates not only the possible future of the mill villages, but also where too they might seek to grow. These areas, formerly endowed with community-based schools and a virtually guaranteed place of employment, now need to evolve in a way that offers the best educational and employment opportunities to its younger residents.

Moreover, the report indicates ways in which the surrounding community might lend a hand in shaping the collective future of these places: the mill villages of old are remembered as places where the upkeep of the homes and the general appearance of the neighborhood were eagerly maintained by both the mill and its residents. It is odd then that this report should so frequently conclude by urging that a mill village community take heed of its appearance and historical status in an effort to fend off deterioration. And yet, given the aforementioned obstacle of education and lower wages, this challenge is formidable.

Spartanburg City Councilwoman Linda Dogan is an outspoken proponent of efforts to restore mill villages to neighborhoods that rival the glory of their former incarnations. Dogan touts efforts like the outpatient surgery building at the former site of Beaumont Mills as a hopeful sign and fair indication of the investments that Spartanburg needs to be making in these locations. She has high hopes for efforts like those under way in Arkwright, where residents are pushing for the cleanup of a former landfill area.

For Dogan, the key is housing. "I talk about housing no matter what I'm talking about," she laughed. Her efforts have been concentrated on seeing the mill village housing within city

Beaumont village, 2002
(Courtesy, Mark Olencki)

limits brought up to city standards. Moreover, she said she wants to move the legacy of the villages forward, having them put on the historic register and recognized as the valuable asset that they have been and can be.

There have been many other changes in the mill villages, too: changes that can't and shouldn't be measured in numbers or translated into data. These changes can best be reckoned by listening to those people who still remain to tell the tale of the mill villages in those early years.

These are changes such as the passing of people like Inman Mills' James Chapman. Robert Boyce, superintendent of the mill's cloth room for 16 years, recalls that if a child rode by on his bike and was dragging his feet, "Mr. Jim" was up and calling after him to stop ruining his shoes that his parents worked so very hard to purchase.

The change is evident in the stories of Johnny Messer, who also grew up in Inman's mill village. He recalls a time when the police chief of Inman Mills patrolled the neighborhood on foot, and the village residents, "almost genuflected when he came around." No more is a time when a mill-sponsored Boy Scout troop travels to Jamborees on the back of a flatbed truck.

What has changed is a time recounted in the pine-knot-hard voice of Drayton's Albert Knight. His memories, polished bright by years of careful handling, reflect the change with an uncommon brilliance. "I might not be able to remember what I did Monday," he said, "but I can tell you everything about how it was back then. You knew who your neighbors were. We had no radio, no telephone, no car. If we went anywhere, we walked. It was all one big family. If someone in the village got in trouble—sick, clothes, groceries—the neighbors took care of it. They didn't go hungry and they didn't go cold."

Yet it is often outsiders, not the tellers of these tales, who dismiss the future of the mill villages and relegate them instead to the past.

On a mild January afternoon in 2002, Knight walked through bramble and brush into a wooded lot. Everywhere Knight looked, specters of the past loomed large. He pointed to a stand of five-story high pines. "That's where our ballpark was, and right there was the grandstands." His eyes followed his finger as it roved over all points: the past locations of pigpens, cow stalls, pastures, and garages, all of them grown over with pine and oak.

Everywhere in the Drayton village, it is the same. Where there was a boardinghouse, now there is an empty lot. An empty lot, too, stands where the school used to. Former recreation fields are a parking lot, and the old Methodist church site now houses an office.

Hearing Knight speak, one gets a broader sense of the reality of the modern day mill village. There is an acknowledgement of what has come and gone: the "super's" house from which at any hour a doctor could be summoned, the rec center that showed cowboy movies once a week. But it's not all about the past. There, too, is pride in the new. After pointing out where the old school once stood, Knight showed off where Houston Elementary stands now. The old fire department is gone, but an area nearby hosts a brand new, top-of-the-line station. The old post office, closed and boarded up, is not far from where Knight proudly pointed out the new postal facility. The past lives in these places, perhaps more so than anywhere else in the county.

Two scenes of Drayton village, 2002 (Courtesy, Mark Olencki)

A Place Called Riverdale

The Riverdale Mill in Enoree closed in August 2001 after 113 years of producing textiles. At the time, it was the longest continually operating mill in Spartanburg County—a record that likely will never be broken.

Collectively, the employees who lost their jobs there represented just a fraction of the 13,000 textile jobs lost in South Carolina in a two-year period. But individually, they each had interesting, and important, stories to tell.

I first saw Riverdale a couple of years before. I was taking pictures at Inman Mills' more modern Mountain Shoals and Ramey plants nearby, and the supervisor who guided me that day offered to give me a tour of Riverdale—four stories of old red brick clinging to a steep hillside along the Enoree River. He was the first to tell me that Riverdale was a special place.

When I began to visit Riverdale after its impending closing was announced, there were maybe 80 people sharing two shifts there left, about a third of the number just two years earlier. As I began talking with people there and in their homes, before and after the closing, a real sense of community, if not family, was evident among them. The mill was at the center of their lives. I met many second- and third-generation Riverdale workers. One man proudly told me his son, now in high school, worked as a second-shift sweeper.

None of the people that I met and talked with seemed bitter. At least no one blamed the owners for the closing of Riverdale. Instead, there seemed to be an almost universal tone of fatalistic acceptance of the plant's demise.

Many of the people who lost their jobs at Riverdale Mills will not find another job in the textile industry. Most of the machines they once tended were sold to factories in Taiwan or Indonesia, or scrapped. The mill itself may be destined for demolition. "We had some really good times. I'll miss this place," a 31-year veteran told me.

These are some of but a few of the special people who ran Riverdale Mill.

—Photographs and text by Mike Corbin

Leland Nelson, 46 years—*Plant Engineer*
Leland is in the bell house on the roof of the Riverdale Mill, working out a plan to remove the old mill bell.

Tommy Sean, 22 years—*Technician, Card Room*
Tommy's grandfather and father both worked at Riverdale. Tommy's son, Justin, worked in the mill as a sweeper.

Don Harrellson, 43 years—*Mechanic, Mill Shop*
Shown here outside the shop, Don started out at Riverdale in the cloth room, served in the U. S. Army for three years, then came back to the mill and worked until he retired when Riverdale closed.

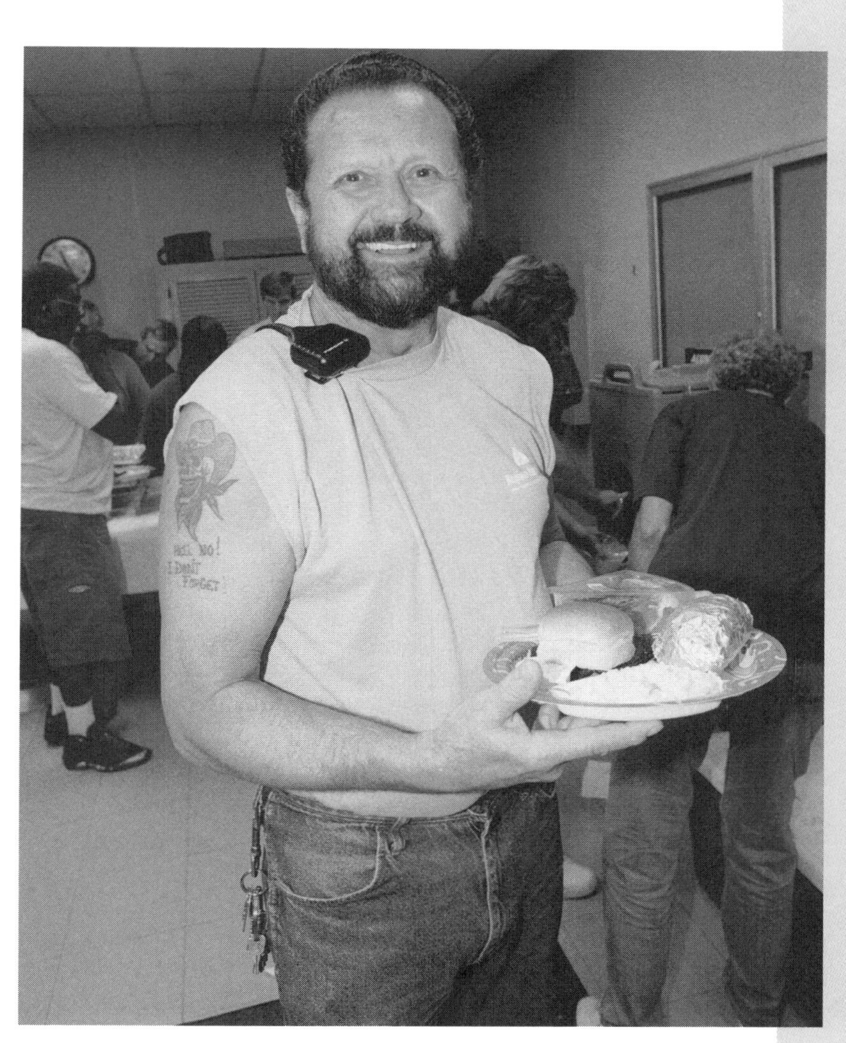

Roger "Rock" Hughes, 30 years—*Head Overhauler*

Rock, shown here at Riverdale's final employee appreciation luncheon, is a 60-year-old mill village kid. "My daddy worked here 'til he was 70 years old. Our family was from a little place up in the mountains, Jerry Creek, Tennessee. We was poor people…recruited by the mill. They sent people up for workers, and the bigger the family was, the more they wanted you. They put all my daddy's family on the train, brought them down here and put them in a new house. They got sent to work the next day."

Dorothy Moates, 22 years & Lesteen Harris, 5 years—*Warper Operators*

Dorothy worked at a variety of jobs at Riverdale before she took over running the warper where her older sister, Lesteen, joined her. They wound up working together for five years before they both "retired" on the same day.

Will "Tree" Dial, 26 years—*Technician, Card Room*

The tallest man at Riverdale, Will planned to go to work for himself when the mill closed. "I fix things. I can build you a deck, shingle a roof, remodel a house, just do about anything you need. I expect I'll be doing more of that when I leave."

Gilbert Kidd, 27 years—*Supply Room*

Gilbert retired from Riverdale in May 2001 to his home in Clinton where he keeps busy around the house. "Oh, I always got something to keep me busy. The garden, the chickens, the bees and the dogs. I've always got something to do."

They Closed Down the Mill
by Matt Rank

Today they closed down the mill at a quarter to four,
Said, "We're losin' too much money,
You don't work here anymore."
The doors have been open for 100 years,
Four generations of hard work, sweat and tears.

Well good-bye to you, my brother, my friend.
They shipped it over the border,
It won't be back again.
The cloth and the mill, they are crumbling now,
Like a moth eaten flag and a rusty ol' plow.

Granddaddy worked there all of his life,
Lived in their mill houses with his children and wife.
And my dad he worked out on the mill floor,
Early in the morning till he grew old and sore.

Well good-bye to you, my brother, my friend.
They shipped it over the border,
It won't be back again.
The cloth and the mill, they are crumbling now,
Like a moth eaten flag and a rusty ol' plow.

Now I'm not political but how fair can it be?
You take away a man's job and ship it over the sea.
If the ones in the big office, and the ones at the bank,
Would push away from the table, we'd have someone to thank.

Matt Rank, a Spartanburg songwriter and band leader, penned this song on May 4, 2001, the day after the sudden shutdown of Spartan International. Rank and his wife Annabelle front the Upstate bands Perry Road and DAM Combo, which perform throughout the Southeast.

Goods made in Pacolet carried the white horse label. (Courtesy, Joe O'Neil)

Textile Town Appendix

produced by Betsy Wakefield Teter, Karen Nutt, and Bill Lynch

Abney Mill. See Woodruff Cotton Mill.

American Fast Print. Established in 1968 by J. S. Fryml, a third-generation Czechoslovakian textile manufacturer who left his homeland when the Communists nationalized industry in 1948 and relocated the business to Canada. American Fast Print was an outgrowth of Montreal Fast Print. Fryml built his small printing and finishing plant along Interstate 85; facility grew tenfold in size by the mid-1990s and had 600 employees. A sister plant, Fryml Fabrics, was built adjacent in 1973. That plant had 120 employees before closing in 1987. Fryml bought Greenville's U.S. Finishing from Cone Mills in 1984. Stung by imports, American Fast Print downsized to 250 workers by 2002.

Apalache Mill (Also, Appalache, Apalachee). One of the earliest mills in Spartanburg County, built between 1836 and 1838 on the South Tyger River two miles north of Greer and originally named South Tyger Manufactory at Cedar Hill. First established by the Rev. Thomas Hutchings. Early investors included Benjamin Wofford, Simpson Bobo, and James McMakin. In 1840, the mill was purchased by McMakin and David W. Moore; five years later, Moore became sole owner. He later sold the factory for $150,000 to Peter Wallace, editor of the *Carolina Spartan*. Wallace filed bankruptcy in 1859, after which Samuel N. Morgan, a local merchant and farmer, secured control and stabilized the company. At the time, the plant operated 800 spindles and employed about 30 people. The mill was known as Cedar Hill Factory from 1880 until 1888, when it became Arlington under president G. T. Walker, and operated 1,300 spindles and 24 looms. After several other changes of ownership, Victor Monaghan Company purchased it in 1911, which began a period of time when the plant experienced strong growth and expansion. The surrounding community was renamed Apalache at this time. J. P. Stevens bought it in 1946, producing cotton yarns, brown sheetings and fancy weaves. At its peak in the 1950s Apalache had 143 mill houses. Stevens closed the plant in April 1981 due to economic conditions. Delta Woodside operated a mill there in the 1990s before closing it in 1996.

Arcadia Mills. See Mayfair Mills.

Arkwright Mills. Located on Fairforest Creek in south Spartanburg. Named for Richard Arkwright, the British inventor of the spinning jenny. Organized in 1896 by Robert Zimmerman Cates (grandson of Glendale stockholder John Zimmerman), H. M. Cleveland, Joseph Walker, J. Boyce Lee, L. J. Simpson and W. F. Smith. With $200,000 in capital and Cates as president, the mill began full operations with a Corlis steam engine in 1898, with 11,000 spindles and 374 looms. The Cates family bought 100 percent of the stock in the 1930s. Mill survived the Depression by selling smoked hams throughout the United States. M. L. Cates became president and was succeeded by his son M. L. (Mac) Cates Jr. Arkwright opened its Camp Croft division in 1948 in a former laundry building for the military base; plant was closed in 1971 when snow collapsed the roof. Built Cateswood Plant in Arkwright in 1968; plant produced 20 percent of the flannel for work gloves in the United States. Original Arkwright Mill closed in 1979 and was demolished. Last remaining plant, Cateswood, was sold to Mt. Vernon Mills of Greenville in 2000.

Arkwright Mill about 1930 (Courtesy, the Spartanburg Herald-Journal)

Arlington. See Apalache.

Armitage. One-story plant built by Milliken & Co. on Interstate 85 in western Spartanburg County in 1968. Opened as an electronics plant and converted to textured polyester yarn. Closed in 1984.

Barksdale Mill. A small, 100-spindle mill organized in 1866 that stood near the site of the Enoree Manufacturing Company and held the machinery formerly used at Hill's Factory on the Tyger. Investors included James Nesbitt Jr., who owned a store in Enoree. About 50 people worked there soon after start-up. Mill was out of business by the time Enoree Manufacturing opened in 1888.

William Barnet & Sons. Founded in Albany, New York in 1898, and began to move south in 1961 with the purchase of a plant in Tryon, North Carolina. Bought the former Draper facility on Hayne Street in Saxon in the early 1970s. Bill Barnet, the fourth generation of his family to run the company, relocated the headquarters to Spartanburg in 1976. Once employed 1,000 people in the Southeast; now has plants in Germany, Belgium, Slovakia, and Italy, and trading companies in Hong Kong, China, and Mexico. Primarily engaged in the processing and trading of fiber forms, now moving toward plastics. Barnet sold the company to his top management in 2001 and was elected mayor of Spartanburg.

The original home of Draper in Saxon, now occupied by Wm. Barnet & Sons LLC (Courtesy, Ron Brown)

Beaumont Manufacturing Company. Located on Chinquapin Creek inside the corporate limits of Spartanburg, it was built in 1890, equipped with 3,072 spindles, 640 twisters, and 40 bag looms. J. H. Sloan (president), John B. Cleveland, Joseph Walker (the original land owner), H. A. Ligon, C. E. Fleming, Vardry McBee, and others filed for the charter. Sloan told the Charleston newspaper his mission was to bring "renewed prosperity and life to Spartanburg." The original building was one-story high and located adjacent to the Richmond & Danville Railroad (later Southern Railroad). A mill village of 15 homes along North Liberty Street was constructed with the mill. The main products in 1900 were carpet warps, seamless bags (for corn and grain), and wrapping twines for the domestic market. These products were made using cotton waste from nearby mills. In 1907, the mill operated with 12,360 spindles and 252 automatic looms. After Sloan's death, Dudley L. Jennings became president. By 1920 there were 142 homes in the village. Walter S. Montgomery Sr. acquired the plant in 1941 when plant equipment was modernized for the production of heavy cotton-duck fabric needed by the armed services for the impending war. Beaumont was the first textile plant in the nation to dedicate its entire production to the war effort. The plant's employees won five coveted Army-Navy "E" awards during World War II. Eventually came under the umbrella of Spartan Mills and employed more than 1,200 people. Closed in 1999 and partially dismantled in 2002.

An aerial photograph of the sprawling Beaumont Mill. At its peak, this mill employed more than 1,800 workers. (Courtesy, Spartanburg Regional Museum)

Bigelow-Sanford. Carpet factory in Landrum opened in 1956. Purchased by Mohawk Carpets in 1996.

Bivingsville. See Glendale.

Blue Ridge Hosiery Company. Organized in Landrum in 1901 for manufacture of hosiery. Located on South Shamrock Avenue in a two-story building. In 1907, capital stock was $39,000, and the mill operated 95 knitting machines. At its height before the Great Depression, the mill employed 150-200 people and manufactured about 350 dozen pairs of hosiery a day. Joseph Lee was the founder, and his son, Roland, ran it until its closing in the early 1930s. The mill was torn down and the brick was used for chimneys in the Landrum area.

Mills often had elaborate corporate logos, such as this one for Landrum's Blue Ridge Mill, which made hosiery. (Courtesy, Bill Lynch)

Buckeye Forest. See Butte Knit.

Burnt Factory. See Weaver's Factory.

Butte Knit. Began operations on Interstate 85 at New Cut Road in 1960. Was the first completely integrated knitted garment manufacturing plant in the United States. In 1967 Butte built Buckeye Forest finishing plant on U. S. Highway 29 West and began a four-building sewing and distribution complex on Sigsbee Road called Butte Industrial Park. Closed all factories in 1980 and 1985. Buckeye Forest was sold to Delta Woodside in the late 1980s, which renovated it for a yarn mill, then sold it to Ameritex Yarns of Burlington, North Carolina.

Camp Croft. See Arkwright.

Cateswood. See Arkwright.

Cedar Hill Factory. See Apalache.

Chesnee Cotton Mill. Two-story mill built in 1910 by John A. Law, founder of Saxon Mill. In 1927 the mill had 20,160 spindles and 440 looms. Bought by Reeves Brothers in 1945. Produced Army twill. Closed in 1997.

Clifton Manufacturing Company. An outgrowth of Glendale, Clifton was incorporated in 1880 with a capital stock of $200,000. Designed by Lockwood Greene, Clifton Mill No. 1 was erected at Hurricane Shoals and began operations in 1881 with 7,000 spindles, 144 looms, and 600 operatives. Mill No. 2 started in 1889, three-quarters of a mile downstream, with 21,512 spindles and 720 looms, later increasing to 27,776 spindles and 861 looms. No. 3, also known as Converse, was built the following year upstream of the other two. It was considered one of the largest mills in the country, with 34,944 spindles and 1,092 looms in 1900. The first president was Edgar Converse, followed by his brother-in-law, Albert H. Twichell, in 1899. Early products for Clifton included standard sheetings, drills, and print cloth. All mills were rebuilt after the flood of 1903. Upon Twichell's death in 1916, J. Choice Evins took over as president. He was replaced in 1945 by Stanley Converse. The company expanded at Converse in 1949 and 1952; it expanded in 1957 at Clifton No. 2. In 1965 company was sold to Dan River Mills, which shut down all mills between 1968 and 1973. Clifton No. 2 operated in a reduced capacity from 1973 to 1983 under lease to Tuscarora Yarns, which purchased the building in 1983 and operated it into the 1990s. The other plants were used for various warehousing operations. A Myrtle Beach company took ownership of No. 1 in the spring of 2002 and began demolition.

Cowpens Manufacturing Company. Organized by a group of Cowpens residents in 1889. R. R. Brown was president and treasurer; directors included A. B. Calvert, S. B. Wilkins, John Dewberry, F. H. Cash, and others. Opened with 75 employees "at a less cost per spindle than any mill in the state." By 1895, the two-story mill operated 3,144 spindles and 200 looms. In 1907, it operated with 17,000 spindles and 406 looms to manufacture sheeting for domestic and export trade. Employed 200 people at that time, with 400 living in the village. Closed and reopened several times between 1930 and 1955, when cotton manufacturing finally ended. Operated as a chemical plant in 1970s. On March 30, 1999, a fire in one of the mill village houses spread to the mill and destroyed it.

Crawfordsville (Also Crawfordville). See Fairmont.

Crescent Knitting Mills. Established in 1904 at the corner of Crescent and Farley Avenues in Spartanburg, the mill produced boys' and women's hosiery for northern markets. The original president was D. D. Little; Ben Mont-

Crescent Manufacturing opened in 1904. (Courtesy, Carol Petty)

gomery and W. W. Lancaster owned and managed the plant for more than 20 years. Closed in 1941 and became a mail order warehouse for Spartan Mills before being demolished during urban renewal.

D. E. Converse Company. See Glendale.

Drayton Mills. Organized in northeast Spartanburg in 1902 and owned by John H. Montgomery, John B. Cleveland, John F. Floyd, W. A. Law, W. E. Burnett, A. C. White, and Arch B. Calvert, who served as president. With lawns as its main product, the mill had 44,800 spindles and 900 plain looms in 1907. Operated by Ben Montgomery in the 1920s. Heavy debt led to a sale to Deering Milliken, its New York sales agent, in 1937. Received coveted "E" Award for war production during World War II. In the 1960s, plant employed 1,200 people. Milliken phased out the plant in 1990 and 1994.

Enoree Manufacturing Company. See Riverdale.

Fairmont Mills. Organized as Crawfordsville by Dr. James Bivings in 1847 on 650 acres he owned on the Middle Tyger River. Mill named in honor of John Crawford who lived nearby. In 1850, it employed 11 people. Sold in 1856 to the firm of Grady, Hawthorne, and Turbeyville. As the Civil War came, mill had 1,000 spindles and 20 looms. After the war, the mill was sold to the Morgan Brothers, then to Harris & Dillard, who expanded it in the late 1880s. Factory later became Fairmont Mills, organized in 1892 by John Bomar Jr., John Zimmerman, and D. E. Converse with $50,000 in capital. In 1890s, mill had 4,000 spindles and 120 looms. Just after 1900, the mill used oil lamps attached to the walls for nighttime production, skylights during the day. Between the 1920s and 1957, the mill had a series of owners, including Guy Harris, Joseph Sulton & Son, and Buffalo Manufacturing Corp. Reeves Brothers owned it from 1957 to 1962, when it was bought by Lyman Printing and Finishing Co. Mill was destroyed by fire in 1977.

Fairforest Finishing. Constructed west of Spartanburg in 1929 as a partnership between Reeves Brothers and the Ligon family, which owned Arcadia Mills. Reeves became sole owner when Arcadia ran into financial trouble in 1935. Arthur Ligon became head of southern operations for Reeves. Mill village, now gone, was called Clevedale. From 1940s to 1960s, plant primarily produced finished fabric for the U. S. military, winning an "E" Award during World War II. At mid-century, it was the sole supplier of water repellant fabric for London Fog raincoats. Plant was expanded in 1942 and 1947. Facility closed in 1974 because of inadequate pollution control facilities. Reopened later by Reeves as a urethane coating plant, creating fabrics for use in aerospace, automotive, and military applications. A second plant was added in 1994 to weave fabric for automotive air bag material.

Fingerville Cotton Factory. Joseph Finger partnered with Gabriel Cannon in 1849 to form the 400-spindle mill 15 miles north of Spartanburg on the North Pacolet River with an investment of $5,000. Original name of the company was Pacolet Manufacturing Company, and the plant started up with 15 operatives. In 1880, there were 1,000 spindles and 15 looms. Fire destroyed original factory building and machinery in 1885. New mill built five years later and incorporated by Belton Liles, T. J. Trimmier, J. A. Henneman, J. A. Lee, A.G. Floyd, and F. P. Simms. With Liles as president, the new yarn mill began operations with 100 employees and 3,000 spindles (increased to 7,000 within two years). Mill was heavily damaged in flood of 1903. Owned in the 1920s by Franklin Process Spinning Mills of Rhode Island, who sold it in 1957 to Indian Head Mills of New York. Last owner was Oneita Knitting Mills, which closed it in 1995.

Fort Prince Spinning Company. See Jackson Mills.

Franklin Process Spinning Mills. See Fingerville.

Fryml Fabrics. See American Fast Print.

Glendale Mills. First factory at Glendale was started by Dr. James Bivings in 1836 on the banks of Lawson's Fork east of Spartanburg. Called Bivingsville Cotton Factory, the mill had 1,200 spindles and 24 looms in its early days. Other investors were Simpson Bobo and Elias C. Leitner. Bivings severed ties with his local investors in 1844 and left to start an ill-fated mill on the Chinquapin. In 1850 Bivingsville employed 52 people and produced $32,000 of goods annually. Leitner, a member of the S. C. House of Representatives, ran it for a time; deeply in debt, he disappeared one day and was never seen again in Spartanburg County. The mill went into bankruptcy about 1855. John Bomar and five investors purchased the property at auction for $19,500 the following year and hired Edgar Converse to run it. The mill in 1860 included 14,435 spindles and 26 looms, and during the Civil War, it produced cloth for the Confederate government. After the war, the Bivingsville Factory was torn down, and a new mill was built about 1870 when company name changed to D. E. Converse and Company. In 1878, at the encouragement of Mrs. Converse, mill's name was changed to Glendale. Employment topped 150 in 1882. Albert Twichell, brother-in-law of Converse, became president in 1899 upon Converse's death. Flood of 1903 nearly destroyed Mill No. 1. In 1907, the rebuilt mill operated with 37,392 spindles, 518 plain looms, and 550 automatic looms. Main products included heavy drills and print cloth sheeting. Later presidents included W. E. Lindsay, Choice Evins, and Jervey Dupre. Mill was sold to Stifel and Sons of Wheeling, West Virginia, in 1947 and to Indian Head Mills in 1957. Closed in 1961.

An aerial photograph of the Glendale Mill and its community. (Courtesy, Spartanburg Regional Museum)

Hill's Factory. Set up as the Industry Cotton Manufacturing Company, the mill was erected in 1819 by brothers Leonard Hill and George Hill, John Clark, and William Sheldon on the banks of the Tyger River, about four miles upstream from the Weaver's Factory. Started with 700 spindles, it was commonly known as Hill and Clark or Hill's Factory. Also locally called "the cotton factory" in its early days as the mill was an attraction for people who came from "far and near" to see it, according to historian J. B. O. Landrum. This mill manufactured "seamless pictorial counterpanes," according to a Hill family journal. Between 1816 and 1830, the year Leonard Hill became sole owner, the building at Hills Factory was twice consumed by fire. In 1825, it contained 432 spindles and eight looms. The factory employed 16 people in 1860. The Hill family sold the machinery to Nesbitt & Wright in 1866, which used it to begin production at the Barksdale Mill on the Enoree River.

Hoechst-Celanese. German-owned company that built a $500 million facility along Interstate 85 to manufacture polyester staple, monofilament, spunbond, and later, plastic products. Complex began in 1965 as Hercules and was later known as Hystron, then Hoechst Fibers. More than 3,000 people were employed in the 1980s. Merged with Celanese Corp. in 1987. Downsized dramatically in the 1990s, then pieces of the business were sold off to other companies. Main production facility eventually was purchased by KoSa, a Mexican-American partnership.

Huckleberry Mill. See Mary Louise Mill.

Industry Cotton Manufacturing Company. See Hill's Factory.

Inman Mills. Organized in 1901 by lawyer James A. Chapman and his brother and uncle, who were Inman cotton farmers. Located along the Spartanburg and Asheville Road and along the railroad right-of-way for a new track being laid from Spartanburg to Asheville. The original plant and mill village of 150 homes were built on a 600-acre family-owned farm. In 1907 spindles totaled 18,336 and looms, 500. Its main product was fine sheeting. A second four-floor addition was added in 1928, increasing production capacity to 48,000 spindles and 1,368 looms. Company purchased Enoree Manufacturing (then called Riverdale) in 1934. Built $5 million Saybrook plant in Inman in 1959, $7.5 million Ramey Plant in Enoree in 1966, and Mountain Shoals Plant in Enoree in 1990. Company's primary products became apparel fabric and cotton yarn. Closed Inman and Riverdale in 2001.

Island Creek Mills. See Mary Louise Mill.

Jackson Mills. Plant originally began as Jordan Manufacturing in 1905, a small mill producing terry towels and quilts. Located two miles east of Wellford, the mill operated 36 looms in 1907, and E. C. Rogers served as president. About the beginning of World War I, the plant was sold to a new company chartered as Wellford Manufacturing Co. That plant was soon bought by H. M. Cleveland, Alfred Moore, and J. R. Snoddy, who named it Fort Prince Spinning Co. In 1922, Jackson Mills, an Anderson County company whose founders included tobacco baron R. J. Reynolds, purchased the plant. Wellford native Alfred Moore, a founder of Tucapau Mills and a board member at Jackson Mills, was named president. At Moore's death in 1940, Cary L. Page became president and began a modernization program, installing 30,840 new spindles and 629 wider looms. Cary L. Page Jr., was elected president in 1970, at a time when the plant employed more than 400 people. The company was purchased in a leveraged buyout by former Milliken executive George Stone in 1983. Closed in 1997.

John H. Montgomery Plant (Chesnee). A modern, one-story, air-conditioned factory opened in 1965 by Spartan Mills at a cost of $7 million. Included 20,000 spindles and 660 looms, primarily producing corduroy fabric. A second plant, Montgomery No. 2 was added in 1970. Closed by 1996.

Jordan Manufacturing Company. See Jackson Mills.

KoSa. See Hoechst Celanese.

Leigh Fibers. Founded by Hans Lehner in Boston in 1922 as an affiliate of a British cotton textile waste company. Had plants in New Bedford and Fall River, Massachusetts, before opening in Wellford in 1960 and gradually consolidating in Spartanburg County. Plant, located along Interstate 85, expanded numerous times. In 2002 Leigh employed 250 people and was the largest processor of textile waste and fiber by-products in North America. The company purchases fiber and fabric waste of all sorts from textile, fiber, carpet, and apparel companies and reprocesses it for sale to such markets as automotive, coarse yarn, and home furnishings.

Lyman Printing and Finishing. See Pacific Mills.

Mary Louise Mill. Built in the Mayo area about 1885 by Buddy Cash, this mill was first known as Island Creek Mill and Huckleberry Mill. It opened with 1,040 spindles. It was later classified as a "private mill" owned by M. and F. Bok. About 1900 it was sold to W. E. "Ball" Watkins who named it for his daughter, Mary Louise. Mostly destroyed in the flood of 1903,

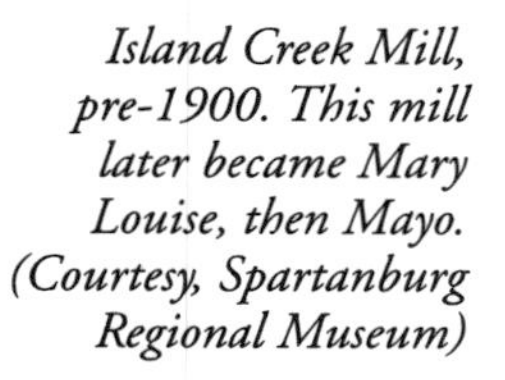

Island Creek Mill, pre-1900. This mill later became Mary Louise, then Mayo. (Courtesy, Spartanburg Regional Museum)

then rebuilt. During the 1920s, the William Whitman Company of Boston owned a controlling interest in this mill, producing high-quality carded yarn. The company went broke and closed in 1932. Mr. Craigler re-opened the plant in 1940 as the Mayo Mill. It was sold again several other times before closing in December 1968. The mill was destroyed by fire February 21, 2000.

Mayfair Mills. Organized in 1903 as Arcadia Mills by local pharmacist Dr. H. A. Ligon with financial backing from the Cleveland and Manning families. Original plant built near Fairforest, just north of area that became known as Westgate. Plant was equipped with 14,624 spindles and 244 looms. In 1923, a second mill was built, known as the Baily plant. Company went into bankruptcy after the stock market crash of 1929 and was sold to Mayfair Mills Inc., a new company organized by Joshua L. Baily & Co., its New York cotton agent. Mayfair purchased Glenwood Cotton Mill in Easley in 1948; purchased Pickens Mill in 1963; opened one of the nation's first open-end spinning plants in Lincolnton, Georgia, in 1974; and purchased plant in Starr, South Carolina, in 1980. Fred Dent, whose mother was a Baily, came south to run the family company in 1947 and was named president. Local employment grew to more than 600 people. Fred's son, Rick Dent, became president in 1989. Company went into bankruptcy in 2001 and closed all its plants.

Model Mill. See Powell Knitting Company.

Mt. Vernon Mills. See Woodruff Cotton Mill.

New Prospect Plant. One-story plant built by Milliken & Co. in northern Spartanburg County in 1960s. Closed in 1985.

Niagra Mill. Opened in 1956 near Saxon, on Williams Street, as Spartan Mills purchased the machinery of the old Niagra Mills of Lockport, New York, and relocated it to Spartanburg. Closed in 1968. This building previously had housed Raycord Sewing Factory, a Spartan Mills facility that first made tire cord fabric in 1945 and converted to shirts and pajamas by 1948.

Olympia Knitting Mills. Built by Andrew Teszler in 1970 on U.S. Highway 221 between Spartanburg and Woodruff. The 400,000-square-foot plant employed 900 people making dyed yarn for dresses. Monsanto bought Olympia in 1974, just as the double-knit business went into a steep decline, and turned it into the world's largest silicon wafer plant in 1979. Silicon plant, later known as MEMC, closed in late 1990s.

ARCADIA MILLS
Fine Converting Cloths

Pacific Mills. Boston-based corporation that built giant manufacturing facility in Lyman between 1924 and 1931. Well known for its high-quality Pacific sheets. Its impressive mill village consisted of parks and recreational facilities, library, modern school buildings, two churches, and 375 mill houses. The complex was purchased in 1955 by M. Lowenstein and Co., which added a research and development center and the Wamsutta Specialty Fabrics Division. Plant was renamed Lyman Printing and Finishing, and the village adopted the name "Mill Town USA." Employed more than 3,400 people in the 1960s and 1970s. Sold to Springs Industries in 1985.

Pacolet Mills. Trough Shoals was the early name for Pacolet Mills. Located on the Pacolet River, Mill No. 1 was chartered in 1882, beginning operations in December of the following year. Organized as the Pacolet Manufacturing Company by John H. Montgomery (president), John B. Cleveland, Joseph Walker,

An aerial photograph of both Pacolet mills and their villages. (Courtesy, Pacolet Elementary School)

Dr. Charles Edward Fleming, and others, with the financial aid of Seth M. Milliken of New York. At the time, the three-story mill had 12,000 spindles and 328 looms, producing drills under the "Buckshead" label. By 1884, village contained 62 cottages and several boarding houses for Pacolet's 500 operatives. In 1888, an adjoining four-story mill was built, and in 1894, a third mill, for a total of 57,000 spindles and 2,190 looms. Company built plant in New Holland, Georgia, in 1902. Flood of 1903 destroyed the great double mill and severely damaged No. 3, all of which which were later reconstructed with financial assistance from Seth Milliken. Mills became known as "No. 3" and "No. 5." By 1907, Pacolet was considered the largest manufacturing complex in Spartanburg County and one of the largest in the South. Operated for many years by Victor Montgomery and Victor Jr., his son. Milliken family took on greater ownership in the 1930s and '40s. Roger Milliken became president in 1947. Pacolet Industries was consolidated into Deering Milliken in 1967. Employment grew to more than 800 people. Milliken shut down weaving in 1981 and spinning in 1983. Later dismantled all Pacolet plants.

Pelham Mills. The Rev. Thomas Hutchings started this mill on the Enoree River in an area known as Buena Vista as early as 1819 and then sold it to Philip C. Lester and Josiah Kilgore around 1830. Also known as Lester's Factory, the mill, which straddled the Greenville-Spartanburg County line, was destroyed by fire in 1853. The partners had no insurance on the property and took a loss of $12,000 as a result of the blaze. Kilgore then sold his interest in the remaining property to Lester. Lester rebuilt the mill, installing about 500 spindles, and enlisted his sons as partners of the new factory, which became known as Lester and Sons. In 1883, Charlestonian Arthur Barnwell and a group of investors from Pelham, New York, incorporated the Pelham Mill Company to purchase the mill. In 1897, a knitting mill was added with five machines, which increased to 70 by 1900, for men's, women's, and children's hosiery. By 1900, the spinning mill operated with 11,000 spindles. At 195 feet, the stack of Pelham Mills was, in 1900, the highest in the state. The mill was later purchased by J. P. Stevens & Co. and burned about 1950. All that remains is the original mill dam.

This cartoon of Enoree Mills, distributed by King Syndicate to newspapers across the country in 1934, spotlighted the fact that Enoree had entrances on all four levels. (Courtesy, Mike Becknell)

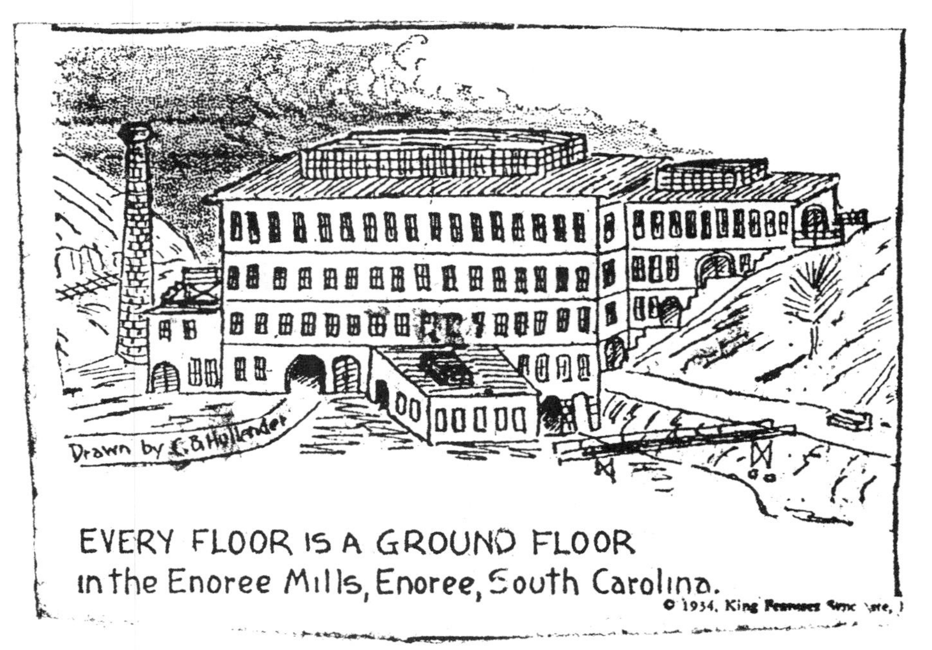

Pequot Mills. See Whitney Mills.

Powell Knitting Company. Originally built around 1919 as the "Model Mill." Operated a number of years with student-workers producing a superior shirting marketed under the name "Character Cloth." The plant, which could not sustain itself, was sold to Powell Knitting Company of Rhode Island in the late 1920s, which used it to manufacture knitted fabrics. In 1958 it was sold to Spartan Mills, which originally used it to manufacture knitted fabrics and socks, and later, geotextiles. Closed by Spartan Mills in the late 1990s. Later used as a warehouse for another Rhode Island company.

Raycord. See Niagra Mills.

Ramey/Mountain Shoals. See Inman Mills.

Riverdale. Incorporated as Enoree Manufacturing Co. by a group of Charlestonians in 1888 and began operations with 10,000 spindles, 320 looms, and about 400 operatives. First president was Grange S. Coffin of Charleston. Unusual mill, built between two hills, had Two Mile Creek running under the middle of the factory and ground-level entrances on all four floors. The original village had about 100 homes, a school, and a church. Closed in 1914 and purchased a year later by W. O. Gay and Co. of Greenville, which renamed it Enoree Mills. Still unprofitable, it was purchased in 1926 by Robert Zimmerman Cates, who renamed it Riverdale. James A. Chapman bought it in 1934. World War II contracts brought strong profits. Mill had 750 employees in the 1950s. Merged with Inman Mills in 1954, the year it became one of the first in the nation to be fully air-conditioned. The addition of new looms in the 1950s made the mill shake so badly that one floor was disconnected from the walls and hung on springs. Once employed more than 600 people. Closed in August 2001. Scheduled for demolition in 2002.

Saxon Mill, shown in 1915, once stood in a wide-open farming area. (Courtesy, Spartanburg Regional Museum)

Saxon Mills. Organized around 1900 by Spartanburg banker John A. Law, this mill was located just outside the Spartanburg city limits. In 1907 it had 5,088 spindles, 300 plain looms and 320 automatic looms. Sold to Reeves Brothers, along with Chesnee Mill, in 1945 and produced print cloth. In the 1950s, Reeves converted Saxon to a polypropylene fiber facility, then sold the entire business to Phillips Fibers Company in 1964. Plant was later sold to Amoco Fibers Company and Drake Extrusion. Closed briefly in 2001, then reopened in 2002 by a Georgia partnership, Saxonfibers LLC, which hired 50 people and planned to grow to 75.

Saybrook. See Inman Mills.

Shamrock Mills. Organized in 1914 by J. E. Mallory, Kye Spears, and others interested in damask tablecloths. Located on Shamrock Avenue in Landrum, the mill was later managed by Mallory's brother-in-law, Ansel J. Mitchell. Product changed from damask material to jacquard bedspreads. Mill had its own village of about 20 homes. Eventually sold to a North Carolina manufacturer of furniture upholstery materials, which closed it in the middle 1960s. Later, part of the operations of South Carolina Elastic Company (formerly Rhode Island Textile Company) moved into the Shamrock Mills building.

South Carolina Manufactory. See Weaver's Factory.

South Tyger Manufactory at Cedar Hill. See Apalache Mill.

Spartan Mills. Organized in 1888 by John H. Montgomery, homebuilder W. E. Burnett, and others. Company was formed in a consolidation with Whitfield Mills of Newburyport, Massachusetts, in the "first direct movement southward of New England capital." Mill No. 1 began operating in 1890 with the Massachusetts machinery—30,000 spindles and 1,100 looms—becoming the first mill in the Spartanburg city limits. At the time, it was the largest mill in the state, consuming one third of Spartanburg County's cotton crop. At opening, Spartan had 150 mill houses in a village called Montgomeryville. A second mill was added in 1896. By 1900, the mills had 74,000 spindles and 2,450 looms. Walter S. Montgomery became president in 1902, and his son Walter, became president in 1929. Company eventually absorbed Tucapau, Beaumont, Powell Mill, Niagra, and Whitney, and spun off Raycord Sewing Factory. Built John H. Montgomery Plant in Chesnee and Rosemont in Union. Walter Montgomery Jr. became president in 1972. Company grew to

This photograph of Spartan Mills shows Dixon Street in the foreground. (Courtesy, Ed and Mary Ann Hall)

13 plants and 5,000 employees in the 1970s. Employment at downtown factories peaked at 1,300. Company renamed Spartan International in 2000 and closed in 2001. Demolition of both mills began in the spring of 2002.

Springs Industries. See Pacific Mills.

Startex Mills. See Tucapau Mills.

Tietex: Founded by British citizen Arno Wildeman in 1974 on North Blackstock Road west of Spartanburg. With 600 employees, it is the largest textile company still in operation in Spartanburg County in 2002. In 1 million square feet of production space, Tietex weaves, prints, and finishes a variety of unique products, including home furnishings for beddings and window coverings, designer fabrics for indoor and outdoor residential upholstery, and specialty fabrics for roofing, shoes, and automotive uses. After Arno Wildeman's death in 1987, his son, Martin Wildeman, continued to operate the company, now known as Tietex International Ltd. A sister plant was added in the mid-1990s in Thailand to serve Asian markets.

TNS Green Plants: A trio of highly automated plants constructed along Interstate 85 near Greer by Japanese-owned Tzusuki Spinning Co. in 1992-1993. Totaled more than 500,000 square feet of manufacturing space. A weaving facility for geotextiles, known as the TNS Advanced Technologies Plant, was added last. All plants closed in late 2001.

TNS Spartanburg: A large spinning and weaving complex opened by Japanese-owned Tzusuki Spinning Co. in mid-1980s in a former Butte Knit factory north of Spartanburg on Interstate 85. Expanded several times in 1990s. Closed in late 2001.

Tucapau. Originally called Tucapaw (and sometimes known as Tucaupau, Tucapah, or "Turkey-Paw"), this mill was built west of Spartanburg on the Middle Tyger River in 1896 by Alfred Moore, Dr. C. E. Fleming, John H. Montgomery, Edgar Converse, and John B. Cleveland. With Moore as president, it opened as a calico mill with 10,000 spindles and 320 looms. The automatic Draper looms were the first of their kind ever installed in a Southern factory, and people came from all over to watch them work. The original village had 33 tenements, and the water supply came from two wells on each side of the river. In 1900, Cleveland was president of the company when the mill included 16,656 spindles and 468 looms. In 1907, the mill employed 580 people and included 64,744 spindles and 1,696 looms to produce print cloths. A second plant for towel manufacturing was built in 1923. Company ran into financial difficulty in the 1920s and had a series of owners including Lockwood Greene and a group of New England creditors. A bleachery and finishing plant were added in 1930. A new company, headed by Walter S. Montgomery Sr., bought the mill in 1936, made renovations and changed the name to Startex Mills (after the Startex brand name of household textiles). Eventually came under the umbrella of Spartan Mills and employed 1,100 people in the 1960s. Plant was closed in 1997 and 1998 and dismantled in 1999.

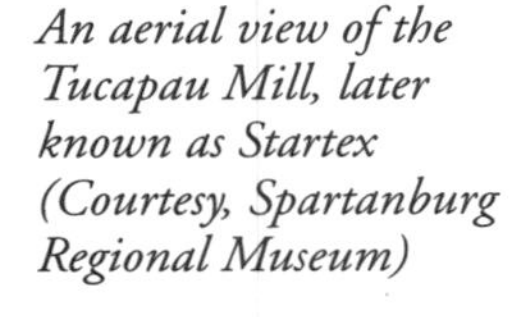

An aerial view of the Tucapau Mill, later known as Startex (Courtesy, Spartanburg Regional Museum)

Tyger Cotton Mill. See Fairmont Mills.

Valley Falls. Small factory built on Lawson's Fork five miles north of Spartanburg in 1857 by William D. McMakin and John Weaver (a member of the Weaver family that founded the first Spartanburg County mill). Community originally was known as Lolo. Subsequently owned and operated by Henry White and William Finger, then by George S. Turner. Employed roughly 15 people between 1860 and 1890. Although the original yarn mill survived the Civil War, it was destroyed by lightning in 1891 when it was owned by F. H. Cash. Rebuilt by 1900 by Trimmier and Company, along with J. P. Stevens. W. P. Roof was president in 1907 when the mill had 7,000 spindles, 200 looms, and about 110 employees. Plant was purchased in 1911 by Martel Mills; in 1957 by Burlington Industries; in 1963 by Charles Jones; and in 1976 by Milliken & Co. Closed in 1992.

Valley Falls Mill. This photo was taken in the 1930s. (Courtesy, Herald-Journal Willis Collection, Spartanburg County Public Libraries)

Victor Manufacturing Company. Located in Greer and owned by the Victor Monaghan Co. of Greenville. Began 1896 with 5,000 spindles and 125 looms. By 1907, the mill had 52,800 spindles and 1,347 automatic looms, producing sheetings, print cloths, and fancy dress goods. At that time, the mill had 662 employees and 1,303 residents of its mill village. Employment peaked in the 1960s at nearly 1,000 people. Purchased by J. P. Stevens after World War II. Closed in 1982. In 2002, a church and day care operated there.

W. S. Gray Mills. Chartered in Woodruff in 1907 as a yarn mill with W. S. Gray as president. Later acquired by Mills Mill of Greenville. After its purchase by Reeves Brothers in 1923, weaving equipment was added, eventually totaling 44,000 spindles and 474 looms. More than 500 worked there in the 1960s. Closed in 1995.

Wadsworth Mills. Yarn mill that operated in the former Camp Wadsworth "sawtooth" laundry building on Blackstock Road from 1922 until the 1940s. Operated by James Marion Trout, grandson of Joseph Finger, founder of Fingerville. Mill made combed cotton yarns on 11,392 spindles. Mill became sewing factory in 1970s, then burned in 1997. The location became the site of the westside Post Office.

Warrior Duck Mill. A small mill that opened on Daniel Morgan Avenue in 1941 to manufacture canvas for military tents. Owned by Reeves Brothers. Closed in 1965. The building later became home of Harbison Furniture and a mattress store.

Wallace Factory. See Apalache Mill.

Weaver's Factory. Built around 1816 on the Tyger River six miles from Cross Keys by the Weaver brothers—Philip, Lindsay, and Wilbur (brother John joined later), who secured a loan from Benjamin Wofford. Started with 489 spindles and 60 acres of land. Between 1816 and 1826, the Weavers operated two small factories at different points on the Tyger, both of which were destroyed by fire. Also known as Burnt Factory and South Carolina Cotton Manufactory.

Wellford Manufacturing Company. See Jackson Mills.

Whitney Manufacturing Company. Organized in 1888 by Dr. C. E. Fleming (president), John Montgomery, John B. Cleveland, and other Spartanburg businessmen. Named for Eli Whitney, the inventor of the cotton gin. Mill designed by Lockwood Greene and located on Lawson's Fork Creek, three miles northeast of Spartanburg. Deering Milliken was exclusive selling agent. By 1890, there were 200 operatives and 40 mill houses. In 1907, the mill had 20,572 spindles and 740 automatic looms, producing mainly standard sheeting. Plant was run for many years by Victor Montgomery and his son Victor Jr. as part of Pacolet Manufacturing Co. The plant ran into problems of debt in the 1940s and was owned briefly by Spartan Mills, Beaumont Manufacturing, then Crescent Corporation, a Charlotte textile machinery dealer. In 1949, it was purchased by Naumkeag Steam Cotton Co. of Salem, Massachusetts, which changed the name to Pequot Mills (much to the consternation of locals). Textron Co. of New York owned it next, calling the plant Indian Head Mills. In 1964 it was repurchased by the Montgomery family as a division of Spartan Mills and renamed Whitney. That operation ended in early 1996, and the plant closed. Plant was dismantled in 2000.

Woodruff Cotton Mill. Organized in 1900 by Aug W. Smith. Investors included J. B. Kilgore and W. W. Simpson. In 1907, the mill had 37,000 spindles and 750 automatic looms, producing shade and print cloths. In 1928, Woodruff Mills consolidated with Brandon Mills and Poinsett Mills in Greenville to form the Brandon Corporation. In 1946, the Brandon Corporation was purchased by Abney Mills of Greenwood. Eventually owned by Mt. Vernon Mills. Closed in 1995 and destroyed by fire in 2001.

Illustration from Spartanburg Journal, *September 1906*
(Courtesy, Spartanburg Regional Museum)

A Textile Glossary

Battery: A round, barrel-like device on a loom that holds spare bobbins of thread. Battery fillers moved up and down weave room aisles replacing empty bobbins with full ones. Workers with this job were often called bat fillers.

Bleachery: Part of the factory where fabric is whitened before dyeing.

Bobbin: A polished wooden spool where yarn is wound

Bossline: A row of houses where the supervisors lived, such as the area surrounding Victor Park in Pacolet

Buck Weaver: A term applied to a man who marries a woman in a mill village, then leaves her and marries someone else in another mill village. South Carolina marriage laws were lax in the early 20th century and required no record of marriage. The woman left behind was called a "grass widow."

Carding: Pulling cotton fibers between a comb-like device with steel points to clean them, make them parallel and ready for further manufacture.

Cheese Truck: A wagon that held full bobbins of thread to be loaded onto a loom's battery.

Come Out: To go on strike, as in, "The weave room is going to come out tomorrow."

Crash Mill: A mill that made fabric for towels, such as the one in Startex.

Creel: A framework arranged to hold slivers, rovings, or yarns so that the ends won't tangle.

Doff: To remove. Doffers removed full spindles of thread or bolts of cloth from the machines where they were made.

Dope Wagon: A cart with Coca-Colas and snacks that was pushed around the manufacturing floor. Early Coca-Cola contained small amounts of cocaine, thus the name "dope."

Drawing a Worm: When mill workers received their pay, they were given a small envelope (known as "a ticket") with pay inside. If they owed the company store more than the company owed them, a squiggly line was drawn across the bottom of the envelope. This was called "drawing a worm."

Dummy: A train that ran on a railroad spur to the mill, carrying cotton in and cloth out.

Fixer: A worker who repaired looms. Usually the highest paid in the mill.

Dope wagons provided snacks in the mills (Courtesy, Junior West)

Flying Squadron: A method used to convince textile workers to walk off the job during the general strike of 1934. Hundreds of striking workers would take to their vehicles and drive together in a motorcade to a mill still operating. Upon seeing the mass of striking workers, the people inside the mill would get the courage to stop their machines and shut down the plant.

Heddle: Fine steel wires with an eye through which ends of warp yarn are threaded on a loom.

Inside Token: A brass token given for piece work or performing a specific task. They had a monetary value either at the company store or the paymaster's office.

Jay Flips: Company script, usually made of paper. Also called "jayclips," "tickets," and "chits."

Jitney: A bus, such as the one that ran between Inman and Spartanburg

Lap: What goes into a carding machine. A continuous sheet of fibers rolled under pressure into a cylindrical package weighing about 50 pounds.

Employees of Arcadia Mills, about 1912. (Courtesy, Mayfair Mills)

Linthead: A derogatory name used to describe mill workers with lint dust in their hair.

Loonies: Company script, usually brass coins emblazoned with the company name. They could be used at the company store for merchandise. Also called "floucans," "flucums," and "bobos."

Mill Hill: Where the houses sat. This was a term coined by the early mill workers who came from the mountains of North Carolina and Tennessee. They were used to terms such as "hills" and "valleys" but were unfamiliar with words like "community" or "village." One lived *on* a mill village, not *in* a mill village.

Opening Room: Where the first step of the manufacturing process begins. A place where operatives strip bales of cotton of their bags and ties and drop small layers of each bale into a machine for cleaning.

Operative: Early name for employee

Picker: Someone who opens and cleans staple, making a lap that goes into a carding machine.

Red Apple: A person in favor of the union during a strike

Roving: A rope-like mass of cotton fibers sent from the card room to the spinning room and placed above the spinning machines in large coils.

Scab: People who ignored a strike and worked in the mill anyway

Stretchout: A name coined to describe the system of extending operatives' labor and responsibility over a greater number of machines. This practice led to the General Strike of 1934.

Shuttle: A canoe-shaped device with a metal tip that holds the bobbin of filling yarn and carries it back and forth across a loom in the weaving process.

Slasher: A worker who operated slashing machines, which joined multiple warp beams of yarn into a single, full-loom beam in preparation for weaving.

Sliver: A continuous untwisted strand of cotton fibers, about one inch in diameter.

Slubber: A machine that reduces the sliver and inserts the first twist before spinning.

Side: One side of a row of spinning machines. A worker responsible for eight "sides" was actually operating four full machines.

Spindle: A small steel rod, about 14 inches in height, on which the bobbin is placed. When it revolves at great speed, it gives yarn its twist.

Warp: Threads that are wound onto a loom before weaving. They run lengthwise.

Waterhouse: Bathroom in the mill

Winder: A machine that transfers yarn from one package to another.

Textile Town Authors

Ross K. Baker is professor of political science at Rutgers University and a regular contributor to the editorial pages of *The Los Angeles Times* and *Newsday*. His commentaries have appeared in *The New York Times, The Washington Post, The Chicago Tribune,* and he is heard on the National Public Radio Program "All Things Considered." Before going to Rutgers, he served on the staffs of Senators Walter F. Mondale, Birch Bayh, and Frank Church and was a research assistant at the Brookings Institute. He is the author of several books on American political history.

Doyle Boggs is the executive director of communications at Wofford College. He earned his B.A. at Wofford in 1970, where he was named a Woodrow Wilson Fellow. He completed his M.A. and Ph.D. in history at the University of South Carolina. He is past president of the Spartanburg County Historical Association and a member of the executive council of the South Carolina Confederation of Local Historical Societies. He is a retired South Carolina Army National Guard lieutenant colonel.

Robert Botsch is a professor of political science at USC Aiken, where he has taught for the past 24 years. He was a CASE professor of the year in 1987-88 for South Carolina and was named one of the top ten college teachers in the nation. He has authored two books on southern workers, *We Shall Not Overcome* (UNC Press, 1980) and *Organizing the Breathless* (University of Kentucky Press, 1993), a book that focuses on the Carolina Brown Lung Association. He is also a co-author of *African-Americans and the Palmetto State*, written for the state Department of Education.

A freelance writer and graphic designer, **JoAnn Mitchell Brasington** lives in Woodruff, South Carolina, not far from what's left of the textile mills where her grandmother and 12 great aunts and great uncles worked. Her grandmother grew up on mill villages in Upstate South Carolina and North Georgia and always had stories to tell. JoAnn is a graduate of Woodruff High School, Wofford College, and the University of South Carolina, where her master's thesis related to the perceptions of Spartanburg County textile workers during World War II.

Glenn Bridges, a Spartanburg resident, is associate editor of *GSA Business Journal* in Greenville where he has written numerous articles about the textile industry and its impact on the Upstate. He holds an associate degree from Spartanburg Methodist College and a B.A. in communication from Coker College. He previously served 10 years as director of public information at SMC, which was founded in 1911 as Textile Industrial Institute to serve the educational needs of textile workers.

Allen Buie, AAIA, is a Spartanburg native who was raised within a mile of the Clifton community. A graduate of Clemson University and a member of the American Institute of Architects, he is pursuing a career in architecture and urban planning with the goal of reclaiming the landscape compromised by suburban sprawl.

Katherine D. Cann seemed fated to spend her life in the shadows of textile mills. Born in Greer, a textile town, she attended Lander College in Greenwood, home of Greenwood and Abney Mills. After brief sojourns in the "southern part of heaven" in Chapel Hill and amid the big city lights of Columbia where she earned graduate degrees, she returned to the Upstate to teach history at Spartanburg Methodist College and live in Greenwood. In 2001, she moved to Spartanburg, *the* textile town.

David L. Carlton grew up in Saxon, where his father worked as a mill clerk before moving into middle management at the Fairforest Finishing Company; an uncle grew up in Arcadia. A 1966 graduate of Spartanburg High School, he received his higher education at Amherst College and

Yale University. Since 1983 he has taught history at Vanderbilt University in Nashville, Tennessee. A specialist in the history of southern industrialization, he is the author of *Mill and Town in South Carolina, 1880-1920* (LSU Press, 1982), a contributor to *Hub City Anthology 2*, and a lot of other stuff nobody reads.

Gladys Coker is a native of Spartanburg County who grew up in Chesnee. Her aunt and uncle are Alaree and Alfred Dawkins, who integrated the Reeves Brothers textile mill in Chesnee. She attended the University of South Carolina at Spartanburg and works for the Greenville County Library System.

Laura Hendrix Corbin, associate director of communications at Wofford College, is a lifelong resident of Spartanburg. She is a freelance writer, specializing in business and economic development issues as well as feature writing. She spent 20 years at the Spartanburg *Herald-Journal* as a reporter/editor and was communications manager for the Spartanburg Area Chamber of Commerce for six years.

Mike Corbin, a native of Ohio, moved to Spartanburg County in 1976. He is a teacher of art and photography at Carver Junior High School. His work is featured in the Hub City title, *Family Trees: The Peach Culture of the Piedmont* (1998). More recently, he worked with the Spartanburg Regional Medical Center, creating a permanent collection of photographs for the Gibbs Cancer Treatment Center.

James Dunlap is associate professor of history at Limestone College. His maternal grandmother, Alta Wood Genoble (1900-1995), was a cloth inspector at Arkwright Mills. James received a Ph.D. in history at the University of South Carolina and wrote his dissertation on the textile industry in Greenville, South Carolina, especially the rise and eventual decline of Woodside Cotton Mill and Parker High School.

Susan Willis Dunlap is the director of development at the Spartanburg Day School. Her senior thesis at the University of South Carolina's Honor's College was "A Mill Town in 1934," an oral history of union activity at the Piedmont Manufacturing Company. Susan is the granddaughter of Clyde and Louise Gilreath, lifelong textile mill employees. She and her family live in the historic Hampton Heights neighborhood.

Jim DuPlessis has covered the textile industry as a reporter for *The Greenville* (S.C) *News* from 1988 to 1992 and has written about the general textile strike of 1934 for *The Knoxville* (Tenn.) *Journal, Southern Exposure* magazine and *The Greenville News*. He was the associate producer for the Honea Path segment of the 1995 documentary film, "The Uprising of '34." He's now a business reporter for *The State* newspaper in Columbia, South Carolina.

Born and raised in New Jersey, **Bruce W. Eelman** holds a Ph.D. in history from the University of Maryland, College Park, where he wrote a dissertation in 2000 entitled "Progress and Community from Old South to New South: Spartanburg County, 1845-1880." Bruce has taught at the United States Naval Academy, the University of Maryland, and is currently an assistant professor of history at Siena College in Loudonville, New York. He continues to research and write on the Spartanburg region and on industrial development in the nineteenth-century South.

Terry A. Ferguson is a practicing geoarchaeologist and currently director of the Wofford College Instructional Technology Laboratory. He has taught geology and anthropology at Wofford since 1984. He graduated from Wofford College and earned M.A. and Ph.D. degrees from the University of Tennessee. He is currently chairman of the board of the South Carolina Heritage Trust.

John Fowler is a traditional musician, storyteller, song collector, and a member of the South Carolina Community Scholars Institute. He shares his work with thousands in the Southeast as a performing artist. He lives in northern Spartanburg County with his wife Kathy and son Taylor.

Tanya Bordeaux Hamm is a freelance writer and editor who works from her home near downtown Spartanburg. She earned B.A. degrees in English and sociology from Wofford College in 1989, received an M.A. degree in journalism from the University of South Carolina in 1991, and worked for The Associated Press, *The Lancaster* (S.C.) *News* and in public relations. She and her husband, Benjy, also a journalist, have two young daughters, Emily and Callie.

Mike Hembree, a newspaper journalist in the Upstate for 30 years, has written or co-written eight books and has been a contributor to several others. His work includes books on the Clifton and Glendale textile villages and Hub City's *The Seasons of Harold Hatcher*. Mike's parents, Wilburn and Ethel Hembree, worked in the Clifton and Beaumont textile mills.

Alice Hatcher Henderson is Distinguished Professor Emerita of History at the University of South Carolina at Spartanburg, where she continues to teach part-time. Her specialty is nineteenth and twentieth century social and cultural history. She and her husband Don are authors of a book on German composer Carl Maria von Weber.

Gary Henderson is a general assignment reporter for the Spartanburg *Herald-Journal* and was named "Journalist of the Year" by the South Carolina Press Association for 2000. He is an essayist, appearing in *Hub City Anthology* and *Hub City Christmas*, and author of *Nine Days in Union, The Search for Alex and Michael Smith*. Gary is the co-author of Hub City's *The Lawson's Fork: Headwaters to Confluence*. His maternal grandfather, John M. Johnson Sr. (1894-1968) began work in the mills along the Pacolet River and Lawson's Fork when he was nine years old.

John Lane is one of the founders of the Hub City Writers Project. A widely published writer and poet, he teaches at Wofford College. His mother's family worked in Spartanburg County cotton mills for three generations. From his office suite at Wofford College he can see the smokestack of Spartan Mills. Just below it his great-grandparents, straight off the farm, met in a mill boarding house and were married at Arch Street Baptist Church.

Jennifer Griffith Langham is a Spartanburg writer who currently writes and designs communications pieces for a local software company. She holds a M.A. degree in English from the University of Tennessee.

Baker Maultsby grew up in Spartanburg and graduated from Wofford College. He is a staff writer for the Spartanburg *Herald-Journal* and in his free time enjoys playing music. With his pal Peter Cooper, Baker co-wrote a song about southern food—"Fatback and Egg on a Bun," a favorite selection of Hall's Grill, formerly located in Una, now located in Arcadia.

Ray Merlock is a professor of communications-journalism at the University of South Carolina Spartanburg, where he teaches courses in mass media, film studies, and video production. He has written film reviews for newspapers, has been a film reviewer on weekly television series, both in Oklahoma and South Carolina, and continues with the "Film Fanatic" series on "The Bill Drake Show" on WSPA-AM Radio. He has also written numerous scholarly articles on film.

Born and reared at Inman Mills, **John Messer** received a college scholarship from the Inman-Riverdale Foundation and became a textiles graduate at Clemson A&M. He was a member of the management team at Riverdale Mills for 14 years and at Hoechst-Celanese for 25 years, retiring in 1992. John is a contributing writer at the *Inman Times* and a volunteer at Spartanburg Regional Medical Center and Mobile Meals. He is a Mason, a Shriner, a Gideon, and a substitute teacher.

A native of Spartanburg, **Harold Miller** began writing as a student of the legendary Louvenia Barksdale at Carver High School in the mid-1940s. He spent nearly 40 years in the food service business at The Charcoal House, then as maitre d' at The Piedmont Club. In recent years, he has been serving as a bailiff at the Spartanburg County Courthouse.

Toby Moore, a former reporter for *The Greenville News*, now lives in Washington, D.C., and works for the U.S. Justice Department. He received a Ph.D. in geography from the University of Iowa where he wrote a thesis entitled, "The Unmaking of a Cotton Mill World: Place, Politics and the Dismantling of the South's Mill Village System."

Karen Nutt is a former newspaper reporter and magazine editor with 16 years' experience in feature writing and one published book, *And Baby Makes Four* (1994). She has a regular column for the Spartanburg *Herald-Journal* called "If You Ask Me." She also writes memoirs, interviewing people and compiling their thoughts into books, which are given to family and friends as keepsakes.

Mark Olencki is a working artist/photographer with three decades of experience in Spartanburg. In 2000, Mark was selected for a Spartanburg artists' exchange program with Winterthur, Switzerland, and exhibited his photographs in both cities. His photographs have been included in the permanent collection of the state of South Carolina, as well as the private collections of many businesses and individuals. Mark's photography work has been featured in several Hub City books, and he has been the designer of all 15 Hub City titles.

Thomas Perry is a fourth-generation linthead, now personnel manager of American Fiber & Finishing Inc. in Newberry, South Carolina. He received a B.A. and M.A. in English at Wake Forest University. He published *Textile League Baseball* (1993) and *The Southern Textile League Basketball Tournament: A History* (1997) and recently finished the manuscript of a historical novel on the wife of Shoeless Joe Jackson. He is married to Donna Adams and they have a daughter, Meghan.

Mickey Pierce is a native of Spartanburg whose grandfather was a local mill worker and participated in the General Strike of 1934. He is an aspiring fiction writer and a member of the board of directors of the Hub City Writers Project.

Norman Powers, who lives in Landrum, had a 23-year career in television and film production in New York City. He was the winner of the 1999 Hub City Hardegree Prize in creative non-fiction and was a contributing author to Hub City's *In Morgan's Shadow* (2001).

Phil Racine was born in Maine where several family members worked in textile mills. He graduated from Bowdoin College and earned M.A. and Ph.D. degrees from Emory University. He is the William R. Kenan Jr. Professor of History at Wofford College where he has taught since 1969. He has edited and written several books on Spartanburg and Southern history, including Hub City's *Seeing Spartanburg* (1999). Phil and his wife, Frances, have two children, Russell and Ali.

Lisa Caston Richie is a freelance writer in Spartanburg. All four of her grandparents retired from textile mills, and her parents were raised in the mill villages of Spartan and Jackson Mills.

Danny Shelton is a graduate of Spartanburg High School and Wofford College. He served Spartanburg's Share the Vision for two years, now is director of admissions at the Spartanburg Day School and is a member of the board of directors of the Hub City Writers Project.

Born in Spartanburg in 1942, **Allen Stokes** graduated from Wofford College, which he attended with a scholarship for the sons and daughters of Spartan Mills employees. He received M.A. and Ph.D. degrees from the University of South Carolina. After serving in the U.S. Army he began working for the South Caroliniana Library where he currently serves as director. His father worked in the shipping office at Spartan Mills and Beaumont. Allen was a spinning room sweeper and yarn tester at Beaumont for several summers.

Philip Stone, a native of Spartanburg, graduated from Wofford College in 1994 and holds a master's degree in history from the University of Georgia. A candidate for the Ph.D. in history at the University of South Carolina, Stone has been the archivist at Wofford since 1999. His dissertation explores the politics of economic development in South Carolina in the mid-twentieth century.

Betsy Wakefield Teter, a native of Spartanburg, received a B.A. degree in history from Wake Forest University in 1980. She had a 17-year journalism career with several newspapers in South Carolina, including serving as business editor of the Spartanburg *Herald-Journal*. Betsy is a founder and executive director of the Hub City Writers Project and was a contributor to *Hub City Anthology* and *Hub City Christmas*. She is the mother of two sons, Rob and Russell.

Diane Vecchio, history professor at Furman University, has published articles on immigrant factory workers in the Northeast. She has a book chapter forthcoming: "Gender, Domestic Values and Italian Working Women in Milwaukee" in *Foreign, Female and Fighting Back: Women, Work and the Italian Diaspora*. She recently completed a manuscript titled, *Work, Family and Tradition: Italian Migrant Women in Urban America, 1900-1935*.

G. C. Waldrep III was born in Halifax County, Virginia, in 1968, where his father had moved as an employee of Burlington Industries. His grandfather and namesake moved from an Upstate cotton farm to the Judson Mills village of Greenville County in the early 1930s, seeking work; his paternal grandmother's family had already settled in Judson, having moved there in the 1920s. Waldrep, who received his doctorate in history from Duke University, is the author of *Southern Workers and the Search for Community, Spartanburg County, South Carolina* (University of Illinois Press, 2000).

Thomas Webster is an elementary school teacher and freelance writer. Although not of mill stock himself, he has the good fortune of being married to a woman who was born and raised on "the Union mill hill." He, his wife, and their two children now make their home in Inman.

Jeffrey R. Willis is the Andrew Helmus Distinguished Professor of History at Converse College, where he has taught since 1967. He is the current president of the Spartanburg County Historical Association and editor of the *Papers and Proceedings of the Greenville County Historical Association* and *The New Greenville Mountaineer*. He is the author of *A Postcard History of Spartanburg, South Carolina* (Arcadia Publishing, 1999) and *Converse College: A Pictorial History* (Arcadia, 2001). He was a contributor to *Hub City Anthology 2*.

Bibliography

Articles:

Arsenault, Raymond. “The End of the Long Hot Summer: The Air Conditioner and Southern Culture,” *Journal of Southern History 50*, November 1984.

Blanchard, Paul. “Communism in Southern Cotton Mills,” *The Nation* 128, No. 3329, April 24, 1929.

—“One Hundred Per Cent Americans on Strike,” *The Nation* 128, No. 3331, May 8, 1929.

Caulkin, Simon, “The Road to Peerless Wigan: Revered by Gurus, Bestowed with Honours, Milliken Leads the Way on the Quality Crusade,” *Management Today*, March 1994.

Converse College Bulletin, “Farewell, Mr. Walter,” Summer 1996.

Lizza, Ryan. “Silent Partner: The man behind the anti-free trade revolt.” *The New Republic,* Jan. 10, 2000.

Phillips, Jim. “Milliken Does it Right,” *Textile Industries*, June 2001.

Shaffer, E.T. H. “Southern Mill People,” *Yale Review*, October 1930.

Shaw, Russell. “Roger Milliken,” *Delta Sky*, December 1985.

Books:

Anderson, Cynthia. *The Social Consequences of Economic Restructuring on the Textile Industry: Change in a Southern Mill Village*. New York: Garland Publishing Inc., 2000.

Andrews, Mildred Gwin. *The Men and the Mills: A History of the Southern Textile Industry*. Macon, GA: Mercer University Press, 1987.

Ayers, Edward L. *The Promise of the New South: Life After Reconstruction*. New York, 1992.

Beardsley, Edward H. *A History of Neglect: Health Care for Blacks and Mill Workers in the Twentieth-Century South*. Knoxville, TN: University of Tennessee Press, 1987.

Bilanchone, Linda, ed., *The Lives They Lived: A Look at Women in the History of Spartanburg*. Spartanburg, SC: Spartanburg Sesquicentennial Focus on Women Committee, 1981.

Biographical Directory of the American Congress, 1774-1996, Congressional Quarterly, 1997.

Blicksilver, Jack. *Cotton Manufacturing in the Southeast: An Historical Analysis*. Atlanta: Bureau of Business and Economic Research School of Business Administration Georgia State College of Business Administration, 1959.

Botsch, Robert E. *Organizing the Breathless: Cotton Dust, Southern Politics, and the Brown Lung Association*. Lexington, KY: University Press of Kentucky, 1993.

Brown, Jimmie Lou Bishop. *The Early Years of Inman, South Carolina*. Greenwood, SC: Bagpipe Press, 1983.

Browning, Wilt. *Linthead: Growing up in a Carolina Cotton Mill*. Asheboro, NC: Down Home Press, 1980.

Camak, David English. *Human Gold from Southern Hills*. Parthenon Press, 1960.

Calcott, W.H., ed. *South Carolina: Economic and Social Conditions in 1944*. Columbia, SC: The University of South Carolina Press, 1945.

Carlton, David. *Mill and Town in South Carolina 1880-1920*. Baton Rouge, LA: Louisiana State University Press, 1982.

Cash, W. J. *The Mind of the South*. New York: Alfred Knoph, 1941.

Chase, William, *Five Generations of Loom Builders: A story of loom building from the days of the craftsmanship of the hand loom weaver to the modern automatic loom of Draper Corporation*. Hopedale, MA: Draper Corporation, 1951.

Coggeshall, John M. *Carolina Piedmont Country (Folklore in the South)*. Oxford, MS: University of Mississippi Press, 1996.

Conway, Mimi. *Rise Gonna Rise: A Portrait of Southern Textile Workers*. Garden City, NY: Anchor Press/Doubleday, 1979.

Cooper, Peter. *Hub City Music Makers: One Southern Town's Popular Music Legacy*. Spartanburg, SC: Holocene Publishing/Hub City Writers Project, 1997.

Corbin, Mike. *Family Trees: The Peach Culture of the Piedmont*. Spartanburg, SC: Hub City Writers Project. 1998.

Crawford, Margaret. *Building the Workingman's Paradise: The Design of American Company Towns*. London: Verso, 1995.

Davis, Bob and David Wessell. *The Coming 20-Year Boom and What it Means to You*. New York: Times Books, 1988.

Davis, William C., ed. *A Fire-Eater Remembers: The Confederate Memoir of Robert Barnwell Rhett*. Columbia, SC: The University of South Carolina Press, 2000.

Dunn, Robert and Jack Hardy. *Labor and Textile: A Study of Cotton and Wool Manufacturing*. New York: International Publishers, 1931.

Edgar, Walter. *South Carolina: A History*. Columbia, SC: University of South Carolina Press, 1998

Escott, Paul D. *After Secession: Jefferson Davis and the Failure of Confederate Nationalism*. Baton Rouge, LA: Louisiana State University Press, 1978.

Etheridge, Elizabeth W., *The Butterfly Caste: A Social History of Pellagra in the South*. Westport, CT: Greenwood Publishing Company, 1972.

Ezell, Helen H. *A Brief History of the First Fifty Years of Chesnee, South Carolina, 1911-1961*.

Ferguson, Terry A. and Thomas A. Cowan. "Iron Plantations and the Eighteenth- and Nineteenth-Century Landscape of the Northwestern South Carolina Piedmont." In *Carolina's Historical Landscapes: Archaeological Perspectives*, ed. by Linda F. Stines et. al. Knoxville, TN: University of Tennessee Press, 1997.

Fitzsimmons, Frank. *From The Banks Of The Oklawaha*. Hendersonville, NC: Golden Glow Press, 1976.

Fleming, Willie. *The History of Pacolet, Volumes I* and *II*. Printel, 1983.

Foster, Vernon. *Spartanburg: Facts, Reminiscences, Folklore*. Spartanburg, SC: The Reprint Company, Publishers, 1998.

Galenson, Alice. *The Migration of the Cotton Textile Industry from New England to the South, 1880-1930*. New York and London: Garland Publishing, 1985

Hall, Jacquelyn Dowd et al. *Like a Family: The Making of a Southern Cotton Mill World.* Chapel Hill, NC: University of North Carolina Press, 1987.

Hembree, Michael and Paul Crocker. *Glendale: A Pictorial History*, 1994.

Hembree, Michael and David Moore. *A Place Called Clifton.* Clinton, SC: Jacobs Press, 1987

—*A River of Memories.* Clinton, SC: Jacobs Press, 1987

Hembree, Michael. *Seasons of Harold Hatcher.* Spartanburg, SC: Hub City Writers Project, 2000.

Hemphill, J. C. *Men of Mark in South Carolina*, four vols. Washington: Men of Mark Publishing Co., 1907-1909.

Hodges, James A. *New Deal Labor Policy and the Southern Cotton Textile Industry, 1933-1941.* Knoxville, TN: The University of Tennessee Press, 1986.

Huff, Archie Vernon, Jr. *Greenville: The History of the City and County in the South Carolina Piedmont.* Columbia, SC: University of South Carolina Press, 1995.

Huss, John E. *Senator for the South.* New York: Doubleday, 1961.

Irby, Hannah B. *Woodruff: An Historical View.* 1974.

Irons, Janet. *Testing the New Deal.* Urbana and Chicago: University of Illinois Press, 2000.

Jacobs, William P. *Problems of the Cotton Manufacturer on South Carolina.* Jacobs & Co. Press, 1932.

Kalogerides, Carla. "Milliken in Motion, A Pursuit of Excellence," *Textile World*, December 1990.

Kanter, Rosabeth Moss. *World Class: Thriving Locally in the World Economy.* New York: Simon & Schuster, 1995.

Kohn, August. *The Cotton Mills of South Carolina.* Columbia, SC: South Carolina Department of Agriculture, Commerce and Immigration, 1907.

Lander, Ernest. "Slave Labor in South Carolina Textile Mills." *The Journal of Negro History*, April 1983.

—*The Textile Industry in Antebellum South Carolina.* Baton Rouge, LA: Louisiana State University Press, 1965.

Landrum, J. B. O. *History of Spartanburg County.* Atlanta: The Franklin Printing and Publishing Company, 1900.

Lane, Herbert. *The Cotton Mill Worker.* New York: Farrar & Rinehart Inc., 1944.

Lawrence, James W. *The Who and Why of Landrum, S.C., 29369.* Kearney, NE: Morris Publishing, 2000.

—*They Called it Inman.* Kearney, NE: Morris Publishing, 2001.

Leonard, Michael. *Our Heritage: A Community History of Spartanburg, S.C.* Spartanburg: Band & White, 1983.

Leitner, Jeffrey. Michael Schulman and Rhonda Zingraf, *Hanging by a Thread: Social Change in Southern Textiles.* Ithaca, N.Y.: ILR Press, 1991.

Lemert, Ben F. *The Cotton Textile Industry of the Southern Appalachian Piedmont.* Chapel Hill, NC: University of North Carolina Press, 1933.

Marshall, F. Ray. *Labor in the South, Wertheim Publications in Industrial Relations.* Cambridge, Massachusetts: Harvard University Press, 1967.

McLaurin, Melton. *Paternalism and Protest: Southern Mill Workers and Organized Labor, 1875-1905.* Westport, CT: Greenwood Press, 1971.

McMillan, Montague. *Limestone College, a History: 1845-1970.* Gaffney, SC: Limestone College, 1970.

Miller, Anthony Berry. *Palmetto Politicians: The Early Political Career of Olin D. Johnston.* Chapel Hill, NC: University of North Carolina Press, 1976.

Minchin, Timothy. *Hiring the Black Worker: The Racial Integration of the Southern Textile Industry, 1960-1980.* Chapel Hill, NC: University of North Carolina Press, 1999.

—*What Do We Need a Union For?: The TWUA in the South, 1945-1955.* Chapel Hill, NC: University of North Carolina Press, 1997.

Mitchell, Broadus. *The Rise of Cotton Mills in the South.* Baltimore: The Johns Hopkins Press, 1921 [Reprinted, Columbia, SC: University of South Carolina Press, 2000].

Morland, John. *Millways of Kent.* Chapel Hill, NC: University of North Carolina, 1958.

Murchison, Claudius T. *King Cotton is Sick.* Chapel Hill, NC: The University of North Carolina Press, 1930.

Perry, Thomas. *Textile League Baseball: South Carolina's Mill Teams, 1880-1955.* Jefferson, NC: McFarland and Co., 1993.

—*The Southern Textile League Basketball Tournament: A History*, Jefferson, NC: McFarland and Co., 1997.

Pictorial Field-Book of the Revolution Vol. 2. New York: Harper & Brothers Publishers, 1859.

Polk County Historical Society. *Polk County, N.C. History* (2nd Ed.). Spartanburg, SC: The Reprint Company, 1999.

Pope, Liston. *Millhands and Preachers.* New Haven: Yale University Press, 1942.

Potwin, Marjorie. *Cotton Mill People of the Piedmont.* New York: Columbia University Press, 1927.

Racine, Phillip. *Seeing Spartanburg: A History in Images.* Spartanburg, SC: Hub City Writers Project, 1999.

—, ed. *Piedmont Farmer: The Journals of David Golightly Harris, 1855-1870.* Knoxville, TN: The University of Tennessee Press, 1990.

Rowan, Richard and Robert Barr. *Employee Relations Trends and Practices in the Textile Industry.* Philadelphia: University of Pennsylvania, 1997.

Russel, Robert R. *Economic Aspects of Southern Sectionalism, 1840-1861.* New York: Russell & Russell, 1960.

Salmond, John R. *Gastonia 1929.* Chapel Hill, NC: The University of North Carolina Press, 1995.

Simon, Bryan. *A Fabric of Defeat: The Politics of South Carolina Millhands, 1910-1948.* Chapel Hill: University of North Carolina Press, 1998.

Simpson, William Hays. *Life in Mill Communities.* Clinton, SC: PC Press, 1941.

—*Some Aspects of America's Textile Industry.* Columbia, SC: University of South Carolina Press, 1966.

Sloan, Cliff and Bob Hall. "It's Good to be Home in Greenville," Chapter 23 of *Working Lives: The Southern Exposure History of Labor in the South*, Marc S. Miller, ed., New York: Pantheon Books, 1980.

Tang, Anthony M. *Economic Development in the Southern Piedmont, 1860 - 1950.* Chapel Hill, NC: University of North Carolina Press, 1958.

Taylor, Linda Dearbury. *History of Cowpens, South Carolina*. Intercollegiate Press, 1982.

Taylor, Walter Carroll. *History of Limestone College*, Gaffney, SC: Limestone College, 1937.

Tippett, Tom. *When Southern Labor Stirs*. New York: Jonathan Cope and Harrison Smith, 1931

Tompkins, D. A. *Cotton Mill, Commercial Features*. Charlotte, NC: Self-published, 1899.

Tucker, Barbara M. *Samuel Slater and the Origins of the American Textile Industry, 1790-1860*. Ithaca, NY: Cornell University Press, 1984.

Tullos, Allen. *Habits of Industry: White Culture and the Transformation of the Carolina Piedmont*. Chapel Hill, NC: The University of North Carolina Press, 1989.

W.P.A., *A History of Spartanburg County*. Spartanburg, SC: The Reprint Co., 1977 [1940].

Williams, George Croft. *A Social Interpretation of South Carolina*. Columbia, SC: The University of South Carolina Press, 1946.

Wallace, David Duncan. *The History of Wofford College, 1854-1949*. Nashville, TN: Vanderbilt University Press, 1951.

Watson, Ruth Trowell. *A Village Called Trough*. Self-published, 1998.

Williamson, Gustavus Galloway Jr., *Cotton Manufacturing in South Carolina*. Ph.D. dissertation, Johns Hopkins University, 1954.

Woodward, C. Vann. *Origins of the New South 1877-1913*. Baton Rouge, LA, 1951.

Young, M.W. *Textile Leaders of the South*. Columbia, SC: The R.L. Bryan Co., 1963.

Files and Papers:

Files, letters, and interviews from the Brown Lung Association and its organizers and members collected by Robert E. Botsch.

Files of James F. Byrnes, Special Collections, Robert M. Cooper Library, Clemson University.

W.C. Coker Papers, Darlington County Historical Commission.

Converse Family Papers, Converse College Archives.

Alester Furman Papers, Special Collections, Robert M. Cooper Library, Clemson University.

J. J. Legare Folder, Letter to Brigadier General Thomas Jordan, October 30, 1862, South Caroliniana Library, University of South Carolina.

Lipscomb Family Papers, Elizabeth Lipscomb letter to her sister, October 23, 1863, Southern Historical Collection, University of North Carolina, Chapel Hill.

Capt. John H. Montgomery Letterbook, March 8, 1899-June 4, 1901. Special Collections, Robert M. Cooper Library, Clemson University.

Olin D. Johnston Papers, Modern Political Collections, South Caroliniana Library.
Lalla Pelot Papers, William R. Perkins Library, Duke University.

Files and letters of the National Association for the Advancement of Colored People, Thomas Cooper Library, University of South Carolina.

Records of The Presidents' Committee on Equal Employment Opportunity, Discrimination Complaint Files, National Archives at College Park, Maryland.

Records of the National Recovery Administration (NRA), National Archives, Washington, D.C.

Southern Labor Archives, Special Collections Department, Pullen Library, Georgia State University, Atlanta.

Newspapers:

Beaumont "E" (October 1942-November 1945)
The Carolina Spartan
The Daily News Record
The Free Lance (1902-1905)
The Georgia Cracker
The Greenville News
The Inman Times
Manufacturers' Record
The (Charleston) News & Courier
The (Una) News-Review
The Pacolet Neigh
The (Columbia) State
The Spartanburg Herald-Journal
The Spartanburg Journal
Textile News
The Textile Tribune
The Wall Street Journal
Women's Wear Daily
The Yorkville Enquirer

Pamphlets:

Coe, Roger. *The Textile Industry in South Carolina: A Vocational Guidance Study Prepared by the National Youth Administration for South Carolina.* Columbia, 1939.

Community Service Council. *Bulletin of all Educational and Social Work for the Building of Better Citizenship in Pacolet Mills Community.* Spartanburg, SC: Band & White Printers, publication date unknown.

Cotton Manufacturers Association. "The Truth About the Cotton Mills of South Carolina," 1929.

Crafted With Pride in USA Council Inc., Washington, D.C. "Our History."

Southern Summer School, 1936. *Scrapbook.*

Hatcher, Harold O. *The Textile Primer.* New York: Council for Social Action, April 1936.

Textile Voices, Spartan, Beaumont, and Startex Mills, 1943.

Charleston News & Courier. *A Story of Spartan Push: The Greatest Cotton Manufacturing Centre in the South, Spartanburg, South Carolina, and its Resources*, July 1890.

The Exposition, Vo. 1, No. 5. *Spartanburg, South Carolina: A Great Manufacturing Center.* Charleston, April 1901.

Reports and Government Documents:

Annual Report of the Commission of Agriculture, Commerce and Industry of the State of South Carolina, 1925.

Congressional Record, December 15, 1977.

Harlan, John Marshall, 2d, U.S. Supreme Court Justice. U.S. Supreme Court ruling March 29, 1965 in Textile Workers Union of America v. Darlington Manufacturing Co. et al, 380 U.S. 263 (1965).

South Carolina Department of Agriculture, *South Carolina: Resources and Population, Institutions and Industries.* Charleston, 1883.

South Carolina Credit Report Ledger, R. G. Dun and Company Collection, Baker Library, Harvard University.

South Carolina General Assembly, Reports and Resolutions, Report of the Commissioner of Agriculture, Commerce and Industries, 1910.

South Carolina General Assembly, Reports and Resolutions, Report of the Commissioner of Agriculture, Commerce and Industries, 1915.

Unpublished manuscripts:

Billings, Susan B. *Beaumont: A Community of Friends and Workers.*

Brasington, JoAnn M., *Textiles Go to War: The Story of Spartanburg, South Carolina's Mill Villages During World War II As Seen in the Pages of the Spartanburg Herald.* M.A. Thesis, University of South Carolina, 2000.

Cann, Mary Katherine Davis. *The Morning After: South Carolina in the Jazz Age.* Ph.D. dissertation, University of South Carolina, 1984.

Clark, Corry. *Mill Operatives and Families at Pacolet Manufacturing Co., 1881-1920: Social Issues.*

DeLorme, Charles DuBose Jr. *Development of the Textile Industry in Spartanburg County from 1816 to 1900.* Unpublished manuscript. Columbia, SC: University of South Carolina, 1963.

Eelman, Bruce W. *Progress and Community from Old South to New South: Spartanburg County, South Carolina, 1845-1880.* Ph.D. thesis. College Park, MD: University of Maryland, 2000.

Grigsby, David. *Breaking the Bonds: Spartanburg County During Reconstruction.* M.A. thesis. Chapel Hill, NC: University of North Carolina, 1991.

Jackson, Bobby Dean. *Textiles in the South Carolina Piedmont: A Case Study of the Inman Mills.* M.A. thesis. Auburn University, 1968.

Moore, Toby Harper. *The Unmaking of a Cotton Mill World: Place, Politics and the Dismantling of the South's Mill Village System.* Ph.D. thesis. University of Iowa, 1999.

Porritt, Edward. "The Cotton Mill in the South," Caroliniana Library, 1895.

Stokes, Allen H. Jr. *John H. Montgomery: A Pioneer Southern Industrialist.* M.A. thesis, University of South Carolina, 1967.

—-*Black and White Labor and the Development of the Southern Textile Industry.* Ph.D. dissertation, University of South Carolina, 1977.

Videos:

Uprising of '34, a film by George Stoney, Judith Helfland and Susanne Rostock

Upstate Memories, a series produced by S.C. Educational Television and the Spartanburg County Commission on Aging

Acknowledgments

Textile Town is truly a community history. More than 100 people contributed to its creation by donating photographs, offering anecdotes, giving interviews, and filling in the historical gaps. Special thanks go to the enthusiastic team of writers and historians who saw the value of this project, made all their deadlines, and produced some dynamic work. More than two dozen new writers came into the Hub City fold with this project.

Thanks go to librarians across South Carolina for their assistance in locating primary source documents, especially: Tom Johnson and Allen Stokes at the Caroliniana Library in Columbia, Philip Stone at Wofford College, Jim Harrison at Converse College, and the staff at the Cooper Library at Clemson University. The staff at the Kennedy Room at the Spartanburg County Public Library provided invaluable assistance to many of the writers involved in this project, as did the kind people at the Spartanburg County Regional Museum. Walter Hill at the National Archives went beyond the call of duty to retrieve the 40-year-old testimony of African-American job seekers. The Spartanburg *Herald-Journal* opened its photo archives for this project, and we are grateful to Benjy Hamm, Chris Winston, and Les Duggins for their patience with us as we collected.

Thanks go to Laura Burris for typing parts of the manuscript; to Nancy Ogle for sharing her library of ETV videotapes with us; and to Sheri Haire for transcribing some of those tapes.

Textile Town was a massive proof reading project, and we are grateful to Jill McBurney, Lisa Isenhower, Carolyn Creal, Winston Hardegree, Doyle Boggs, and Jim Barrett for reviewing it in its raw form. Hub City office assistant Melissa Timmons took on some of Betsy Teter's work so that she could spend nearly a year with her hands in this book. Mark Olencki spent three months of his life designing this book, and Winnie Walsh and Susan Thoms of the Spartanburg County Public Library provided the handy index.

Specials thanks go to key residents of Spartanburg County's mill communities: Elaine Harris and Betty Littlejohn of Pacolet; Don Bramblett of Clifton; John Messer of Inman; Sarah Koon of Beaumont; Lynn Sellars of Una. Hub City also expresses its appreciation to Dr. Carmen Harris of the University of South Carolina at Spartanburg for her guidance in reporting African-American history.

Finally, we are most grateful to the textile people who opened up their lives to us and allowed us to record their words for posterity. These people are the true heroes of *Textile Town*.

INDEX

"Hunk" Gossett's Store, 98
"Pug" Guyton's Barber Shop, 98
(Charleston) Courier, 20
13th Regiment, South Carolina Volunteers, 24, 76
18th Regiment, South Carolina Volunteers, 68
1903 Flood, 71, 77–79, 98, 231
1916 Federal Child Labor Law, 70
1929 stock market crash, 73
1934 General Strike, 129
42nd Division (Rainbow Division), 137

–A–

A Fabric of Defeat (book), 202
A League of Their Own (movie), 205
A Passion for Excellence (book), 289
A Street, 199
Abbeville Plant, 268
Abbeville, S.C., 268
Abney Mills, 307, 318
Acker, Fredda, 205
Act III, 251
Acuff, Roy, 194
Adair, Janet, 269
Adams, John Quincy, 19
Adidas America, 255
AFL-CIO, 219, 241
African Americans, 70, 81–83, 99–101, 126, 139, 158, 168, 181, 197, 210–211, 221–223, 225, 232, 238, 246–248, 274, 298
Agilon, 214
Aiken County, S.C., 81
Aiken District, S.C., 21–22
Air Force Academy, 200
Alabama, 17, 201, 247, 294
Alamo Polymer, 230
Albany State College, 182
Albany, N.Y., 308
Alcoa Fujikura, 255, 274
Aldersgate Methodist Church, 128
Alfred Moore Foundation, 298
All-American Girls Baseball League, 205
Allen, Joe, 166
Allen, Mrs. _____, 178
Allison, William, 28
Alma Plant, 268
Amalgamated Clothing and Textile Workers Union, 289
Ambassador Corp. Inc., 294
American Civil Liberties Union, 249
American Cotton Manufacturers Institute, 292
American Cotton Manufacturers' Association, 90
American Enka, 248
American Fast Print, 307
American Federal of Labor, 172, 173
American Federation of Labor, 151
American Jewish Committee, 249
American Legion, 136, 168
American Textile History Museum, 22
American Textile Manufacturers Institute, 243–244, 268, 277, 289
American Textile Workers Union, 271
American Wool and Cotton Reporter (book), 148
Ameritex Yarns, 309
Amigo Corp Inc., 294
Amoco Fibers Co., 315
Anacona Co., 148
Anderson County, S.C., 73, 137, 312
Anderson, Dwayne, 237
Anderson, S.C., 92, 160, 205
Andrew Teszler Learning Adjustment School, 253
Anti-Defamation League of B'nai B'rith, 249
Apalache Mills, 17, 57, 89, 212, 247, 307, 315, 317
Apalache, S.C., 30, 274
Appleton Mills, 148
Appomattox, Va., 24, 76
Aragon Mill (song), 261
Arcadia Foundation, 297, 298
Arcadia Mills, 41, 96–97, 99, 135, 147, 160–161, 197, 212, 221, 307, 310, 313
Arcadia, S.C., 133, 158, 160, 224, 250–251, 257–259, 274, 279–280
Arkansas, 252
Arkwright Baptist Church, 234
Arkwright Foundation, 297, 298
Arkwright Fund, 298
Arkwright Mills, 40, 98, 102, 135–136, 215–216, 234, 246, 266, 268, 277, 279, 307, 309
Arkwright, Richard, 18, 266, 307
Arkwright, S.C., 153, 157, 234, 299
Arlington Mills, 17, 40, 147, 307
Armitage Plant, 308
Armory Hall, 184–185
Army Air Corps, 199
Arts Partnership of Greater Spartanburg, 260, 298
Asheville Highway, 205
Asheville, N.C., 39, 83, 143, 311
Asia, 194
Athens, Greece, 73
Atkins, Robert, 42–43
Atlanta & Charlotte Air Line, 39
Atlanta, Ga., 73, 80, 106, 221, 245, 254, 257
Atlantic City Auditorium, 73
Atlantic Ocean, 66
Atlantic Seaboard, 235
Aug. W. Smith Co., 77
Augusta, Ga., 52, 81, 278
Australia, 268
Austria, 42, 255
Avalon Plant, 268

–B–

Back Line, 98
Babies in the Mill (song), 35
Badgett, Thomas, 28
Bailey, H.E., 250
Bailey, Willie, 197
Baily Plant, 97, 313
Baker, Ross K., 249
Baldwin Mills, 196
Ball, Lucille, 226
Ballad of Spartan Mills (song), 207
Balmer, Hans, 255
Bamberg County, S.C., 297
Barber, Ruth, 242
Barksdale Mill, 17, 25, 28, 308, 311
Barnet Foundation, 298
Barnet Southern, 258
Barnet, Bill, 308
Barnwell Plant, 268
Barnwell, Arthur, 314
Barnwell, S.C., 268
Barr, Daisy, 225
Barrett, Jim, 297
Barry, Thomas B., 73, 74
Barton, Max, 272
Battle of Cowpens, 76
Bay State Cotton Corp., 148
Beachley, R.G., 140
Beardsley, Edward, 86, 112
Beattie, S.M., 156, 199
Beaumont Avenue, 188
Beaumont Co., 170
Beaumont E (newspaper), 171, 172, 175
Beaumont Lassies, 205
Beaumont Methodist Church, 249, 297
Beaumont Mills, 40, 43, 57, 102, 135, 162–163, 166, 170–173, 188, 194, 212, 244, 246, 249–250, 265, 267, 277–279, 297, 299, 308, 315, 318
Beaumont, S.C., 178, 188, 220, 299
Belgium, 42, 255, 268, 308
Belton, S.C., 268
Belue, Earl, 152
Beresford Box, 255
Berry Field, 205, 206
Best Baby Inc., 294
Best Industries Inc., 294
Beverage-Air Corp., 229
Bi-Lo, 86
Big Hungry River, 131
Big Thicketty Creek, 32
Bigelow-Sanford, 308
Biggerstaff, Julius, 80

Bile Them Cabbages Down (song), 195
Bird, William, 83
Bishop, Barney, 151
Bivings, James D., 19–22, 25, 27, 31, 310–311
Bivings, Susan Elizabeth, 20
Bivings, Susan Von Storre, 19
Bivingsville Cotton Manufacturing Co., 19, 22–24, 31–32, 311
Bivingsville, S.C., 21, 23–26, 266
Bivins, Lillie Lee, 102
Black Lung Movement, 291
Blacksburg, S.C., 159, 268
Blackstock Road, 317
Blackwell, Gordon, 129
Blackwood, Ibra, 137, 160
Blankenship, Harry, 212
Blanshard, Paul, 196
Blease, Cole L., 59, 137
Blechman, Frank, 291
Blowers' Dray, 60
Blue Ridge Hosiery Co., 308–309
Blue Ridge Mountains, 30, 69, 83
Blue Ridge Power Co., 131
BMG Entertainment, 255
BMW Manufacturing Corp., 109, 255
Board of Missions, Methodist Episcopal Church, South, 130
Bobo Funeral Chapel, 192
Bobo, Sandra, 274
Bobo, Simpson, 17, 19, 23, 25, 33, 307, 311
Boggs, Doyle, 199
Boiling Springs, S.C., 57
Bok, M. and F., 312
Bolt, 135
Bomar & Co., 25
Bomar, Converse & Zimmerman, 25
Bomar, John, 23, 25, 32, 311
Bomar, John Jr., 310
Bomar, Thomas, 76
Bomar, Thomas M., 39–40
Bonner's Store, 98
Borden, 274
Bostic, N.C., 268
Boston, Mass., 33, 73, 145, 312–313
Botsch, Robert E., 290
Boy Scouts, 165, 243, 300
Boyce, Robert, 300
Boys and Girls Club, 298
Brackett, Alvie Westmoreland, 194
Brackett, Leonard, 195
Brackett, Thomas "Ted," 195
Bradley, Jennie, 116
Bramblett, Don, 65, 283
Brandon Corp., 116, 318
Brandon Mills, 146, 318
Branham, William M., 77
Brasington, JoAnn Mitchell, 191–192, 235, 294
Brazil, 268
Brennan, William, 250
Brevard Music Center, 260
Brevard, N.C., 192
Brezhnev, Leonid, 260
Brian Lyttle Inc., 255
Brice, Robert, 67
Bridges, Glenn, 72, 257
Bright, Adam, 44
Britain, 18, 167–168
Britton, W.J., 46, 89
Broad River, 30, 77, 80
Brooks, Henry, 192
Brooks, Judson L., 192–193
Brookshire, L.E., 161, 162
Broome High School, 81
Brown Lung Association, 291–292
Brown Street, 126
Brown's Chapel Baptist Church, 98
Brown, Albert G., 23
Brown, Permelia Twichell, 23
Brown, R.R., 309
Brown, Ron, 258–259
Browning, Wilt, 84
Bruckner Machinery, 255
Bruere, Robert W., 155
Bryn Mawr College, 186, 187
Buchanan, Pat, 289
Buck Creek, 77
Buckeye Forest Mill, 252, 309
Buckley, William F., 213
Buena Vista, 314
Buffalo Manufacturing Corp., 310
Buie, Allen, 74
Bungalow Town, 98
Bureau of East-West Trade, 259
Burlington Industries, 265, 271, 277, 297, 317
Burlington, N.C., 309
Burnett Street, 98
Burnett, Vernon R., 157
Burnett, W.E., 41, 310, 315
Burnt Factory, 309, 317
Burrell, Homer, 100
Bush, George, 289
Business Council, 260
Butte Industrial Park, 252, 309
Butte Knitting Mills, 228–229, 236–237, 251–253, 268, 309, 316
Buy American, 276
Byrnes, James F., 144, 155, 167, 203
Byssinosis, 290–291

–C–

Cadillac, 47
Caldwell, Frances, 236, 237
Caldwell, Toy, 263
Calhoun, John C., 18
California, 266, 294
Calvert, Arch B., 309–310
Calvert, Felix, 80
Calvert, Henry, 47
Calvert, Lou, 80
Calvert, Mary Petty, 76
Calvert, Mr. and Mrs. Augustus, 80
Camak, David English, 62, 108, 129–130, 296
Cambridge University, 199
Camby, Clyde, 177
Camby, Don, 167, 222
Camby, Jennings, 167
Camby, Thurl, 167
Cameron Street, 167
Camp Croft, 168–169, 204, 227–228, 297, 309
Camp David, 260
Camp Wadsworth, 92
Campbell Soup Co., 73
Campbell Trophy, 293
Campbell, J. Luther, 152
Campbell, William Henry, 161
Campobello, S.C., 77, 78
Can Holler, S.C., 98
Canada, 23, 255, 307
Canaday, Mrs. _____, 60
Cane Creek, 15
Cann, Katherine, 89, 130, 146
Cannon Mills, 271, 293
Cannon's Shoals, 30
Cannon, Gabriel, 20, 310
Cannon's Campground, 177
Cape May, N.J., 260
Caribbean, 276
Carlton, David L., 55, 209
Carolina Association, 135
Carolina Brown Lung Association, 264
Carolina Cash, 269
Carolina Scenic Coach Lines, 177
Carolina Spartan (newspaper), 20, 22, 24, 32–33, 38–40, 46, 49, 52, 56, 74, 76, 307
Carolina, Clinchfield & Ohio, 39
Carpet tiles, 214
Carr, Betty Hughes, 242
Carr, June, 80
Carson, Fiddlin' John, 221
Carter Family, 194
Carter, John, 193
Cash, Buddy, 312
Cash, F.H., 309, 317
Cates family, 211, 277, 297, 307
Cates, MacFarlane L. Jr., 218, 234, 268, 276, 297, 307
Cates, Robert Zimmerman, 307, 314
Cateswood Plant, 215, 277, 307, 309
Catlett, Gibson, 41, 56
Cauthen, Henry, 295
Cedar Hill Factory, 17, 23, 25, 30, 40, 268, 307, 309
Cedar Hill, S.C., 307
Celanese Corp., 257, 311
Center for Accelerated Technology Training, 255
Central Bank, 64
Central Labor Union, 186, 192, 201
Central Labor Union Hall, 161
Central Methodist Church, 17, 76
Central Park, 146
Chace Mills, 148
Channing, Carol, 268
Chapman family, 211, 266, 275, 292
Chapman, Elizabeth, 275
Chapman, Alfred Foster, 292
Chapman, James A. (1), 47, 61, 128, 292, 311
Chapman, James A. (2), 60–61, 106, 292–293, 300, 314
Chapman, James A. Jr. (3), 218, 231, 276, 292–293
Chapman, Joseph W., 293
Chapman, Marshall, 268, 275, 293

Chapman, Norman H., 293
Chapman, Robert H., III, 293
Chapman, Robert Hett, 80, 293
Chapman, Robert Hett Jr., 60, 293
Chapman, W.H., 182
Character cloth, 130, 314
Charles Lea Center, 253, 298
Charleston, 38
Charleston Cotton Mills, 81, 82
Charleston County, S.C., 137
Charleston, S.C., 15–16, 18–19, 24, 38, 41, 51, 53, 55, 76, 80–81, 83, 131, 212, 237, 308, 314
Charlotte, N.C., 82, 92, 145, 163, 215, 247–248, 257, 318
Cherokee County, S.C., 34, 126
Cherokee Falls Manufacturing Co., 40
Cherokee Falls, S.C., 34
Cherokee Ford, 30
Chesnee Mills, 41, 64, 104, 138, 161, 212, 232, 309, 315
Chesnee, S.C., 112, 158, 209–210, 215, 244, 265, 267, 271, 274, 276–279, 312, 315
Chestnut Street, 118
Cheves, Langdon, 18
Chicago Bloomer Girls, 135
Chicago Match Queens, 206
Chicago, Ill., 71, 248
Chicopee Manufacturing Co., 146
Chicopee, Ga., 146
Child, Bill, 65
China, 48, 98, 260, 308
Chinquapin Creek, 19, 20, 57, 308
Chiquola Mill, 137
Christian Endeavor, 64
Christian Science Monitor (newspaper), 267
Church of God, 129
Church Street, 47
Cincinnati Reds, 248
Cindy (song), 195
CIO, 172, 173, 241
City of Boerne v. Flores, 250
Civil Rights Act of 1964, 225, 246
Civil Rights era, 210
Civil War, 20–22, 24–28, 31, 34, 37–38, 42, 68, 84, 128, 131, 222, 310–311, 317
Civilian Conservation Corps, 168
Clark family, 28
Clark, John, 16, 311
Clark, Shelton, 103
Clay, Tompkins, 56
Clement Lumber Co., 132
Clemmons, Otis, 235
Clemson College, 157, 293
Clemson University, 82, 205, 248, 250
Clemson, S.C., 245, 268, 286
Clerimond, 266
Clevedale, S.C., 274, 310
Cleveland County, N.C., 221
Cleveland family, 96, 313
Cleveland Hotel, 245
Cleveland Junior High School, 72
Cleveland Street, 98
Cleveland, H.M., 307, 312
Cleveland, Jesse, 20
Cleveland, Jesse Franklin, 240
Cleveland, John B., 20, 41, 49, 240, 308, 310, 313, 316–317
Cliburn, Van, 263
Clifton Avenue, 234
Clifton Manufacturing Co., 33, 38–39, 42, 46, 48, 52, 66–68, 70, 72–77, 80, 85–86, 89–90, 103–104, 106, 119, 127, 129, 150, 163, 165, 172, 175, 192, 194, 221, 241–243, 263, 275, 292
Clifton Mill No. 1, 30, 38, 40, 46–47, 66, 78, 96, 155, 166, 231, 240, 242, 309
Clifton Mill No. 2, 30, 40, 48, 66, 75, 77–81, 98, 165–166, 240, 242, 309
Clifton Mill No. 3 (Converse), 77–78, 80, 309
Clifton Mill Store No. 2, 221
Clifton No. 1 Baptist Church, 194
Clifton, S.C., 38, 42, 51–52, 66–67, 72, 75–77, 80, 85–86, 89–90, 103, 127, 129, 150, 165, 175, 192, 194, 221, 242–243, 275
Clinton, S.C., 244, 268, 303
Coates, Kenneth, 200–202
Cobb, _____, 60
Cobb, J.L.H., 41
Coffin, Grange S., 314
Cogdell, P.A., 171
Coggeshall, John M., 93
Cohoes, N.Y., 23
Coker, Gladys, 233
Cold War, 211, 223, 245
Coleman, Frank, 28–29
Coleman, Ray, 152
College Park, Md., 246
College Street, 143
Collins & Aikman, 265
Columbia Record (newspaper), 100
Columbia University, 138, 182
Columbia, S.C., 25, 80–81, 100, 163, 199, 201, 206, 248, 291
Columbus, Ga., 287
Columbus, N.C., 132, 268
Commission on an All-Volunteer Army, 259
Commission on Human Relations, 253
Committee on Equal Employment Opportunity, 224–225, 246
Community Cash, 192, 263
Conanicut Mills, 148
Cone Mills, 307
Congregational Church, 164
Congress of Industrial Organization, 216
Congressional Record (book), 100
Connecticut, 118, 138–139, 266
Consultex, 255
Converse College, 27, 64, 68, 139, 243–244, 263, 296
Converse family, 211
Converse Heights, 215, 234, 293
Converse Mills, 40, 48, 59, 77–78, 80, 106, 135, 197, 242, 248, 282, 309
Converse, Dexter Edgar, 23–27, 30, 32–33, 38–39, 47, 49, 52, 66–68, 74–76, 128, 296, 309–311, 316
Converse, Helen Antoinette (Nellie), 26, 266
Converse, Olin, 23
Converse, S.C., 77–78, 127, 155
Converse, Stanley, 106, 172, 175, 231, 242, 309
Cooleemee, N.C., 200, 201
Cooper, Bill, 28
Cooper-Limestone Institute, 55
Coopersville Iron Works, 31–32
Coopertown, S.C., 98
Corbin, Laura Hendrix, 254
Corbin, Mike, 285, 301
Cornell University, 293
Cornelson, Rose Bailey, 244
Cotton Ball Dance, 204
Cotton Blossom Plant, 268
Cotton Manufacturers Association of South Carolina, 114, 139
Cotton Mill Association of South Carolina, 180
Cotton Mill Colic (song), 87
Cotton Mill People of the Piedmont (book), 138
Cotton Mill, Commercial Features (book), 145
Cotton Mills in the South (article), 31
Cotton Mills of South Carolina (book), 43, 128
Cotton Mill People of the Piedmont (book), 139
Cotton Textile Code, 197
Cotton Textile Labor Relations Board, 139, 155–157, 159, 161, 172
Cotton Textile Queen, 203
Council for Social Action, 164
Court of Common Pleas for Spartanburg County, 249
Cousin Bud and the Hillbilly Hit Parade (radio show), 194
Cousin, B.F., 167
Cowpens Manufacturing Co., 40, 149, 159, 163, 309
Cowpens, S.C., 89, 93, 160, 249, 265, 309
Cox, James L., 53
Craft Drugs, 263
Crafted with Pride in the U.S.A., 268, 289
Crafted With Pride in the U.S.A. Council, 289
Craig, Bob, 210
Craig, John, 16
Craigler, _____, 313
Cramer, Stuart W., 215
Crawford, John, 20, 310
Crawfordsville Mill, 20, 22–23, 32
Crawfordsville, S.C., 20, 30
Createx, 245
Crescent Corp., 318
Crescent Knitting Mills, 93, 158, 309, 310
Crescent Avenue, 309
Crocker, Clarence, 211
Crocker, Clyde, 55
Crompton, Jeff, 85–86
Crompton, Lillian, 85–86
Crosby, Philip, 270
Cross Keys, S.C., 15, 317
Crossroads of the New South, 210, 251
Cumming Street School, 102
Cunningham, Billy, 248
Cushman Plant, 268
Cypress Plant, 268

–D–

D.E. Converse and Co., 24–26, 310–311
Dacron, 256
Daily News Record (newspaper), 275
Dale Carnegie course, 275
Dallas, Texas, 73
Dalton, Ga., 163
Dan River Mills, 227, 231, 309
Dandy, Beatrice, 181
Daniel Construction Co., 216, 287
Daniel Morgan Avenue, 317
Daniel, Charles, 287
Danielson, John W., 41
Danville, Va., 159, 231
Dare Foods, 255
Darlington Manufacturing Co., 288
Darlington, S.C., 35, 217, 288
Davidson College, 248, 293
Davis, Daisy, 126
Davis, Jefferson, 32
Davis, Jimmy, 194
Davis, Julia, 76
Davis, L.J., 44
Davis, Sammy, 268
Davis, Tom, 44
Dawkins, Alaree, 232–33
Dawkins, Alfred, 232–33
Dayton University, 248
De Havilland, Olivia, 179
DeBow's Review, 21
Deering Milliken, 71, 213, 216, 230, 246, 265, 287–288, 310, 314, 317
Deering Milliken Research Corp., 214, 229
Deering Milliken Research Trust, 214
Deering Milliken Service Corp., 245
Deering, William, 71
DeFore Plant, 268
Delta Woodside Industries, 272, 307, 309
Deming Way, 270
Deming, Charles, 270
Democratic Party, 138, 288
Denmark, 268
Dent family, 212
Dent, Frederick B., 212, 218, 227, 251, 259–260, 313
Dent, Frederick B. Jr., 260, 279, 313
Depression, 64, 70–71, 93, 102, 129, 136, 151, 154, 159–160, 163–164, 172, 177, 192, 195, 210, 215, 234–235, 265, 309
Detroit, Mich., 236
Dewberry, John, 309
Dewey Plant, 268
DeYoung, Robert, 193, 207
Dial, Will "Tree," 303
Dill, Iris, 276
Dillard, Richard, 246
Dillon, S.C., 200
Dixie Jamboree (radio show), 194–195
Dixie Shirt Co., 244
Dixiecrats, 138
Dixon Brothers, 221
Dixon family, 123
Dixon Street, 315
Dixon, Dorsey, 35
Dogan, Linda, 299
Doggette, R.L., 133
Dolline's Restaurant, 86
Dominican Republic, 276
Dorsey, Pvt., 171
Double X Boys, 194
Douglas, Paul, 139
Drake Extrusion, 315
Draper Corp., 72, 215, 229, 239, 257–259, 273
Draper, Earle S., 106, 146
Draper-Texmaco, 259
Drapour family, 257
Drayton Black Dragons, 197
Drayton Darlings, 205, 206
Drayton Mills, 41, 49, 68, 71, 93, 135–136, 152, 170, 197, 212, 238, 241, 246, 258, 266, 276, 310
Drayton, John, 15–16
Drayton, S.C., 98, 120, 127, 163, 177, 179, 181, 249, 298, 300
Drayton, William Henry, 266
Dual Lane Highway, 212
Duke Power Co., 132
Duke, James B., 216
Duncan Memorial Methodist Church, 44
Duncan Park, 168, 197
Duncan Park Stadium, 203
Duncan Stewart Plant, 268
Duncan, D.R., 41
Duncan, Donnie, 220
Duncan, S.C., 274
Dunean Mills, 162, 248
Dunean, S.C., 146
Dunlap, James, 128
Dunlap, Susan Willis, 198
Dunn, Robert, 96
DuPlessis, Jim, 286
DuPont, 256
DuPre, A. Mason, 201
Dupre, Jervey, 311
Durham, N.C., 159
Dutchman's Creek, 15
Dye, J.N., 44

–E–

E Award, 173, 308, 310
E Street, 98
Earnhardt, Dale, 221
Easley, S.C., 84, 313
East Main Street, 47
Eastern Carolina League, 135–136
Eastern Illinois University, 200
Edwards, James, 288
Eelman, Bruce W., 24, 31, 33
Egerton, John, 85
Eight Month's Strike (essay), 186
Electric Railway, 79, 81
Elizabethton, Tenn., 196
Elm City Plant, 268
Embargo Act of 1807, 18
Emory, Bud, 80
England, 199, 255, 266
English Manufacturing Co., 33
Enoree Manufacturing Co., 36, 40, 59, 104, 156, 157, 285, 308, 310, 312, 314
Enoree Methodist Church, 29
Enoree River, 17, 25, 30–31, 89, 266, 301, 311, 314
Enoree, S.C., 28, 47, 98, 107, 111, 115, 179, 220, 264, 275, 284, 292–293, 299, 301, 312
Enro Shirt Co., 190, 236, 275
Ensley, Elizabeth, 104
Enterprise Plant, 268
Episcopal Church of the Advent, 250
Epting, J.C., 133
Equal Employment Opportunity Commission, 225
Erwin, Arthur, 224
Europe, 92, 151, 171, 194, 243, 251
Evins, J. Choice, 106, 112, 309, 311
Evins, S.N., 23, 25
Excelsior Plant, 268
Exposition, 44, 53
Ezell, Frank, 181

–F–

F Street, 122
Fair Employment Practices Committee, 223
Fairforest Creek, 307
Fairforest Finishing Co., 72, 99, 170, 212, 274, 310
Fairforest, S.C., 96, 106, 147, 313
Fairmont Methodist Church, 20, 27
Fairmont Mills, 20, 30, 40, 172, 229, 310, 309, 317
Fairmont, S.C., 223, 274, 280
Fairview Heights, S.C., 177
Fall River, Mass., 148, 312
Fallout Shelter Incentive Payment Plan, 245
Farley Avenue, 309
Farley, H.L., 33
Farley, William, 271
Farnsworth Mill, 71
Federal Emergency Relief Administration, 186
Federal Housing Administration, 250, 251
Federal Trade Commission, 257
Federal Writers Project, 91
Ferguson Creek, 15
Ferguson, Joe, 236
Ferguson, Terry, 34
Fiber Industries, 256
Fieldcrest, 271
Findley, Mr. and Mrs. J.B., 78
Finger, Joseph, 20–22, 30, 266, 310, 317
Finger, William, 22, 317
Fingerville Cotton Factory, 20, 25, 40, 136, 158, 266, 310, 317
Fingerville, S.C., 21–23, 77, 298
Finley, Mrs. J.R., 80
Firestone Steel Products Co., 229
First Amendment, 249
First Baptist Church, 68
First Presbyterian Church, 25, 64, 72, 131, 298
Fiske-Carter Construction, 263
Fleming, Charles Edward, 49, 308, 314, 316–317
Fleming, Willie, 80

Flood of 1903, 59, 77, 80, 310–314
Florence, S.C., 257
Florida, 252–253, 280
Floyd's Mortuary, 192
Floyd, A.G., 310
Floyd, John F., 310
Fluor Daniel, 288
Flynn, Dewey, 212
Flynn, Hall, 212
Foerster, Paul, 255–257
Fonda, Henry, 179
Forbes (magazine), 286
Ford Motor Co., 228
Ford, Gerald, 259–260, 276
Forest Street, 113, 143
Fort Prince Spinning Co., 310, 312
Foster, Fay, 99
Foster, Flora, 99
Foster, Vernon, 46, 168–169, 244
Fowler, Art, 248
Fowler, Bessie Holland, 194
Fowler, J.C., 156
Fowler, Jesse Dean, 169
Fowler, John Thomas, 193
France, 137, 167–168, 255, 268
Francis Marion College, 248
Franklin Process Spinning Mills, 310
Franklin, Gene, 194, 195
Fredrick, Zam, 248
Free Lance (newspaper), 57
Fryml Fabrics, 307
Fryml, J.S., 307
Fuller, Belle, 141
Furman Co., 250–251
Furman College, 196
Furman University, 129
Furman, Alester, 250–251

–G–

G Street, 122
G.E. Capital Corp., 279
Gadsden, Al., 159
Gaffney Manufacturing Co., 40, 71, 244
Gaffney Mill No. 2, 40
Gaffney, Michael, 18
Gaffney, S.C., 76, 80, 159, 172
Gainesville, Ga., 68, 71, 82–83, 146, 268
Garland, Hank, 194
Garner, Eloise, 110
Garnett, Kevin, 248
Garrick, Tom, 294
Gaston County, N.C., 34, 149
Gastonia, N.C., 92, 149, 196
Gates, Christine, 190–191
Gault, Mary Irene Lavender, 120–21
Gayley Plant, 268
General Baking Co., 228
General Strike, 146, 161, 177, 184, 187, 192, 207
Gentry's pasture, 57
Gentry, Margie, 179
Georgia, 18, 20–21, 29, 118, 193, 247, 263, 271, 275, 280, 294, 315
Georgia Tech, 287
Gerber, 294
Germany, 167, 255–256, 268, 308
Gerrish Milliken Plant, 268
Gershon, Charles, 143–144
Gettysburg, Penn., 76
Gibbs International, 274
Gibbs, Jimmy, 273
Gibson, Betty, 85
Gillespie Plant, 268
Gilliamtown, S.C., 126
Gilliland Plant, 268
Gilreath, Clyde, 198–199
Gilreath, Hovey "Mutt," 199
Gilreath, Louise Porter, 199
Gilreath, Mary Lee, 199
Ginn, Heath & Co., 76
Glendale Baptist Church, 195
Glendale Mills, 24, 34, 38, 47–48, 52, 91, 135, 194, 211, 217–218, 227, 231, 240, 266, 292, 307–311
Glendale Shoals, 30
Glendale, S.C., 26–27, 30, 45, 47–48, 51, 59, 68, 75, 77, 79, 81, 98, 105, 113, 178, 193, 195, 246, 275, 280, 282
Glendalyn Avenue, 234
Glenwood Cotton Milll, 313
Globe Yarn Co., 148
Gobbler's Knob, 98
Godfrey, H.C., 116
Gold Seal Merchant, 269
Goldberger, Joseph, 85, 143–144
Goldberger, Joseph Jr., 144
Goldberger, Mary Farrar, 144
Golden Valley Plant, 268
Goldwater, Barry, 213, 288
Golightly, Grover, 194
Golightly, Richard, 67
Gorman, Francis, 161, 207
Gosa, Fleetia, 80
Gossett, John Wesley, 197
Goudelock, Dawsey, 188
Goudelock, Walter, 188
Grable, Betty, 204
Graded School, Pacolet Mills, 51
Grady, Hawthorne and Turbeyville, 20, 310
Graf Metallic of America, 255
Grand Ole Opry (radio show), 194, 221
Granite Street, 98
Graniteville, S.C., 21–22, 145
Gray, Cliff (Farmer), 194
Great Britain, 268, 277
Green River, 132
Green River Valley, 131
Greene, Blanche Kirby, 169
Greene, Stephen, 41, 72–73, 82
Greensboro, N.C., 159, 291
Greenville Air Base, 197
Greenville County, S.C., 26, 57, 89, 160, 289
Greenville District, S.C., 17, 32
Greenville Old Pros, 248
Greenville United Textile Workers, 198
Greenville, S.C., 52, 57, 81, 92, 116, 146, 156–157, 159, 161–162, 197, 210, 212, 216, 228, 245, 247–248, 250, 293, 307, 314, 317–318
Greenville-Spartanburg Airport, 215
Greenville-Spartanburg Airport Commission, 287
Greenville-Spartanburg International Airport, 287
Greenwich Village, 212
Greenwich, Conn., 260
Greenwood Mills, 276
Greenwood, S.C., 197, 318
Greer Mills, 161, 212
Greer, Ola, 89
Greer, S.C., 57, 160, 247, 307, 316–317
Gregg, William, 20–22
Gregory, Alvin, 237
Gregory, Jessie Mae, 267
Grier, Mr. ______, 80
Groton School, 286
Guest, C.M., 205
Guilford Mills, 277, 297
Guy Harris, 310

–H–

H Street, 199
Hagerstown, Md., 140
Haggar, 276
Hair, Buster, 136
Haiti, 276
Hall, Ben, 237
Hall, Ella, 80
Hall, Jimmie, 80
Hall, Lola, 80
Hall, Mr. and Mrs. Joel H., 80
Hamilton, Arthur, 135
Hamm, Tanya Bordeaux, 245, 252, 259
Hammett, Henry P., 26
Hammond Brown Jennings, 192
Hampton Avenue, 64, 65
Hamrick family, 172
Hamrick Mills, 270
Hamrick, Jim, 270
Hamrick, John, 277
Handbook of South Carolina (book), 89, 119
Hanes Underwear, 297
Hanes, Jim, 297
Happy Hollow, 98
Harbison Furniture, 317
Harden, William, 77
Harding, Warren G., 144
Hargraves Mills, 148
Harrellson, Don, 302
Harris & Dillard, 310
Harris, David Golightly, 32
Harris, Elaine, 128, 130
Harris, Lesteen, 302
Harrison, Mildred Carrington, 260
Harrison, William Henry, 19
Hartwell, Ga., 268
Harvard Business School, 255
Harvard Law School, 292
Harvard University, 274
Hastoc School for Boys, 244
Hatch Plant, 268
Hatcher Garden and Woodland Preserve, 164
Hatcher, Harold O., 164
Hatchett, Harold, 168
Hawkins, Ernest "Powerhouse," 136

Hayne Street, 308
Hayne, S.C., 224
Hearon Circle, 228
Heath, Minnie, 247
Hebron Methodist Church, 16
Heinitsh Drugs, 192
Helms, Jesse, 289
Hembree, Mike, 48, 66, 240
Henderson County, N.C., 132
Henderson, Alice Hatcher, 68, 138
Henderson, Gary, 143, 203
Henderson, Mrs. _____, 80
Henderson, Oveida, 247
Hendersonville, N.C., 83, 131–132, 140
Henneman, J.A., 310
Henry, Charles H., 76
Henry, James E., 21
Henry, Sam, 160, 184–186
Henson, Mrs. _____, 80
Hercules Powder Co., 230, 256–257, 311
Herring, Harriet, 177
Hetzel, Fred, 248
Heyward, Duncan C., 80
Hickory, N.C., 136
High Bridge, 132
High Shoals, 34
Hill family, 28
Hills Factory, 15–17, 19, 23, 25, 28, 30–31, 308, 311
Hill, Frank, 86
Hill, George, 16, 44, 311
Hill, Herbert, 247
Hill, James, 31
Hill, James Leonard, 28
Hill, John Leonard, 16
Hill, Leonard, 15–17, 28, 311
Hill, Polly, 65
Hill, William, 34
Hillcrest, 253
Hillcrest-Sommer Plant, 268
Hillside Plant, 268
Hines, Geneva, 246
History of Pacolet, Volume II (book), 80
History of Spartanburg County (book), 23
Hitler, Adolph, 167
Hobbysville, S.C., 67–68
Hobourn Area Components, 255
Hobourn Sales, 255
Hodge, Olin, 158
Hoechst Fibers, 230–231, 257, 311
Hoechst-Celanese, 255–256, 311
Hoerger, E.L., 201
Holcombe, David, 67
Holcombe, Susan A., 67
Holden, Kathryn Mabry, 240, 241
Holder, Betty Brown, 206
Holland, Charlie, 134
Hollings, Ernest, 138
Hollis, L.P., 247
Holyoke, Mass., 33
Honea Path Plant, 268
Honea Path, S.C., 138, 268
Hong Kong, 308
Hookworm, 109
Hoover, Brenda, 237
Hope, Bob, 268
Horse Creek Valley, 81
Houston Elementary School, 300
Howard Johnson's, 210
Howard, Winslow, 146–147, 184
Howe, C.E., 44
Hub City, 128, 131, 206, 228
Hub City Courts, 168
Hub City Gents, 197
Hub City Lunch, 192
Hub City Writers Project, 279
Huckleberry Mill, 311–312
Hughes, Roger "Rock," 302
Human Gold from Southern Hills (book), 62, 129
Humphrey Plant, 268
Humphrey, Hubert H., 138
Huntsville, Al., 163
Hurricane Shoals, 30, 66, 76, 309
Hutchings, Rev. Thomas, 15, 17, 307, 314
Hystron Fibers, 257, 311

–I–

Iannazzone, Ralph, 253
Ike's Corner Grille, 188
Imai, 270
India, 98, 211
Indian Head Mills, 310–311, 318
Indiana, 164
Indianapolis Clowns, 197
Indonesia, 301
Industrial Commission, 58
Industrial Revolution, 18, 34, 258
Industry Cotton Manufacturing Co., 16, 311
Inglis, Bob, 289
Inman Mills, 41, 56, 60, 93, 106, 122, 128, 135, 151, 159, 161, 172–173, 181, 194, 199, 208, 215–217, 221, 223, 228, 230–231, 264, 267–268, 270–271, 275, 277, 279, 284–285, 292–293, 297, 301, 311–312, 314–315
Inman Mills Baptist Church, 128
Inman Mills, S.C., 61, 98, 193, 195, 300
Inman, S.C., 48, 57, 61, 133, 210, 265, 268, 284, 312
Inman-Riverdale Foundation, 292–293, 297, 298
Internal Revenue Service, 275
International Brotherhood of Pottery and Allied Workers, 228
International Harvester Company, 71
International Ladies Garment Workers Union, 219, 228–229, 236, 252, 268
Interstate 26, 73, 251, 257
Interstate 85, 73, 214, 216, 229–230, 251–253, 257, 307–309, 312, 316
Irby, W.H., 29
Ireland, 67, 255
Irons, Janet, 163
Ishikawa, 270
Island Creek Mill, 40, 312
Isom, John B., 193
Israel, Boyd, 182–3
Italy, 272, 308

–J–

J.E. Sirrine, 293
J.E. Sirrine Textile Foundation, 293
J.P. Stevens and Co., 212, 227, 265, 271, 297, 307, 314, 317
Jack and Roland, 194–195
Jack, James, 211
Jackson Mills, 41, 72, 165, 277, 298, 310, 312, 317
Jackson, Andrew, 19
Jackson, Bobby Dean, 151
Jacobs, William P., 98
James Creek, 15
Japan, 98, 176, 211, 252, 255, 268, 270, 276, 288
Jaya Mills, 259
Jefferson, N.C., 278
Jennings, Dudley L., 308
Jerry Creek, Tenn., 302
Jim Crow law, 197, 224
Jimmy and Marsha Gibbs Foundation, 298
John A. Walker Co., 77
John Bomar and Co., 23
John H. Montgomery Plant, 209, 215, 244, 265, 274, 312, 315
Johnson and Johnson, 146
Johnson, _____, 135
Johnson, B.W., 133
Johnson, Ben, 77
Johnson, George Dean Jr., 270
Johnson, Junior, 221
Johnson, Lyndon B., 224, 246
Johnson, Mrs. B.S., 80
Johnson, Oliver, 80
Johnson, Paul, 152
Johnson, Roscoe, 80
Johnson, Vance, 133
Johnson, Willie, 181
Johnston Industries, 287
Johnston Plant, 268
Johnston, Gladys Atkinson, 137
Johnston, Olin D., 129, 136, 138, 224, 288–289
Johnston, Rose, 244
Johnston, S.C., 268
Jonathan Logan Inc., 219, 228, 251, 253
Jones, Charles, 317
Jones, Leola, 225
Jones, Paul, 231
Jonesville Black Tigers, 197
Jonesville, S.C., 244, 266, 268
Jordan Manufacturing, 312
Joseph Sulton & Son, 310
Joshua L. Baily & Co., 212, 259, 313
Journal (magazine), 200
Journal and Carolina Spartan (newspaper), 76
Judson Plant, 268
June Freshet of 1903, 77
Junior Order of the American Mechanics, 108
Justice Grocery, 192
Justice, Una Mae Abbott, 192
Justice, Warren, 192
Justice, Wilburn P., 192

–K–

Kahn, Si, 261
Kannapolis, N.C., 293
Kanter, Rosabeth Moss, 255, 274
Kearns, Tommy, 248
Keg Town, 98
Kelly, Betty, 294
Kennedy Center, 260
Kennedy Free Library, 69, 139
Kennedy, John F., 224, 246
Kentucky, 277
Kestler, Henry, 20
Kex Plant, 268
Kidd, Gilbert, 303
Kilgore, J.B., 96, 318
Kilgore, Josiah, 314
Kinard, Terry, 248
King Tut, 197–198
King's Mountain Iron Co., 34
Kingsley Plant, 268
Kingsport, Tenn., 145
Kingstree Plant, 268
Kingstree, S.C., 268
Kirby, Amanda Brown, 85
Kirby, Leo, 154
Kirby, Maggie, 80
Kirby, Mrs. William, 80
Kirby, Nora Bell, 181
Kirby, Robert, 181
Kmart, 271
Knight, Albert, 300
Knights of Labor, 52, 73
Knights of Pythias, 108
Knightson, S.E., 157
Knuckles, Mary Brown, 126
Kohler Co., 228, 273
Kohler, Wis., 228
Kohn, August, 43, 51, 55, 59, 128
Korean Conflict, 211, 223, 282
KoSa, 230, 257, 311
Ku Klux Klan, 224
Kusters Corp., 230, 271

–L–

LaGrange, Ga., 268, 288
Lake Adger, 132
Lake Bowen, 30
Lake Lure, 145
Lake Summit, 131–132, 139–140
Lake Summit Co., 140
Lake Toxaway, 81
Lancaster, Maynard, 129
Lancaster, W.W., 310
Lander, Ernest, 20
Landrum, J.B.O., 67, 128, 311
Landrum, S.C., 70, 99, 119, 308–309, 315
Lane, John, 30
Lane, Mary, 70, 86, 203–204
Lanford, Nellie Maude, 203
Langdon Avenue, 188
Langford, Mrs. Carl, 156
Langham, Jennifer Griffith, 125
Latham, A.M., 32
Laurens County, S.C., 28
Laurens, Henry, 18
Laurens, S.C., 29, 268
Lavonia, Ga., 268
Law family, 132, 211
Law, Carolyn Leonard, 64–65
Law, John A., 64–65, 112, 129, 131, 138–139, 203, 212, 266, 296, 309, 315
Law, John A. Jr., 64
Law, Margaret, 132
Law, Marjorie Potwin (see also Potwin, Marjorie), 140
Law, W.A., 310
Lawrence, Mass., 147
Lawson's Fork Creek, 19, 22, 25, 30, 57, 78, 231, 311, 317
Lawson's Fork Factory, 25
Lee, 276
Lee, J. Boyce, 307
Lee, J.A., 310
Lee, Joseph, 309
Lee, Robert E., 24
Lee, Roland, 309
Lehner, Hans, 312
Leicester College of Technology, 253
Leigh Fibers, 273, 312
Leitner, Elias C., 19, 23, 24, 311
Leitner, George, 23
Leonard, Barry, 278
Leonard, Michael, 226
Lester and Sons, 314
Lester's Factory, 314
Lester, Philip C., 314
Lewis, J. Woodrow, 249
Lewiston, Maine, 41
Life (magazine), 203
Life in the Mill Communities (book), 134
Ligon family, 310
Ligon, Arthur, 310
Ligon, H.A., 308, 313
Liles, Belton, 310
Limestone College, 55, 68
Limestone Springs, 76
Limestone, S.C., 76
Lincoln, Samuel B., 73
Lincolnton Cotton Factory, 19
Lincolnton, Ga., 313
Lincolnton, N.C., 19–20, 23
Lindbergh, Charles A., 116
Linder's Shoals, 31
Linder, Mrs. C.W., 78
Lindsay, W.E., 311
Lipscomb, Elisabeth, 32
Lisbon, Maine, 71
Little, D.D., 309
Live Oak Plant, 268
Local 581, 237
Local 1881, 282
Local 2070, 163, 165
Lockhart Mills, 68, 71, 82
Lockhart, S.C., 86
Lockport, N.Y., 313
Lockwood Greene & Co., 38–39, 72–73, 309, 316–317
Lockwood Greene Engineers Inc., 72–73
Lockwood, Amos, 72–73
Lodge, L.D., 68
Loftis, Lee, 122–23
Logan, Henry, 248
Lolo, S.C., 317
London, England, 266
Long Island, N.Y., 287
Long, John D., 212
Long, Mr. and Mrs. Garland, 80
Longe, Shirley, 116
Loray Mill, 149, 196
Los Angeles, Calif., 272
Louin family, 80
Louisiana, 17
Louisville, Ky., 236
Lowcountry, 21, 34, 55, 82, 137
Lowell, Mass., 148, 289
Lumpkin, Pickett, 201
Lyles, Wayne, 194
Lyman Blues, 197
Lyman Printing and Finishing Co., 136, 161–163, 197, 212, 229, 267, 310, 312–313
Lyman, S.C., 72, 106, 127, 145–147, 160, 168, 184, 193, 198, 313
Lyttle, Brian, 255

–M–

M. Lowenstein and Co., 212, 271, 313
Made in the U.S.A., 268, 269, 294
Magnolia Finishing Plant, 268
Mahlo America, 255
Maid of Cotton, 226
Main Street, 120, 161
Main Street Mall, 263
Maine, 70, 71
Malcolm Baldridge National Quality Award, 287
Mallory, J.E., 315
Malone, Tom, 287
Management Design (magazine), 275
Manhattan, N.Y., 71
Manning family, 96, 313
Manning, Jack, 274
Manufacturers Power Co., 131
Manufacturers' Association of the Confederate States, 20
Marietta, Ga., 268
Marion, N.C., 196
Markham, Edwin, 200
Marshall Plan, 244
Marshall Tucker Band, 263
Marshall, F. Ray, 218
Martel Mills, 96, 277, 317
Mary Black Clinic, 72
Mary Louise Mill, 119, 158, 266, 311–312
Marysville School, 126
Marysville, S.C., 126–127
Mason Machine Works, 148
Masons, 108
Massachusetts, 22, 23, 38, 96, 118, 147, 247, 286
Massey, Mrs. _____, 80
Mauldin High School, 248
Maultsby, Baker, 85
Maybank, Burnet, 288

Mayberry, _____, 135
Mayfair Mills, 95, 160, 212, 230, 250, 257, 259–260, 266, 268–269, 271, 277–279, 297–298, 307, 313
Mayo Mill, 313
Mayo, S.C., 119, 312
Maysville, S.C., 126
McAbee, Charles W., 159
McBee, Vardry, 23, 25, 308
McCarley, Marguerite Cooper, 29
McCarn, Dave, 87, 221
McClure, Carl, 246
McCormick Plant, 268
McCormick, S.C., 268
McCrary, E.R.W., 67
McDowel, J.B., 32
McDowell County, N.C., 66
McGee, Elizabeth "Libba," 226
McGraw Hill Publishing Co., 73
McMakin, James, 307
McMakin, William D., 22, 317
McMillan, S.C., 20
McSwain, W.A., 135
Mechanics Mills, 148
Meek, Martin, 28–29
Melton, Ellen, 192
MEMC, 313
Memorial Day, 204
Memphis, Tenn., 253
Menzel Inc., 255
Mercantile Stores, 287
Merchant, John, 78
Merlock, Ray, 294
Merriman, John, 68
Messer, John, 133, 300
Messner, Ike, 55
Mexico, 308
Meynardie, Rev. J. Simmons, 52
Michelin Tire, 255, 263
Michener, John, 94
Middle Tyger River, 20, 310, 316
Middlesboro, Ky., 292
Midnight Massacre, 230
Midnight Ramblers, 195
Midway Plant, 268
Midwest, 25
Mikro, 248
Mill architecture, 74
Mill Mother's Lament (song), 149
Mill Spring, N.C., 132
Mill Town USA, 313
Miller, Ben, 102
Miller, Deborah, 102
Miller, Grace, 102
Miller, Harold, 102
Miller, Isaac, 102
Miller, J.O., 44
Miller, Jean, 274
Miller, Jonathan, 102
Miller, Susan, 102
Millhands and Preachers (book), 129
Milliken, 120, 229, 246, 257
Milliken & Co., 71, 265, 267–268, 270–272, 275–277, 279, 286–287, 289, 308, 310, 312–313, 317
Milliken Design, 268
Milliken family, 151, 172, 212, 296
Milliken Fine Arts Center, 296
Milliken Foundation, 297
Milliken Packaging, 268
Milliken Research Center, 214
Milliken Research Corp., 245, 263, 268
Milliken Research Group, 287
Milliken Research Park, 270
Milliken Sales, 268
Milliken Science Center, 297
Milliken Street, 104, 169
Milliken, Gerrish, 71, 286, 287
Milliken, Roger, 212, 215, 217–218, 227, 230, 243, 245, 254–255, 259, 265, 270, 272, 275–277, 286–290, 298, 314
Milliken, Seth, 38–39, 41, 52, 55, 68, 70–72, 81–82, 212, 222, 314
Millitron dyeing machine, 214
Mills Mill, 158, 170, 203, 212, 317
Mills, J.M., 167
Mills, Robert, 16
Minchin, Timothy, 213, 221, 224–225, 274
Minot, Maine, 71
Mississippi, 17, 144
Mitchell, Ansel J., 315
Mitchell, Broadus, 89
Mitchell, Clarence, 246
Moates, Dorothy, 302
Mobile Meals of Spartanburg, 297–298
Model Mill, 96, 109, 130, 313, 314
Mohawk Carpets, 308
Monaghan Mill, 247
Monarch Plant, 268
Monsanto, 274, 313
Montgomery & Crawford Hardware Co., 243
Montgomery Building, 72
Montgomery College, 55
Montgomery family, 132, 151, 172, 180, 211–212, 296–297, 318
Montgomery Industries, 244
Montgomery Memorial Methodist Church, 128
Montgomery Plant No. 2, 312
Montgomery, Ben, 79, 93, 120–121, 310
Montgomery, Benjamin Franklin, 67
Montgomery, Harriet Moss, 67
Montgomery, John, 67
Montgomery, John H., 30, 39–41, 43–45, 52–53, 55, 58, 67–68, 70–72, 74, 81–83, 93, 128, 131, 139, 212, 222, 310, 313, 315–316, 317
Montgomery, Mrs. Ben, 121
Montgomery, Rose, 266
Montgomery, Victor, 67, 79, 107, 115, 126, 128, 156, 275, 296, 314, 318
Montgomery, Victor Jr., 314, 318
Montgomery, Walter S. (1), 44, 46, 79, 93, 106, 130–131, 171, 218, 243, 296, 315
Montgomery, Walter S. Jr. (3), 244, 270, 278–279, 316
Montgomery, Walter S. Sr. (2), 173, 188, 212, 244, 250, 263–264, 266–268, 276–278, 282, 294–295, 297, 308, 315–316
Montgomeryville, S.C., 43–44, 315
Montreal Fast Print, 307
Moody, Peter Richard, 199–202
Moore, Alfred, 312, 316
Moore, David W., 307
Moore, G. Walter, 159, 163, 165
Moore, Pearl, 248
Moore, Toby, 27, 92, 145, 195
Morgan Brothers, 310
Morgan Square, 161, 268–269
Morgan Stanley, 271
Morgan, Samuel N., 307
Morland, John, 104, 155, 217, 219–220
Morris, Mike, 284–285
Moscow, 260
Mostel, Zero, 204
Motlow Creek, 77
Mountain Shoals, 17, 30
Mountain Shoals Plant, 293, 301, 312, 314
Mountain Shoals Plantation, 28
Mountainview Home, 168
Mr. Cotton Boll, 95
Mt. Vernon Mills, 276–277, 307, 313, 318
Mullwee, Claude, 103
Multi-Fiber Arrangement, 276
Murrell, A.B., 46
Murry, E.P., 293
Myers Park, 145
Myrtle Beach, S.C., 309

–N–

Nader, Ralph, 290–291
Naismith, James, 247
Nance, Larry, 248
Nashville, Tenn., 221
Nation (magazine), 196
Nation's Health (magazine), 140
National Archives, 246
National Association for the Advancement of Colored People, 224, 246–247
National Business Hall of Fame, 287
National Cotton Council, 226
National Defense Appropriation Act of 1942, 297
National Distillers, 230
National Energy Conservation Council, 260
National Guard, 160, 168, 184, 196
National Industrial Recovery Act, 156–158
National Labor Relations Board, 172, 198, 288
National Recovery Act, 159, 192
National Review (magazine), 213
National Union of Textile Workers, 81
National War Labor Board, 170, 172–173, 178
Naumkeag Steam Cotton Co., 318
Neal, Ellerbe "Big Daddy," 248
Negro American League, 197
Negro League, 197–198
Nelson, Leland, 301
Nelson, Maude, 135
Nesbitt & Wright, 17, 311
Nesbitt House, 28
Nesbitt Iron Manufacturing Co., 34
Nesbitt's Shoals, 30
Nesbitt, Caroline Burton, 28
Nesbitt, James, 67
Nesbitt, James Jr., 28–29, 308

Neville Holcombe Distinguished Citizenship Award, 287
New Bedford, Mass., 312
New Cut Road, 252, 309
New Deal, 129, 137–138, 156, 158, 175, 190, 197
New England, 15, 17–19, 23, 28, 33, 37, 41, 53, 56, 69, 72–73, 75, 82, 89, 93, 96, 116, 139, 146–148, 180, 211, 213, 216, 229, 257, 273, 315
New Holland Plant, 268
New Holland, Ga., 71, 314
New Jersey, 73, 116
New Mexico, 151
New Montana, Canada, 73
New Orleans, La., 27
New Prospect Plant, 313
New Prospect, S.C., 20, 122
New Republic (magazine), 289
New Town, S.C., 118
New York, 73, 96, 280, 310, 318
New York Giants, 248
New York Stock Exchange, 258
New York Times (newspaper), 119, 144, 226, 231, 246, 289
New York, N.Y., 27, 38–39, 41, 64, 68, 71, 76, 89, 146–147, 164, 182, 210, 212, 215, 219, 228, 247, 252, 259, 268, 272, 280, 286–287, 292, 314
Newberry, S.C., 33, 82, 159, 197
Newburyport, Mass., 41, 315
Newman, Marvin, 194
News & Courier (newspaper), 82
Newsview (magazine), 163
Newton Plant, 268
Nguyan, Am, 279
Niagra Mills, 244, 246–247, 313–315
Nichols, Ga., 268
Nichols, Montague, 135
Ninety Six Mill, 248
Nix, _____, 135
Nixon, Richard M., 259–260, 265, 288
Noble Trees Foundation, 298
Noblitt's Store, 194
Nolen, John, 145
North American Free Trade Agreement, 276, 282, 289
North Blackstock Road, 316
North Carolina, 18, 21, 23, 42–43, 66, 83, 102, 104, 118, 131–132, 139, 145, 200, 216, 241, 247, 260, 271, 289, 291, 294, 315
North Carolina State University, 253
North Church Street, 20, 204
North Liberty Street, 188, 308
North Pacolet River, 20–21, 30, 77
North Pine Street, 214
North Tyger River, 31, 212
Northeast, 21, 25
Northern Textile Association, 287
Northrop loom, 50
Norton, Beatrice, 280–281, 291
Nothing Could Be Finer (song), 275
Nutt, Karen L., 67, 70, 205

–O–

O Beulah Land (song), 194
O'Dell, 135
Oak Ridge, Tenn., 73
Oakwood Cemetery, 76
Oatman, L.T., 74
Obed Creek, 77
Ohio, 186
Ohio River Valley, 39
Oklahoma, 73
Old Gold & Black (newspaper), 200
Old Joe Clark (song), 195
Old Timer's Reunion, 248
Olmstead, Frederick Law, 145
Olympia Knitting Mills, 253, 313
Oneita Knitting Mills, 310
Operation Dixie, 216, 288
Opportunity School, 108
Order of the Palmetto, 244
Orson, 266
OSHA, 291
Ott's Shoals, 31
Otto Zollinger Inc., 255
Owens, Lunette, 209
Owens, Mrs. John, 80
Owens, Roy, 80

–P–

Pace, Ira Parker, 124–25
Pacific Mills, 72, 96, 109, 145–147, 184, 193, 198, 212–213, 312–313, 316
Pacific Ocean, 171, 260
Pacolet Black Trojans, 197
Pacolet Flood, 59, 77, 80, 310–314
Pacolet Industries, 314
Pacolet Manufacturing Co., 20, 23, 39–41, 43, 45, 48, 53, 59, 67–68, 71, 74, 77, 79, 82–83, 93, 124, 128, 131, 134, 136, 140–141, 156, 165–167, 310, 314, 318
Pacolet Mill No. 1, 79, 313
Pacolet Mill No. 2, 40, 79
Pacolet Mill No. 3, 77, 79
Pacolet Mill No. 5, 71
Pacolet Mills Methodist Church, 128
Pacolet Mills, S.C., 85, 88, 98, 116, 126, 142, 146, 154, 169, 177, 222, 275
Pacolet River, 25, 31, 33, 38–39, 66–67, 70–71, 75–78, 81, 230–231, 240, 256, 266, 313
Pacolet Station, S.C., 222
Pacolet Valley, 77–78, 81, 230
Pacolet, S.C., 30, 43, 47, 50–51, 68, 72, 76–77, 80, 104, 107–108, 115, 129–130, 188, 192, 197, 267, 275
Page, Cary L., 312
Page, Cary L. Jr., 312
Palmetto Council Boy Scouts of America, 293
Panic of 1873, 26
Paris, France, 116
Parker, L.W., 57
Paterson, N.J., 19
Patterson, Elizabeth Johnston, 138, 269, 289
Pawtucket, R.I., 18
PBS, 295
Pearl Harbor, 172, 190
Pearson, David, 221
Peck, _____, 182
Peel, John, 198
Peerless Plant, 268
Peete, Alfred, 33
Pelham Mill Co., 40, 103, 314
Pelham, N.Y., 314
Pelham, S.C., 89
Pellagra, 69, 85, 109, 143–144
Pellagra Hospital, 143, 296
Pelot, Lalla, 33
Pelzer Manufacturing, 52, 73, 248
Pendleton Plant, 268
Pendleton, S.C., 214, 216, 268
Pennell, Dick, 270
Penney's, 210
Pennsylvania, 19, 67, 186, 252, 272
Pequot Mills, 182, 226, 314, 318
Perkasa Engineering, 259
Perkins, Frances, 163
Perot, Ross, 289
Perry, Thomas, 135, 166, 197, 220, 247
Peters, Tom, 270, 288–289
Petty, Charles, 40, 76
Petty, James, 76
Petty, Jim, 194
Petty, Paul, 76
Petty, Richard, 222
Petty, Ruth Cannon, 76
Phi Psi Fraternity, 253
Phifer, Mary H., 171, 175
Philadelphia Press (newspaper), 73
Philadelphia Stars, 197
Phillips Fibers Co., 29, 315
Phillips Petroleum, 29, 230
Phillips, LaBama Lyda, 240
Pickens Mill, 313
Pickens, Andrew, 32
Piedmont, 21, 23, 27, 34, 69, 75, 82, 128, 131, 140, 146, 148, 216, 221, 229, 265
Piedmont & Northern Railroad, 139, 198
Piedmont Club, 297
Piedmont Industry Defense Institute, 245
Piedmont Interstate Fair, 177
Piedmont Manufacturing Co., 26, 198, 248
Piedmont, S.C., 198
Pierce, Mickey, 121
Piggly Wiggly, 101
Pine Mountain Plant, 268
Pine Mountain, Ga., 268
Pocasset Manufacturing, 148
Poinsett Mills, 318
Poland, 167, 168
Polk County, N.C., 43, 132
Polydeck Screen, 255
Polysindo, 259
Ponder, Wililam Fred, 201
Poole, Charlie, 221
Pope, Liston, 129
Porritt, Edward, 31, 38, 59
Porter, Louise, 198
Portland, Maine, 71

Portugal, 268
Post Office, 81
Poteat, Betty, 205–206
Potwin, Marjorie (see also Law, Marjorie Potwin), 70, 129, 132, 138–139, 187
POV (television show), 295
Powell Knitting Co., 96, 109, 130, 163, 244, 276, 313–315
Powell Mill, S.C., 106
Power and Light Co., 131
Powers, Norman, 131
Pressley, Dewey, 241
Pressley, Rosell Suttles, 240–241
Preston Street, 136
Prince, Cecil, 194
Pringle, Mrs. A.F., 76
Progressive Movement, 56, 59, 90
Prototype Mill, 245
Proust, Adrien, 290
Providence, R.H., 41
Prysock, Henry, 246
Publix, 86
Punxsutawney, Penn., 29

–Q–

Quartermaster Corps, 169
Quick Response, 271
Quinn, Eldred, 241

–R–

R.R. Donnelly, 274
Racine, Philip, 37
Radford, Mary, 298
Rainbows and Lollipops Inc., 294
Rainsford, Bettis, 272
Ramey Plant, 215, 264, 275, 293, 301, 312
Ramey/Mountain Shoals, 314
Randolph, A. Philiip, 223
Rank, Matt, 305
Ravan, Lottie, 204
Ravenel, H.E., 49
Ray, Dovie, 181
Raycord Sewing Factory, 276, 313–315
Reagan, Ronald, 268–269, 276, 288–289, 291
Reconstruction Era, 25–26, 34, 38, 68
Red Cross, 246
Red Egypt, 98
Red Men, 108
Reeves Brothers Inc., 96, 99, 147, 170, 190, 212, 230, 232–233, 247, 267, 271, 276–277, 309–310, 315, 317
Reform Party, 289
Relative Industries Inc., 294
Religious Freedom Restoration Act, 250
Religious Land-Use and Institutionalized Persons Act of 2000, 250
Reno, Don, 221
Republican Party, 213, 288
Revolutionary War, 266
Reynolds, R.J., 312
Rhett, Robert Barnwell Jr., 32
Rhode Island, 15–19, 28, 30, 96, 109, 130, 310, 314
Rhode Island Textile Co., 315
Rice, Bernice, 274
Rice, Ivory Jr., 274
Richardson, Tommy, 277
Richie, Lisa Caston, 243, 292
Richmond & Danville Railroad, 308
Riddle, Vernon, 194, 195
Riegel Manufacturing, 294
Rieter AG, 229, 254–255
Right-to-Work law, 217
Riverdale, 293, 310, 312, 314
Riverdale Mills, 156, 216, 266, 279, 284–285, 292–293, 301–302, 302–303
Rivers, Eugene, 102
Robbs, Mr. and Mrs. Ed, 80
Roberts, Ethalia, 85
Rochester, England, 255
Rock Hill, S.C., 104
Rockwell Automation, 258–259
Rockwell Standard, 258
Rodgers, Isham, 188
Rodgers, Jimmy, 194
Rodgers, Laura Goudelock, 188
Rogers, E.C., 312
Rogers, Furman B., 157
Rogers, Ruby, 267
Roof, W.P., 317
Roosevelt, Franklin Delano, 138, 155–156, 158, 161–162, 165, 168, 170, 175, 192, 223, 269
Rosa, D.D., 76
Roseland Presbyterian Church, 297
Rosemont Plant, 244, 266, 315
Route 176, 132
Rural Electricification Administration, 138
Rutherford County, N.C., 159

S

S&S Manufacturing, 276
S.C. Business Hall of Fame, 244
S.C. Chamber of Commerce, 293
S.C. Commissioner of Agriculture, Commerce, and Industries, 109
S.C. Commissioner of Labor, 201
S.C. Cotton Manufacturers Association, 112, 292
S.C. Department of Agriculture, Commerce and Industry, 118
S.C. Educational Television, 294–295
S.C. Employment Security Commission, 249
S.C. Fast-Pitch Hall of Fame, 205
S.C. Federation of Textile Workers, 159, 192
S.C. General Assembly, 69, 90, 97, 101–103, 111, 117, 137, 201
S.C. Hall of Fame, 287
S.C. Highway 9, 20
S.C. House of Representatives, 52, 137, 201, 311
S.C. Industrial Commission, 290
S.C. Institute, 22
S.C. Interstate and West Indian Exposition, 41, 53
S.C. Legislature, 70
S.C. Methodist Conference, 130
S.C. School for the Deaf and Blind, 298
S.C. Senate, 53
S.C. State Board of Health, 109
S.C. State Department of Health, 187
S.C. State Federation of Labor, 118
S.C. State Penitentiary, 187
S.C. State Superintendent of Education, 51, 107
S.C. Supreme Court, 249
S.C. Textile Manufacturers Association, 156, 243, 269, 295
S.C. Workers' Compensation Commission, 290
SACM Textiles, 255
Salee, Jack, 248
Salem, Mass., 318
Salisbury, N.C., 200, 257
Saluda Grade, 132
Saluda Plant, 268
Saluda, N.C., 132
Saluda, S.C., 268
Salvation Army, 298
Sampson, O.H., 82
Sams, Charles, 282–283
San Fernando, Calif., 294
Sandburg, Carl, 200
Sandlapper (magazine), 81
Sandor Teszler Library, 253
Sanford Spinning, 148
Santuck, S.C., 77–78, 81
Satterfield, "Grampa," 128
Satterfield, Maggie, 212
Saxon Avenue, 85
Saxon Baptist Church, 106, 129, 193
Saxon Heights, S.C., 64
Saxon Local Union No. 1882, 167
Saxon Methodist Episcopal Church, 128
Saxon Mill Village, 55, 99, 104, 115
Saxon Mills, 41, 43, 50, 58, 64, 69–70, 84, 112, 117, 129, 131, 135–136, 138–140, 151, 163, 165, 167, 186–187, 203, 212, 219, 230, 243, 246–247, 266, 309, 315
Saxon Mills Methodist Church, 129
Saxon, S.C., 65, 108, 111, 130, 165–166, 203, 224, 273, 280, 308
Saxonfibers LLC, 315
Saybrook Plant, 215, 217, 265–266, 293, 312, 315
SCETV Commission, 294
Schenck, Michael, 19
Schick, 271
Schwartz, David, 251
Schwartz, Richard, 252
Scotland, 67
Scott, Willie, 248
Scruggs, Earl, 221
Seals, Georgia, 243
Sean, Justin, 302
Sean, Tommy, 302
Sears, Roebuck and Co., 210, 271, 294
Segregation Act of 1915, 84
Sessions Court, 187
Seventh-Day Adventist Church, 249
Shake Rag Hill, 98
Shamrock Avenue, 315
Shamrock Damask Mill, 70, 99, 119, 315
Sharon Plant, 268
Shearn, Mrs. C.J., 76

Shecut, John, 18
Shelby, N.C., 257
Sheldon family, 28
Sheldon, William, 16, 311
Shelton, Danny, 239
Sherbert v. Verner, 249–250
Sherbert, Adell Hoppes, 249–250
Sherbert, Frank, 249
Sherman, William Tecumseh, 32
Shook, Mark, 235
Short Line, 167
Siberia, 53
Sibley Plant, 268
Sibley Street, 193
Sigsbee Road, 252, 309
Silver Springs, Fla., 205
Simms, F.P., 310
Simon, Bryant, 152, 202
Simpson, L.J., 307
Simpson, W.W., 96, 318
Simpson, William Hays, 134
Simpsonville, S.C., 268
Sims, Genoble, 80
Sims, Novie D., 80
Singapore, 268
Sirrine, J.E., 157
Sisk, E.M., 167
Sixth Avenue, 212
Slack, Thomas, 15
Slater, Samuel, 15, 18
Slaughter, Jim, 248
Sloan, J.H., 308
Slovakia, 308
Smith v. Employment Division, 250
Smith, Arthur, 194
Smith, Aug. W., 318
Smith, Doll, 91
Smith, Ellison D., 138
Smith, Lesesne, 132
Smith, Ola, 126–127
Smith, Rosa, 225
Smith, Susie, 91
Smith, W.F., 307
Smith, Walt, 91
Smith-Lever Act, 138
Smithfield's, 235
Smyth, Ellison, 52, 58
Snake Pit (movie), 203
Snoddy, J.R., 312
Snyder, _____, 120
Snyder, Henry Nelson, 201–202
Snyder, Nora, 116
Social Security, 69
Soil Conservation Service, 168
Sorrell, Robert, 248
Soto, Victor, 294
Soto, Walter, 294
South Atlantic States Music Festival, 25
South Carolina, 21, 23–24
South Carolina Cotton Manufactory, 15, 317
South Carolina Elastic Co., 315
South Carolina Homespun Co., 18
South Carolina Manufactory, 315
South Carolina Manufacturers Institute, 260
South Carolina Manufacturing Co., 31, 34
South Carolina Volunteers, 24
South Liberty Street, 168
South Main Street, Tucapau, 180
South Pine Street, 47, 215, 229, 257–259
South Shamrock Avenue, 308
South Tyger Manufactory, 17, 23, 307, 315
South Tyger River, 17, 25, 307
Southampton, 287
Southeast, 66, 135, 164, 257, 297, 308
Southeastern Regional Tournament, 205
Southern Avenue, 246
Southern Conference, 248
Southern Organizing Drive, 216
Southern Pediatrics Seminar, 132
Southern Railroad, 39, 308
Southern Railway, 72, 78, 81, 92
Southern Street, 250
Southern Textile Association, 292–293
Southern Textile Basketball Tournament, 247, 249
Southern Textile League, 198
Southside Baptist Church, 138
Soviet Union, 260
Spartan Automotive Service, 192
Spartan Fund, 296, 298
Spartan Herald, 46
Spartan International, 279, 305, 316
Spartan Mill No. 1, 244, 282–283
Spartan Mill No. 2, 40
Spartan Mill Village, 43, 46, 62, 130, 220, 298
Spartan Mills, 39–41, 43–44, 46–48, 53, 60, 67–68, 70–72, 76, 89, 91, 93, 95, 98, 100, 106, 113, 127, 131, 134–136, 143, 151, 155, 157–158, 161, 163, 165, 169–170, 172, 174, 178, 181, 195, 207, 212, 215–216, 220, 228, 238–239, 241, 243–244, 246–249, 257, 263–264, 268, 271–272, 276–278, 282, 294, 296, 308, 310, 312, 313–315, 318
Spartan Mills Community Band, 136, 168
Spartan Regiment, 266
Spartan Shield (newspaper), 228
Spartan Square, 263
Spartan Undies, 268
Spartanbug Development Association, 228
Spartanburg & Union Railroad, 39
Spartanburg and Asheville Road, 311
Spartanburg Area Chamber of Commerce, 254–255, 287
Spartanburg Baby Hospital, 132
Spartanburg Chamber of Commerce, 119, 144, 243, 256
Spartanburg City Council, 299
Spartanburg County Colored League, 197, 220
Spartanburg County Foundation, 243–244, 292, 297
Spartanburg County League, 136, 220
Spartanburg County Medical Society, 143
Spartanburg County Mill Villages (book), 299
Spartanburg County Planning and Development Commission, 260, 299
Spartanburg County Public Library, 294
Spartanburg County Regional Museum of History, 130, 173
Spartanburg County, S.C., 23, 25
Spartanburg Day School, 260, 298
Spartanburg Development Association, 218, 224
Spartanburg District 3, 230
Spartanburg District, S.C., 15–17, 19–23, 31, 40
Spartanburg Female College, 76
Spartanburg General Hospital, 72, 253, 297
Spartanburg Herald (newspaper), 38, 44, 53, 56, 137, 139, 154, 168, 201
Spartanburg Herald-Journal (newspaper), 144, 173, 180, 204, 206, 210, 247, 250, 253, 263, 295
Spartanburg Journal (newspaper), 31, 57, 76, 137, 151, 159, 161–162, 168
Spartanburg Junior College, 131
Spartanburg Manufacturing Co., 72, 76
Spartanburg Memorial Auditorium, 72
Spartanburg Methodist College, 62, 130–131, 296, 298
Spartanburg Music Festival, 25
Spartanburg Rail Gas and Electric Co., 131
Spartanburg Regional Medical Center, 243–244, 260, 298
Spartanburg Rotary Club, 60
Spartanburg School District 6, 251
Spartanburg Technical College, 260, 298
Spearman, James B., 207
Spears, Kye, 315
Spears, Molly, 70, 99
Spivey's Creek, 77
Spring Street, 120
Springfield College, 247
Springfield Republican (newspaper), 74
Springs Industries, 267, 271, 276–277, 293, 313, 316
St. John's College, 76
St. Jude's Hospital, 253
St. Louis Cardinals, 136
St. Luke's Free Medical Clinic, 296, 298
Stafford, T.I., 242
Stamford, Conn., 214
Stapleton, Elbert, 243
Starr, S.C., 313
Startex Finishing, 276
Startex Mills, 163, 165, 172, 180, 212, 243, 268, 278–279, 316
Startex, S.C., 85, 178, 180, 220, 225, 266, 275
Stateburg, S.C., 18
Statistics of South Carolina (book), 16
Staubli, 255
Steadman, _____, 111
Stearns Plant, 268
Stearns, Ky., 268
Stifel and Sons, 311
Stokes, Allen, 76, 81
Stone, George, 312
Stone, J.H., 162
Stone, R. Phillip, II, 136
Stoney, George C., 295
Stribling, Hicks, 77, 78
Strickler, Marshall H., 245
Strike of 1934, 294
Stumptown, 98
Sullivan, Harold, 136

Sulzer AG, 229
Sulzer Brothers, 229
Sulzer Ruti Inc., 255
Summer School for Women Workers, 186
Sumter District, 18
Sunset Baseball League, 142
Sunset Memorial Park, 193
Sunshine Cleaners, 248
Suzaki, 270
Swanton, Vt., 23
Swearingen, Mr. and Mrs. Samuel, 80
Switzer, S.C., 274
Switzerland, 229
Sycamore Plant, 268
Symtech, 255

–T–

Taft-Hartley Act, 138
Taguchi, 270
Tailored Baby Inc., 294
Taiwan, 301
Tate, Frances, 193
Taunton, Mass., 148
Taylor, Frank E., 83
Taylor, Frederick, 195
Taylor, Hobart, 247
Taylorism (scientific management), 195–196
Technical Education Center, 230
Tecumseh Mills, 148
Temple B'nai Israel, 253
Tennessee, 42–43, 66, 83, 149, 280, 294
Teszler, Andrew, 228, 251–252
Teszler, David, 254
Teszler, Sandor, 253
Teter, Betsy Wakefield, 73, 123, 146, 185, 188, 246, 250–251, 256, 263, 281, 283, 285, 296
Tetra Pak, 268
Texas, 17, 29, 67, 144, 250
Texmaco, 259
Texpa America, 255
Textile Code, 157
Textile Council, 192
Textile Hall, 248
Textile Industrial Institute, 62, 96, 108, 111, 129–131, 137, 296, 314
Textile Industrial League, 220
Textile Labor Relations Board, 161
Textile Leader of the Century, 287
Textile Primer (book), 164
Textile Town, 209
Textile Transit Co., 170
Textile Tribune (newspaper), 176, 192–193, 200
Textile Week, 269
Textile Workers Union of America, 231, 288
Textile Workers Union of America Local 325, 241–243
Textile World (magazine), 287
Textiles Go To War, 203
Textiles: Employment and Advancement for Minorities, 225
Textron Co., 318
Thailand, 278, 316
The 500, 98
The Idler, 57
The State (newspaper), 101
They Closed Down the Mill (song), 305
Thomas, Aaron, 294
Thomaston, Ga., 268
Thompkins, D.A., 145
Thompson, Waddy, 19
Thompson, William W. Jr., 42, 47
Thompson-McFadden Commission, 143
Thurmond, J. Strom, 138, 224, 259, 287, 292
Tientsin, China, 73
Tietex International Ltd., 316
Tight Wad, 98
Tillotson, Jean, 212
Time (magazine), 201
TNS, 252, 267, 316
TNS Green Plants, 316
TNS Spartanburg, 316
To a Cotton Mill Worker (poem), 200, 202
Tobe Hartwell Courts, 168
Toccoa, Ga., 268
Toddle House, 210
Tonight Show with Johnny Carson, 86
Travelers Rest, S.C., 89
Trent, Buck, 221
Trent, Connie, 85–86
Trevira, 256
Trimite Powders, 255
Trimmier and Co., 317
Trimmier, F.M., 76
Trimmier, T.J., 310
Trough Shoals, 30, 56, 68, 76, 128, 175, 313
Trough, S.C., 142
Trout, James Marion, 317
True, Dan, 71
Truman, Harry S, 138
Tryon, N.C., 77, 308
Tuberculosis, 69
Tucapau Local 2070, 244
Tucapau Mills, 40, 44, 49, 52, 72–73, 103, 135, 147, 161, 163, 165, 196, 212, 243, 266, 312, 315–316
Tucapau, S.C., 30, 34, 98, 110, 118, 164–166, 180, 192, 280
Tucker, Rosalie, 236–237
Tukey, Jenny, 254
Tukey, Richard E., 254–256, 272
Turner, George S., 317
Tuscaloosa, Ala., 253
Tuscarora Yarns, 309
Tuxedo, N.C., 132
Twichell Auditorium, 263
Twichell, Albert, 24–26, 38, 47, 52, 292, 309, 311
Twichell, Helen Antoinette (Nellie), 23–24
Twichell, Winslow, 23
Twitty, Bobby, 247
Two Mile Creek, 314
TWUA-CIO, 241
Tyger Cotton Mills, 317
Tyger River, 15, 25–26, 30, 308, 311, 317
Tzusuki Spinning Co., 316

–U–

U.S. Army, 172
U.S. Bureau of Labor Statistics, 116, 118
U.S. Chamber of Commerce, 139
U.S. Congress, 250, 268, 277
U.S. Defense Department, 247
U.S. Department of Labor, 70, 116
U.S. Finishing, 307
U.S. Highway 221, 20, 252, 313
U.S. Highway 29, 77, 92, 214, 252, 309
U.S. House of Representatives, 198, 269
U.S. Housing Authority, 168
U.S. Public Health Service, 143
U.S. Secretary of Commerce, 259
U.S. Senate, 287, 289
U.S. Supreme Court, 70, 217, 249–250, 265, 288
U.S. War Department, 169
Una News Review (newspaper), 192–193
Una Northside Baptist Church, 193
Una Post Office, 192
Una Water and Sewer District, 193, 220
Una, S.C., 113, 192–193
Union Camp, 229
Union County, S.C., 67–68, 71, 289
Union District, S.C., 31, 76
Local 581, 237
Union, S.C., 159, 268, 315
United Auto Workers, 228
United Confederate Veterans, 76
United Garment Workers, 159
United Kingdom, 288
United Textile Workers of America, 158–159, 161–162, 187, 196
United Way, 243
Unity Plant, 268
University of Chicago, 138
University of North Carolina, 248
University of South Carolina, 100, 137, 205, 248, 268
University of South Carolina-Spartanburg, 230
Upcountry, 23, 26–27, 151
Upstate of South Carolina, 193, 198, 221, 228, 257, 260, 294–295
USO Center, 204

–V–

Valentine, 266
Valley Falls Mill, 25, 147, 151, 231, 265, 317
Valley Falls, S.C., 22, 41, 89, 127
Valway Plant, 268
Van Patton Shoals, 31
Vass's meadow, 57
Vecchio, Diane, 83
Vermont, 66
Verner, Charlie, 249
Vesta Manufacturing Co., 81–83
Victor Manufacturing Co., 40, 89, 212, 268, 307, 317
Victor Monaghan Co., 307, 317
Vietnam, 260
Vietnam War, 225
Vinegar Hill, 98

Virginia, 18, 23, 76, 91, 247
Virginia Military Institute, 244
Visa, 214
Voiselle, Bill, 248
Volunteers in Service to America (VISTA), 291
Von Goethe, Johann, 74
Voultine, Myrtle, 116
Vox Pop Radio Show, 204

–W–

W.O. Gay and Co., 314
W.S. Gray Cotton Mills, 41, 101, 106, 111, 317
Waddell, Landrum, 80
Wadell, Martha, 80
Wadsworth Mills, 317
Wagner Act, 172, 198, 207, 217, 219
Wal-Mart, 271
Waldrep, G.C., III, 70, 151, 162, 231, 241
Walker and Fleming, 68
Walker Brothers, 115
Walker Street, 98, 177
Walker, E.B., 144
Walker, Fleming & Co., 76
Walker, G.T., 307
Walker, Hattie, 66
Walker, Joseph, 24, 307–308, 313
Walker, Robert, 66
Wall Street, 151, 271
Wall Street Journal (newspaper), 73, 289
Wallace Factory, 317
Wallace House, 76
Wallace, David Duncan, 202
Wallace, Peter, 307
Walsh, Frank, 159
Wamsutta Specialty Fabrics Division, 313
War of 1812, 18
Ware Manufacturing, 22
Ware Shoals, S.C., 160, 196
Warrior Duck Mill, 317
Washington, D.C., 70, 163, 168, 170, 178, 259, 270–271, 281–282
Watergate, 260
Watkins, Mary Louise, 312
Watkins, W.E. "Ball," 312
Watson School, 50–51
Watson, Ruth Trowell, 56, 129
Waynesville, N.C., 83
Wear Cotton Campaign, 97
Weaver, John, 15, 22, 317
Weaver, Lindsay, 15, 17, 317
Weaver, Philip, 15, 17, 317
Weaver, Wilbur, 317
Weavers Factory, 17, 19, 309, 311, 315, 317
Webster, Thomas, 298
Weetamoe Mills, 148
Wellford Manufacturing Co., 312, 317
Wellford, S.C., 49, 163, 273, 312
West Indies, 18
West's Studios, 204
West, Annie Laura, 186–187
West, J.E., 44
West, John C., 253–254, 256
West-Point Pepperell, 271
Western Carolina University, 248
Westgate Mall, 263, 264
Westgate, S.C., 313
What a Friend We Have in Jesus (hymn), 195
Wheeler, George, 143–144
Wheeling, W.Va., 311
When They Ring the Golden Bell (song), 194
Whig Party, 19
White House, 269, 291
White Rose Cemetery, 188
White Stone, S.C., 268
White Trojans, 197
White, A.C., 310
White, Ada, 116
White, Ernie, 136
White, Henry, 22, 317
Whitfield Mills, 315
Whitlock's Store, 98
Whitlock, Sarah R., 278
Whitman, David, 72
Whitney Elementary School, 182
Whitney Mills, 40, 48, 68, 71, 76, 91, 93, 135, 183, 221, 226, 241, 244, 265, 266, 278, 279, 314, 315, 317, 318
Whitney, Eli, 18, 266, 317
Whitney, S.C., 51, 55, 57, 72, 127, 177, 179, 181–182
Wiggins, Ella May, 149
Wilburn, Clarence, 238–239
Wildeman, Arno, 316
Wildeman, Martin, 316
Wilkins, S.B., 309
Wilkinsville, S.C., 76
William Whitman Co., 313
Williams Street, 247, 313
Williams, Dock, 80
Williams, J.H, 82
Williams, Jane, 80
Williams, Judge _____, 80
Williamston, S.C., 268
Willis, Alfred, 154
Willis, George, 78
Wilmington, Del., 256
Wilson, Shirley, 205, 206
Wilson, Stanyarne, 52, 53
Wilson, Woodrow, 171
Winfield Plant, 268
Winfield, Tenn., 268
Wingate, Jerry, 280, 281
Winnsboro, S.C., 130
Winston-Salem, N.C., 297
Wisconsin, 228
WK&B, 294
WLW (radio station), 248
Wm. Barnet & Sons, 273, 308
Wofford College, 20, 64, 76, 130, 135, 137, 139, 143, 166, 199–202, 248, 253, 292–293, 297–298
Wofford Park, 135, 197
Wofford Street (extension), 168
Wofford's Shoals, 31
Wofford, _____, 123
Wofford, Ben, 69
Wofford, Benjamin, 16–17, 307, 317
Wofford, William, 34
Wofford/Berwick Iron Works, 19
Women's Army Corps, 206
Women's Bureau, U.S. Department of Labor, 70
Wood, Alvin, 264
Wood, Charles G., 116
Wood, William, 80
Woodmen of the World, 108
Woodruff Cotton Mill, 41, 96, 116, 313, 318
Woodruff News (newspaper), 216
Woodruff, Janie, 181
Woodruff, Josh, 212
Woodruff, S.C., 29, 31, 41, 89, 92, 101, 111, 117, 158, 161, 170, 190, 212, 253, 271, 276, 294, 313, 317
Woodside Mill, 156
Wooten, Earl, 248
WORD (radio station), 194
Worker's Compensation Act, 137
Works Progress Administration, 23, 165, 168, 186
World Class: Thriving Locally in the Global Economy (book), 255
World War I, 70, 89, 93, 96, 100–102, 104, 108–109, 111, 115, 118, 131–132, 137, 144–145, 147, 192, 195, 312
World War II, 67, 73, 86, 102, 129, 136, 138, 167, 175–177, 190, 194, 199, 204–205, 209–210, 212, 215, 219, 221–223, 226–227, 238, 240–241, 243, 260, 265, 297, 308, 310, 314, 317
World War III, 245
Worth Street, 212
Worthy, Quay, 80
WRET (television station), 294
Wright, Coy, 271
WSM (radio station), 194, 221
WSPA (radio station), 194, 221
Wyatt, Horace, 248
WYFF (television station), 295

–X–

Xerox, Inc., 73

–Y–

Yale University, 213, 260, 270, 286–287
YMCA, 104, 115, 247
York County, S.C., 34
Yorkville Enquirer (newspaper), 82
Yu Yuen Cotton Mill, 73
Yugoslavia, 253
YWCA, 115

–Z–

Zima, 230
Zima/Kusters/EVAC, 255
Zimmer Machinery America, 255
Zimmerli Foundation, 298
Zimmerli, Kurt, 255
Zimmerman, John Conrad, 23, 25, 307, 310
Zirconia, N.C., 132
Zittenfield Inc., 294

The Hub City Writers Project is a non-profit organization whose mission is to foster a sense of community through the literary arts. We do this by publishing books from and about our community; encouraging, mentoring, and advancing the careers of local writers; and seeking to make Spartanburg a center for the literary arts.

Our metaphor of organization purposely looks backward to the nineteenth century when Spartanburg was known as the "hub city," a place where railroads converged and departed.

At the beginning of the twenty-first century, Spartanburg has become a literary hub of South Carolina with an active and nationally celebrated core group of poets, fiction writers, and essayists. We celebrate these writers—and the ones not yet discovered—as one of our community's greatest assets. William R. Ferris, former director of the Center for the Study of Southern Cultures, says of the emerging South, "Our culture is our greatest resource. We can shape an economic base...And it won't be an investment that will disappear."

Hub City Anthology • John Lane & Betsy Teter, editors
Hub City Music Makers • Peter Cooper
Hub City Christmas • John Lane & Betsy Wakefield Teter, editors
New Southern Harmonies • Rosa Shand, Scott Gould, Deno Trakas, George Singleton
The Best of Radio Free Bubba • Meg Barnhouse, Pat Jobe, Kim Taylor, Gary Phillips
Family Trees: The Peach Culture of the Piedmont • Mike Corbin
Seeing Spartanburg: A History in Images • Philip Racine
The Seasons of Harold Hatcher • Mike Hembree
The Lawson's Fork: Headwaters to Confluence • David Taylor, Gary Henderson
Hub City Anthology 2 • Betsy Wakefield Teter, editor
Inheritance • Janette Turner Hospital, editor
In Morgan's Shadow • A Hub City Murder Mystery
Eureka Mill • Ron Rash
The Place I Live • The Children of Spartanburg County

Colophon

The design and production steps of *Textile Town* was akin to the process of cloth manufacturing. Many people, new technologies, and working environments were employed. Image scanning was started at Olencki Graphics, Inc. by Tina Smith, and "mobile" scanning sessions were added by Katherine Wakefield and Mark Olencki. The ole Power Macintosh® 7100/80 and the "newer" G3 were used here. Back in the studio, the other G3 took over the book production from the older twin 7100/80s. Alas, this was the last time those 1991 vintage machines will be a part of HCWP book projects. The new twist came with the introduction of an iMac® G3 400 mhz machine and a LaCie® 60 gig Firewire® external hard drive. This portable production-studio combo went with the designer on his "vacation" to Atlanta and was then at home, in the kitchen, where many late night hours were spent working. *Textile Town* was manufactured in a first printing of 5000 soft-bound copies and a limited edition of 350 case-bound copies. The text type family is AGaramond with a bit of American Typewriter and Kabel. The display faces are Engravers Roman and Serpentine. A constant source of Screaming Yellow Zonkers!® with a variety of diet soft drinks kept the munchies away.